AF701417

Cereals

Novel Uses and Processes

Cereals

Novel Uses and Processes

Edited by

Grant M. Campbell

Colin Webb

and

Stephen L. McKee

Satake Centre for Grain Process Engineering
University of Manchester Institute of Science and Technology
Manchester, United Kingdom

Plenum Press • New York and London

Library of Congress Cataloging-in-Publication Data

Cereals : novel uses and processes / edited by Grant M. Campbell,
Colin Webb, and Stephen L. McKee.
p. cm.
Includes bibliographical references and index.
ISBN 0-306-45583-8
1. Grain--Biotechnology. I. Campbell, Grant M. II. Webb, Colin.
III. McKee, Stephen L.
TP248.27.P55C47 1997
620.1'17--dc21 97-1547
CIP

Proceedings of an international conference on Cereals: Novel Uses and Processes,
held June 4 – 6, 1996, in Manchester, United Kingdom

ISBN 0-306-45583-8

© 1997 Plenum Press, New York
A Division of Plenum Publishing Corporation
233 Spring Street, New York, N. Y. 10013

http://www.plenum.com

All rights reserved

10 9 8 7 6 5 4 3 2 1

No part of this book may be reproduced, stored in a retrieval system, or transmitted in any form or by any means, electronic, mechanical, photocopying, microfilming, recording, or otherwise, without written permission from the Publisher

Printed in the United States of America

PREFACE

"So long as a person is capable of
self renewal they are a living being."
—Amiel

Cereals have been the source of life to the human race, providing nutritional and material needs since the dawn of civilization. As with all dynamic industries, the Cereal industry has renewed itself in the past; as the millennium approaches, it is on the brink of another renewal, in which the versatility and providence of cereals are being rediscovered, but in new and exciting ways.

Cereals are richly diverse; over 10,000 varieties convert minerals and the energy of the sun into a bursting catalog of functional and versatile biomolecules and biopolymers. Processing technology allows these components to be accessed, separated, isolated and purified, while chemical science allows modification for even greater diversity and specificity. The last century has seen the move from cereal- to oil-based chemical and materials industries. But cereals contain a greater variety and functionality of macromolecules than oil. Starch, protein, bran and straw, already diverse across cereal varieties, can be fractionated into more specific elements, modified chemically to enhance function, or used as feedstocks in fermentation-based bioconversion systems, to produce a range of bulk and fine chemicals for industries as diverse as food, pharmaceuticals, plastics, textiles, pulp and paper, transport, composites and boards, adhesives and energy.

There are many incentives and pressures for exploiting this rich catalog of ingredients in ever more beneficial ways. Environmental concerns over renewable resources and biodegradable materials favor cereal products over petrochemicals. Agricultural surpluses, combined with the desire for national self-sufficiency in raw material resources and chemical processing technology, encourage countries world-wide to look at new ways of using their cereals. Cereal processors are pushing the change, seeking to add value to their commodities, while the chemical industries pull the developments as they seek new sources for current and future product ranges. New markets for "smart" materials look to functional biopolymers as the starting point, while functional foods attract increasing interest from the consuming public.

The economics of the process industries requires new approaches to make cereals competitive. In a less competitive past, some components of the crop could be viewed as waste products. This perspective has progressed, through recognizing by-products, to regarding all outputs as co-products, contributing critically to the competitive economic equation. The next stage is to design integrated processes *ab initio*, to utilize the whole crop in an economically optimized system. Increased fractionation will continue to add value to process streams, while co-production of food and non-food products on the same site will coincide with increased mutual technology- and knowledge-transfer between the food and chemical industries.

The shift to cereals will progress, as Incentives give birth to Innovation, then to Improvement in which industry excels, and finally to Economic Competitiveness. What is needed is a critical mass of industrialists, academics and government, with the will and imagination to bring to fruition fresh ideas about novel uses and processes for cereals.

The Satake Centre for Grain Process Engineering was established deliberately in a world class Chemical Engineering department, to encourage just such a fresh approach to cereals. It is fitting that the Centre's first International Conference should have brought together people from over 20 countries, from Australia to Zimbabwe, to focus on "Cereals: Novel Uses and Processes," at a time when industry world-wide is poised to revolutionize the use of cereals.

The editors would like to thank the oral and poster presenters at the conference for the first class presentations which have become the chapters of this book, and the delegates for being such an enthusiastic audience. We thank also our sponsors: the Satake Corporation, the European Commission who supported the event under FAIR-CT96–4811, CPL Scientific Ltd., Kellogg Company of Great Britain Ltd., and Dalgety plc. We are grateful too to our other chairmen, Mr. Eric Audsley of Silsoe Research Institute, U.K. (who also served on the Technical Steering Committee), and Dr. Pauli Kiel of the Institute of Biomass Utilization and Biorefinery, Denmark. The editors also thank the other members of the SCGPE team who were instrumental in organizing the conference and these proceedings: Mr. David Sugden, Miss Paula Whittleworth, and Miss Tracey Donlan.

The editors are also grateful to Professor Bernard Atkinson, who opened the conference with a challenge to the gathered researchers and industrialists to generate a collective momentum which would move the industry forward, in terms of capitalizing on opportunities for benign, biodegradable, cereal-based technologies. This book is part of the response to that challenge.

This book, following the conference structure, firstly overviews the potential of cereals as industrial raw materials for food, feed, and non-food applications. The major cereal components are then considered in Section I: starch, protein, bran, and straw are explored regarding their potential for novel uses, describing research taking place worldwide on these versatile cereal components. Starch provides the raw material for a range of plastics and chemicals, while starch properties are being re-evaluated and cataloged in the quest for specific functionality. Cereal proteins, especially gluten, provide a unique functionality with applications in adhesives and plastic films. Chemical modifications of both starches and proteins offer even greater opportunities for tailoring properties to specific applications.

Bran and straw, traditionally regarded as waste or by-products, also present opportunities for economic advantage. Straw can be burned for energy, or treated to allow fermentation, while harvesting before maturity gives access to the carbohydrates stored in the stems during growth. In addition, the immature seeds co-harvested have potentially interesting nutritional and functional properties. The fractionation of bran follows the trend of increased fractionation generally: flour streams are increasingly fractionated to add value to high quality streams, while protein fractionation enhances specific functionality. In the case of bran, the new fractionation process developed in Australia releases the highly nutritious aleurone cells.

Having considered the cereal components individually, the book brings them together by introducing the Wholecrop Utilization concept in Section II. In a wholecrop system, integrated processes are designed which exploit every part of the crop in an integrated, economically optimized system, producing a range of products, both food and non-food, and including internal energy generation and consumption within the overall economic equation. Such systems increase the productivity of cereals while decreasing the environmental impact of process wastes.

A key technology in integrated wholecrop systems is fermentation. Fermentation allows the benign conversion of biomolecules into a vast range of chemical monomers and polymers. As cereals contain, in a concentrated form, all the nutrients required for microbial life and growth, they offer the ideal medium for fermentation. New fermentation systems based on whole grains as substrates eliminate the need for expensive starch separation and purification, followed by supplementation with vitamins, minerals and a nitrogen source. Internal energy generation from cereal straw completes the total processing concept.

Food uses will continue to dominate cereal usage; Section III considers novel developments in this area. Functional foods, "nutraceuticals", are of increasing interest to consumers and manufacturers; novel processes such as the bran fractionation already mentioned are increasing access to these natural food components. Novel processes are also developing for flour milling, flour usage in crackers and bread, malting and sorghum processing. World-wide, cereals are being re-examined and re-evaluated. The book ends with an account of the shake-up and subsequent revitalization of the New Zealand cereal industry, which has developed into the country's fastest growing export sector. With New Zealand's economic growth into a world leader, this final chapter provides food for thought for the cereal industrialists of every country.

Dean William Inge wrote *"There are two kinds of fool: one says, 'This is old, therefore it is good'; the other says, 'This is new, therefore it is better' "*. The old usage of cereals is no longer good enough. The new does offer prospects for a better way; more effective use of crop components, efficient integrated processes, environmentally friendly functional materials from renewable resources. But the path to the new is not yet defined. Each individual success moves the cereal industry forward. The challenge is for individuals and industries to renew their vision, as they allow cereals to serve the human race into the new millennium.

Grant M. Campbell
Colin Webb
Stephen L. McKee

CONTRIBUTORS

Åkerberg C (Chapter 8) — Department of Chemical Engineering, University of Lund, PO Box 124, S-221 00 Lund, Sweden

Andersen M (Chapter 27) — Institute of Biomass Utilization and Biorefinery, South Jutland University Centre, Industrivej 11, DK-6870 Ølgod, Denmark

ap Rees T (Chapter 3) — Plant Science Department, University of Cambridge, UK

Audsley E (Chapter 24) — Silsoe Research Institute, Wrest Park, Silsoe, Beds. MK45 4HS, UK

Batchelor SE (Chapter 3) — Scottish Agricultural College, Aberdeen, UK

Bekers M (Chapter 21) — Institute of Microbiology and Biotechnology, University of Latvia, Kronvalda boulevard 4, Riga LV 1586, Latvia

Bird MR (Chapter 13) — School of Engineering, Bath University, Claverton Down, BA2 7AY, UK

Bjerre A (Chapter 17) — Environmental Science and Technology Department, Risø National Laboratory, PO Box 49, DK-4000, Roskilde, Denmark

Booth EJ (Chapter 3) — Scottish Agricultural College, Aberdeen, UK

Boudrant J (Chapter 31) — CNRS-LSGC, 2 Avenue de la Forêt de Haye, 54500 Vandoeuvre les Nancy, France

Brock CJ (Chapter 16) — Parascan Technologies Ltd, Unit 8, Padgets Lane, South Moons Moat Industrial Estate, Redditch, Worcs, B98 0RA, UK

Carlsson R (Chapters 11, 20) — Department of Natural Sciences, Kalmar University, PO Box 905, S-391 29 Kalmar, Sweden

Cecchini C (Chapter 18) — Instituto Sperimentale per la Cerealicoltura, via Cassia 176, 00191 Roma, Italy

Cervigni T (Chapter 18) — CRA, via Borgorose 15, 00189 Roma Italy

Cochrane MP (Chapter 10) — Crop Science and Technology Department, SAC, West Mains Road, Edinburgh EH9 3JG, UK

Coombs J (Chapter 1) — CPL Scientific Limited, 43 Kingfisher Court, Newbury RG14 5SJ, UK

Cooper AM (Chapter 10) — Scottish Crop Research Institute, Invergowrie, Dundee DD2 5DA, UK

Corke H (Chapter 12) — Department of Botany, University of Hong Kong, Pokfulam Road, Hong Kong

Corradini C (Chapter 18) — Instituo di Cromatografia del C.N.R. - Area della ricerca di Roma - 00016 Monterotondo (Roma), Italy

Culshaw D (Chapter 19) — ETSU, Harwell, Didcot, Oxfordshire OX11 0RA, UK

D'Egidio MG (Chapter 18) — Istituto Sperimentale per la Cerealicoltura, via Cassia 176, 00191 Roma, Italy

Dale F (Chapter 10) — Scottish Crop Research Institute, Invergowrie, Dundee DD2 5DA, UK

de Graaf LA (Chapter 14) — Agrotechnological Research Institute (ATO-DLO), subdivision Industrial Proteins, PO Box 17, NL-6700 AA Wageningen, The Netherlands

Delatte JL (Chapter 31) — Malteries Soufflet, Quai Sarrail, 10400 Nogent sur Seine, France

Din RA (Chapter 13) — School of Engineering, Bath University, Claverton Down, BA2 7AY, UK

Donini V (Chapter 18) — Istituto di Cromatografia del C.N.R. - Area della ricerca di Roma - 00016 Monterotondo (Roma), Italy

Duffus CM (Chapter 10)	Crop Science and Technology Department, SAC, West Mains Road, Edinburgh EH9 3JG, UK
Ellis RP (Chapter 10)	Scottish Crop Research Institute, Invergowrie, Dundee DD2 5DA, UK
Entwistle G (Chapter 3)	Scottish Agricultural College, Aberdeen, UK
Evers AD (Chapter 16)	Campden & Chorleywood Food Research Association, Chipping Campden, Glos. GL55 6LD, UK
Fliss M (Chapter 31)	Malteries Soufflet, Quai Sarrail, 10400 Nogent sur Seine, France
Forder DE (Chapter 32)	Satake UK Ltd, PO Box 19, Bird Hall Lane, Cheadle Heath, Stockport, SK3 0RX, UK
Gabriel M (Chapter 31)	CRAM IM, Rue Jean Lamour, 54500 Vandoeuvre les Nancy, France
Ghorpade V (Chapters 7, 15)	Industrial Agricultural Product Center, University of Nebraska, Lincoln, NE 68503–0730, USA
Gorton L (Chapter 9)	Department of Analytical Chemistry, Center for Chemistry and Chemical Engineering, University of Lund, PO Box 124, S-221 00 Lund, Sweden
Hacking A (Chapter 3)	Dextra Laboratories, Reading, UK
Hahn-Hägerdal B (Chapter 26)	Department of Applied Microbiology, Lund University of Technology/Lund University, PO Box 124, S-221 00 Lund, Sweden
Hall K (Chapter 1)	CPL Scientific Limited, 43 Kingfisher Court, Newbury RG14 5SJ, UK
Hanna M (Chapters 7, 15)	Industrial Agricultural Product Center, University of Nebraska, Lincoln, NE 68503–0730, USA
Hofvendahl K (Chapter 26)	Department of Applied Microbiology, Lund University of Technology/Lund University, PO Box 124, S-221 00 Lund, Sweden
Howling D (Chapter 2)	Manchester Metropolitan University, Hollings Faculty, Old Hall Lane, Manchester M14 6HR, UK
Hsieh F (Chapter 4)	Department of Biological and Agricultural Engineering, University of Missouri, Columbia MO 65211, USA
Huff H (Chapter 4)	Department of Biological and Agricultural Engineering, University of Missouri, Columbia MO 65211, USA
Kennedy D (Chapter 33)	University of Zimbabwe, Box MP 167, Mt Pleasant, Harare, Zimbabwe
Kiel P (Chapter 27)	Institute of Biomass Utilization and Biorefinery, South Jutland University Centre, Industrivej 11, DK-6870 Ølgod, Denmark
Kolster P (Chapter 14)	Agrotechnological Research Institute (ATO-DLO), subdivision Industrial Proteins, PO Box 17, NL-6700 AA Wageningen, The Netherlands
Larsen NG (Chapter 34)	Crop and Food Research International, PO Box 7, North Ryde, NSW 2113, Australia
Laukevics J (Chapter 21)	Institute of Microbiology and Biotechnology, University of Latvia, Kronvalda boulevard 4, Riga LV 1586, Latvia
Laurell T (Chapter 9)	Department of Electrical Measurements, University of Lund, Lund, Sweden
Lawton JW (Chapter 6)	Plant Polymer Research, National Center for Agricultural Utilization Research, USDA-ARS, 1815 North University Street, Peoria, IL 61604, USA
Lin Y (Chapter 4)	Department of Biological and Agricultural Engineering, University of Missouri, Columbia MO 65211, USA
Lindley TN (Chapter 34)	Grain Foods Research Unit, Crop and Food Research, Private Bag 4704, Christchurch, New Zealand
Lynn A (Chapter 10)	Food Science and Technology Department, SAC, Auchincruive, Ayr KA6 5HW, UK
Mackay GR (Chapter 3)	Scottish Crop Research Institute, Dundee, UK
Marko-Varga G (Chapter 9)	Department of Analytical Chemistry, Centre for Chemistry and Chemical Engineering, University of Lund, PO Box 124, S-221 00 Lund, Sweden
Maurel F (Chapter 31)	Malteries Soufflet, Quai Sarrail, 10400 Nogent sur Seine, France
Moonen H (Chapter 30)	Food Science and Technology Centre, Quest International, PO Box 2, 1400 CA Bussum, Holland
Morrison IM (Chapters 3, 10)	Scottish Crop Research Institute, Invergowrie, Dundee DD2 5DA, UK
O'Brien GS (Chapter 5)	Zeneca Biopolymers, Wilmington, DE, USA

Paterson L (Chapter 10)	Crop Science and Technology Department, SAC, West Mains Road, Edinburgh EH9 3JG, UK
Pignatelli V (Chapter 18)	ENEA INN BIOAG C.R. Casaccia, via Anguillarese 301, 0060 Roma, Italy
Prentice RDM (Chapter 10)	Crop Science and Technology Department, SAC, West Mains Road, Edinburgh EH9 3JG, UK
Ruklisha M (Chapter 21)	Institute of Microbiology and Biotechnology, University of Latvia, Kronvalda boulevard 4, Riga LV 1586, Latvia
Savenkova L (Chapter 21)	Institute of Microbiology and Biotechnology, University of Latvia, Kronvalda boulevard 4, Riga LV 1586, Latvia
Schmidt A (Chapter 17)	Environmental Science and Technology Department, Risø National Laboratory, PO Box 49, DK-4000, Roskilde, Denmark
Sells JE (Chapter 24)	Silsoe Research Institute, Wrest Park, Silsoe, Beds. MK45 4HS, UK
Stenvert NL (Chapter 29)	Goodman Fielder Milling and Baking Group, PO Box 1, Summer Hill, NSW 2130, Australia
Suhner M (Chapter 31)	CRAM IM, Rue Jean Lamour, 54500 Vandoeuvre les Nancy, France
Sun H (Chapter 12)	Institute of Crop Breeding and Cultivation, Chinese Academy of Agricultural Sciences, Beijing 100081, China
Svonja G (Chapter 22)	Barr Rosin, Maidenhead, Berkshire SL6 1BR, UK
Swanston JS (Chapter 10)	Scottish Crop Research Institute, Invergowrie, Dundee DD2 5DA, UK
Tiller SA (Chapter 10)	Scottish Crop Research Institute, Invergowrie, Dundee DD2 5DA, UK
Torto N (Chapter 9)	Department of Chemistry, University of Botswana, P/Bag 0022 Gasborone, Botswana
Trust B (Chapter 33)	University of Zimbabwe, Box MP 167, Mt Pleasant, Harare, Zimbabwe
Vedernikovs N (Chapter 21)	Institute of Wood Chemistry, Dzerbenes Str, 27 Riga LV 1006, Latvia
Walker KC (Chapter 3)	Scottish Agricultural College, Aberdeen, UK
Wang R (Chapter 25)	Satake Centre for Grain Process Engineering, Dept. of Chemical Engineering, UMIST, PO Box 88, Manchester M60 1QD, UK
Webb C (Chapter 25)	Satake Centre for Grain Process Engineering, Dept. of Chemical Engineering, UMIST, PO Box 88, Manchester M60 1QD, UK
Weller C (Chapter 15)	Industrial Agricultural Products Center, University of Nebraska, Lincoln, NE 68583–0730, USA
Whitworth MB (Chapter 16)	Campden & Chorleywood Food Research Association, Chipping Campden, Glos. GL55 6LD, UK
Willett JL (Chapter 5)	National Center for Agricultural Utilization Research, USDA-ARS, Peoria IL, USA
Wood PJ (Chapter 28)	Centre for Food and Animal Research, Agricultural and Agri-Food Canada, Ottawa, Ont K1A 0C6, Canada
Wroe C (Chapter 23)	BP Chemicals Ltd, Britannic Tower, Moor Lane, London, EC2Y 9BU, UK
Wu H (Chapter 12)	Department of Botany, University of Hong Kong, Pokfulam Road, Hong Kong
Yue S (Chapter 12)	Institute of Crop Breeding and Cultivation, Chinese Academy of Agricultural Sciences, Beijing 100081, China
Zacchi G (Chapter 8)	Department of Chemical Engineering, University of Lund, PO Box 124, S-221 00 Lund, Sweden

CONTENTS

SECTION II: WHOLE CROP UTILIZATION

INTEGRATED BIOPROCESSES

FERMENTATION: THE KEY TECHNOLOGY

SECTION III: FOOD PROCESSES

Cereals

Novel Uses and Processes

1

THE POTENTIAL OF CEREALS AS INDUSTRIAL RAW MATERIALS

Legal, Technical, and Commercial Considerations

Jim Coombs and Katy Hall

CPL Scientific Limited
43 Kingfisher Court, Newbury RG14 5SJ, United Kingdom

1. INTRODUCTION

Cereals represent a major component of the human diet worldwide, either directly as baked goods derived from flour, or indirectly as components of animal feed (grain, brans, straws and other residues as appropriate for monogastrics, fowl and ruminants). Global cereal production and trade are dominated by wheat and maize (Table 1). These cereals are also the major raw materials for industrial use, as discussed below. Although supply dropped last year, resulting price increases have led to greater sowing (estimates anticipate 8% increase to 579 Mt (USDA) or 570 Mt (FAO)) with use at 565 Mt - a 3% increase. Maize is expected to be up 11% on last year's poor US harvest, but recent weather suggests this may not be the case, whilst exports are expected to fall reflecting increased US feed demand.

The concept of cereal-based industry can be extended to include flour milling and feed sales. However, in this chapter attention will be paid to those areas where the cereal is subject to fractionation, modification, transformation or formulation prior to sale. Such industrial use covers separation of grains to protein (gluten), flour and oils; the utilisation of by-products of milling; the hydrolysis of starch to sugars; derivation or modification of starch as polymers; the fermentation of sugars for bulk chemicals, fuels, fine chemicals, enzymes, biopesticides and pharmaceuticals; chemical modification of sugars; combustion of straw for heat and power; and the use of straws in composite materials as well as paper, card and board. Current global starch production of around 25 Mt (with over half from the US and EC) is mainly from cereals (77% maize).

The concept of new use also requires similar definition. New use can be through:

- growth of industrial markets using known technology, such as the production of maize-based fuel alcohol in the US

Cereals: Novel Uses and Processes, edited by Campbell *et al.*
Plenum Press, New York, 1997

Table 1. Main producers and traders of major cereals (million tonnes per annum)

	USA	*EU*	*Australia*	*Canada*	*China*	*India*	*Ex-USSR*	*WORLD*
[a]*Wheat*	**59.5**	**86.6**	16.9	25.4	100	57.8	72.2	534
[a]*Maize*	**243.9**	**26.5**	0.2	6.5	104			467
[a]*Rice*			1.7	3.6	116	78.4		348
[b]*Wheat*		3.2			13		9.4	111
[b]*Maize*		2.2					4.2	55
[b]*Rice*		0.65						15
[c]*Wheat*	35.4	18.5	12.8	17.9				106
[c]*Maize*	32.8	2.0		0.1	11.5			55
[c]*Rice*	2.6		0.2	1.2	1.4	0.6	0.1	15

[a]Production, [b]Imports, [c]Exports

- growth of a new market, reflecting novel technology, such as the production of high fructose syrups using isomerase
- growth through substitution of one cereal for another, such as wheat replacing maize in Europe's glucose syrup markets
- growth as a result of consumer pressures or legislative changes, such as the move towards water soluble, solvent (VOC) free adhesives
- growth in the demand for low-calorie or *lite* food products
- growth in a desire for self-sufficiency, or a need to find internal markets for over production
- growth reflecting increasing waste disposal costs or more stringent environmental legislation, resulting in greater use of residues, effluents and by-products as well as the replacement of fossil fuel (coal, gas or oil) based products with biomass-based equivalents
- growth reflecting the ability to introduce new traits into cereals using genetic manipulation.

The extent and rate of growth of such markets does not depend on technical limitations. This may be a surprising conclusion to the practising research scientist. However, as discussed below, if the market conditions (net added value, or profit - depending on the viewpoint) are right then industry will respond - in the same way that man reached the moon and home-based personal computers have evolved to be more powerful than the mainframes of a decade ago. Raw material price, linked to farm (production) costs and support mechanisms remain the main factor affecting greater industrial use of cereals.

2. MARKETS

Cereal components are used in a vast range of food and industrial applications. Chapters 2 and 3 consider the markets for starch, which, along with protein, bran, straw and other cereal components, serves the many industries discussed in the following paragraphs.

Food

Excluding conventional flours and baked goods, growth and innovation have been seen in many categories of ingredients for manufactured foods. The major catagories are modified starches (hydrocolloids), caloric sweeteners (enzyme hydrolysates), *lite* fillers (such as hydrogen peroxide treated bran as a flour substitute), fat substitutes (insoluble fibre, micro-crystalline cellulose, polyglucose, polyols) and functional foods (see Chapters 28 and 29). These include soluble fibre, such as oat β-glucan (including Oatrim, a fat-like gel from enzyme treated oatbran sold as a cholesterol lowering fat replacement, and polyfructans, xylans, etc.). In addition, a number of food enzymes and other ingredients are produced by solid state fermentations based on grain, bran or stover, or submerged fermentation using glucose syrup (vitamins, organic acids, amino acids). Xanthan gum, produced by *Xanthomonas campestris* grown on starch based media, is now a major food hydrocolloid with around 50% of the 30,000 tpa which is used variously as emulsifier, stabiliser, thickener, or gel. Cereal proteins are separated as vital wheat gluten, used for specialty breads, and hydrolysed to produce flavour enhancers (glutamate) or diet supplements (glutamine), while demand for natural anti-oxidants increases.

Cereal starches, and also proteins, vary in properties and functionality, and can be modified chemically to enhance functionality further; Chapters 10 and 12 discuss how the properties of traditional and new starches are being assessed and catalogued with a view to novel food and industrial applications.

Drinks

New formulations of dilutable squashes, and to a lesser extent carbonates, have had a significant impact in Europe on caloric sweetener consumption, whilst the recently introduced alcoholic lemonades provide a novel outlet for grain spirits.

Feed

Over 500 Mt of feed are used worldwide, split fairly evenly between poultry, pigs and cattle. In the past, compound feed has been determined by the price of ingredients on the world market, their nutritional value and the needs of the animals in question (monogastrics, fowl or ruminants), with feed prices moving with the cost of primary ingredients, followed by by-products from corn milling, soya processing, brewing, etc. Composition depends on the age of the animal and the purpose for which it is raised (layers or broilers, beef or milk cattle, etc.). The feed market is again dominated by US and the EU, but with the fastest rate of growth in Asia. The main changes are towards vertical integration of feed growers, animals husbanders, processors and retailers, aiming to establish security of supply and quality, against problems such as *Salmonella* and BSE in the UK. The main influences for growth in the conventional compound feed market are population growth and increased standard of living, with the main determinant of what is used being price. However, on a local basis, changes reflecting a move from hay to silage, the use of inoculants, enzymes and chemical preservation, and modification, as well as a move away from bought-in feed containing animal protein (in the UK in particular), has extended the range of cereals and cereal fractions used, with an accelerating move towards organic and conservation grades.

Cleaners

Include simple or modified APGs and aminosorbitol derivatives, dicarboxylic starch, glucose, sorbitol and methylglucoside peracetates as surface active agents, builders or bleaching agents and indirectly through production of enzymes (alkaline proteases, cellulases) for use in biological washing powders.

Chemicals

A range of solvents (e.g. ethanol, butanol, acetone) and acids (e.g. acetic, propionic, butyric, lactic; see Chapters 8, 9 and 17) can be produced from cereals by fermentation, and aromatics can be produced by hydrolysis or chemical means fairly directly (ferulic acid, vanillin, furfural) or through complex catalytic chemistry starting with ethanol (as in Brazil) or synthesis gas; however, the economics are against this in most countries (see Chapter 23). Excluding food and pharmaceuticals, some products such as itaconic acid are produced in larger volumes by fermentation. However, the main high volume products are modified starches, whilst other chemicals fall into the food, pharmaceutical and fuel markets, considered separately.

Medical and Pharmaceuticals

Starches are used as carriers or binders, as well as raw material for their production (ascorbic acid, fermentation products). Carriers include cyclodextrins, where their structure enables them to entrap the active ingredient. Polyols are also finding increasing use since some are distinguished by their chirality, one of the most rapidly growing areas of medicine.

Personal Care Products

Compounds such as APGs (and other natural products) are increasingly being used in cosmetics, whilst modified starches with high water holding capacity are used for their absorbent properties. Xanthan gum also finds use in liquid soaps, toothpastes, shampoos and other personal care/hygiene applications.

Liquid Fuels and Oxygenates

Conventional yeast-based fermentation of starch hydrolysates followed by azeotropic distillation yields absolute ethanol, which can be added to petroleum-based fuels as an extender, anti-knock (octane enhancer) or oxygenate (see Chapter 23).

Biodegradable Plastics

These include conventional plastics using up to 85% starch fillers - with materials such as polycaprolactone co-polymers with modified starch (polyethylene co-acrylic or co-vinyl alcohol) at one end of the market and polyhydroxyalcanoate, polyhydroxy butyrate and other fermentation-based products (such as polylactic acid) at the other (see Chapters 5, 6 and 26). At the moment, cost is a major constraint, whilst some products which have been marketed are limited by their sensitivity to water. Chapters 6 and 15 discuss coating starch and gluten films, respectively, with polylactic acid and other coatings, to improve mechanical and water barrier properties. Other products include those with a small percentage of starch (biofragmentable products; see Chapter 10). Gluten is also being used again, having in the past served for electrical components such as the rotor cap for Model T Fords. Currently such products account for around 1% of the market.

Loose Fill Packaging

Literally, pop corn is now being used as a substitute for polystyrene beads.

Biopesticides

Whole grain, brans and other fractions may be used as substrates for bacterial or fungal products, and also as carriers and fillers in formulation. Hydrocolloids, derived by fermentation or chemical modification, may also be used for encapsulation.

Pulp and Paper

This market includes products derived from straw and other cereal residues, as well as starches used as fillers or binders (see Chapters 2 and 3). The addition of starch improves the quality of recycled paper. Hence, this is seen as a growth area, increasing from around 3.6 million tonnes, as paper recycling increases.

Composites and Board

In theory, straw can be used for board manufacture. However, the nature (wetability) and fragmentation pattern is such that they are much less suitable than, often low priced, competing materials - especially wood chips which can be derived from mill wastes or off-cuts. In composites, adhesion can also cause problems. Straws can be used in lower density boards.

Textiles

Starch is widely used as a size or stiffener in fabric, especially printed cottons where it can be used to hold materials and prevent diffusion. The choice of starches, both origin and amount of processing or derivatisation, is complex, with cereal starches competing with potato or tapioca on price and performance. In general, historical use and knowledge is greater than present practice, reflecting changes in the fibres used towards synthetics and geographical location towards Asia.

Adhesives

These consist of many ingredients, including solvents, fillers, antifoams, stabilisers and plastifiers, as well as the resin or glue itself. Replacements for organics solvents, which ensure glue remains liquid and evaporates during drying, by water is increasingly occurring to reduce solvent abuse and release of VOCs. Such products may be based on animal products (generally heat softened) or starch (dextrins, acetylated or otherwise modified). Chapter 14 discusses adhesives based on wheat gluten proteins.

Heat and Power

Straw and other crop residues can be used as a fuel for conventional boiler/steam turbine power generation plant in the 0.5 to 10 MW range, or as a component of the total fuel input in larger waste to energy plant (see Chapter 19). Effluents and solid wastes can be dried and burnt, but this may give little net energy gain. An alternative for wet residues is the use of anaerobic digestion to produce biogas, the methane content of which makes it a suitable boiler fuel or for use in internal combustion engines or gas turbines for power generation. In theory, whole crop grain could be grown for combustion in a dedicated power station, giving higher yields than some energy crops.

3. LEGISLATION

Historically, the underlying fiscal policy in the developed world has supported agriculture and taxed industry. As long as cereals were predominately used as food (or as animal feed in the indirect production of food) and were in limited supply, whilst cheap petroleum oil was available in abundance, this arrangement was politically attractive. However, over the last two decades this simple dictum has been overturned by a number of events. These include variable oil prices; creation of surpluses through improved farming techniques; diminishing farm incomes and rural populations; concerns for local and global environments; opportunities arising from advances in biology (genetic engineering); and a move towards global markets (GATT), as the east/west barrier fell and the European Community grew. The impact of these events, and the legislative response, has been significantly different in the US and the EU, resulting in a large, growing maize-based industry in the former which contrasts with technical and market stagnation in Europe. The main areas of contrasting legislation have been as follows:

Production

In the EU, the Common Agricultural Policy (CAP), has changed over the last few years from support-induced surpluses to supported set-aside of land, with compensatory and set-aside payments under the Arable Area Payment Scheme. Although planting for non-food use on set-aside is possible, the terms, including contracts with users, has limited the extent to which this has been adopted. Cereal support, at over £10 billion, represents around 30% of the whole farm budget (whereas in 1989 it was only 15%) and for 1996 is expected to increase by 34%. In the past this was in part due to expansion of the EU to 15 Member States. However, the anticipated rise is in area payments and production refunds, whilst export refunds drop - benefiting from the increase in world grain prices. A major impact has been the 10-fold drop in stocks between 1991 and 1996, partly due to the fact that feed wheat is no longer eligible. The overall impact of price, stocks and costs has resulted in a reduction of set-aside to 10%, which should lead to increased production.

In the US, the new Farm Bill became law in April as the Federal Agricultural Improvement and Reform (FAIR) Act, which represents a major upheaval of previous farm programmes, driven by a need to balance the federal budget, market conditions and political pressures. The previous Acreage Reduction Programme has been discontinued and income support programmes (including loans on stored grain) have been decoupled from market and support prices. Feed grains and wheat are eligible, with payments shared out on the basis of a fixed budget, divided amongst crops and producers, of which maize gets 46% and wheat 26% of an estimated $US 5.6 billion in 1996, decreasing by 30% between 1998 and 2002. In general, the effect should be to bring land back into production, allowing farmers to grow the crops which suit them best. In general, maize production is expected to increase, so prices should drop back.

In the US, trends in cereal use have reflected industrial interest, market pull, investment and favourable legislation, especially in respect to fuel ethanol production. In contrast, in the EU, legislation has been restrictive or (where potentially beneficial) failed to pass into law. The most marked impact has been due to the setting of quotas on the production of enzyme-derived fructose/glucose syrups (known as high fructose corn syrup in the US and as isoglucose in the EU) in the 1970s, which have continued with restriction as each new group of countries have joined the EU. At the same time, attempts to reduce tax on liquid fuels of biological origin (bioethanol) have faltered, although some Member States (France, Italy) have brought in their own laws. This has blocked the two largest potential markets for the industrial use of cereals in the EU, at less than 1 million tonnes per annum.

In other areas, both EU and, in the case of the UK, national legislation has been beneficial. In the food area, the new Sweetener Directive and Ingredients Directives, as well as pending legislation on Novel Foods, Nutritional Claims (medical, nutritional and health) and Quantitative Ingredient Declaration (QUID) and consumer pressures for natural, convenient and/or controlled diet products, has led to an expansion in ingredient markets. One marked effect has been in the soft drinks sector where deregulation (in the UK) of sugar (sucrose) levels and the new Sweetener Directive, linked to concern about tooth decay and young children exceeding the ADI (acceptable daily intake) for saccharin, have resulted in an almost total replacement of sucrose by glucose syrup together with aspartame or acesulfame K.

In contrast, continuing support for fuel alcohol in the US, linked to air quality, has resulted in increased investment in manufacturing capacity, which is expected to grow further. In particular, concerns about urban air pollution have led to the Clean Air Act Amendments of 1990, which require the use of oxygen-containing components in the gasoline used in certain areas where ozone and carbon monoxide levels are high. Ethanol has the advantage that, in addition to being renewable, the required level of additive can be reached with lower amounts than with the alternatives such as MTBE (methyl tertiary butyl ether).

Both in the US and in EU, other environmental concerns have led to legislation covering reduction of waste, encouraging recycling and supporting renewable energy. These aspects are inter-related due to the fact that a large proportion of materials disposed of are packaging, offering opportunities for recycling of metal, glass and plastic as well as composting or combustion in waste to energy plant. The use of biodegradable packaging, fabrics and building materials is also seen as a way of decreasing fossil reserves and contributing to control of carbon dioxide and other emissions which may contribute to climate change, supported by active media and consumer support. However, as discussed in the next section, many people are not willing to pay the higher prices natural raw materials command. Hence, growth depends on taxation and support structures, such as the Non-Fossil Fuel Obligation (NFFO) in the UK which has increased the amount of electricity generated from renewables. However, cost estimates have restricted the number of farm residue plants accepted in the UK, whilst in the US several hundred MW capacity have been installed.

4. COMMERCE

If cereals can be used as a raw material to feed any market where demand is consistent and price is high, then industry will respond, create the product and set up the infrastructure for growth. This is best illustrated by the corn wet-milling industry in the US as shown in Figure 1.

Animal Feed

Demand in the UK has increased slightly (5%) with the main growth in cattle feed, including 6% increase in wheat use in preference to alternatives, reflecting lower 1995 US and Chinese soybean harvests and rising demand. Even so, use of such meals has increased 4% in spite of the illustrated price rises (Figure 2). This trend, possibly driven by BSE, and the resultant ban on meat and bone meal in all compound feeds, may change as consumers move to poultry and pork. However, this would sustain the trend as non-ruminant feed contains more cereals. In the US, demand for meat, as well as meat exports, is growing, in part contributing to the illustrated price rise in feed proteins (Figure 3). This again is an area of increasing use, with over 7 Mt of maize fed to animals which were then slaughtered and exported. Predicting future trends is complicated by the expanding Asian livestock industries which may then consume manioc and rice bran, pushing Europe towards feed wheat.

High Fructose Syrup

Production is more or less static, blocked by legislation in the EU and by market saturation in the US, although several plants are now being built in Asia and feasibility studies have been carried out in a number of countries. Again, the key issue is raw material and final product costs as compared with local sucrose (if available). It is possible that population growth and increased standards of living may be the main factors determining growth in this area. Hence, it is possible that the greatest area of growth would be through consumption of colas in China.

Bioethanol

In 1977 there were no fuel ethanol plants, there are now 70 producing over 6 billion litres, equal to 1% of the market with an investment of $2.5 billion in capital generating over 8,000 jobs. At the same time, technology developments have reduced energy use by

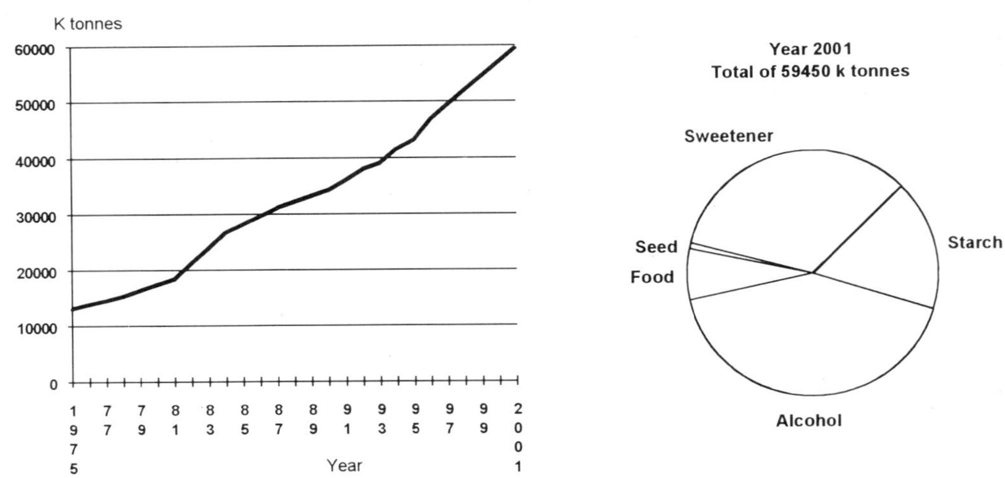

Figure 1. Actual and estimated total food, seed and industrial use of corn, 1975–2001.

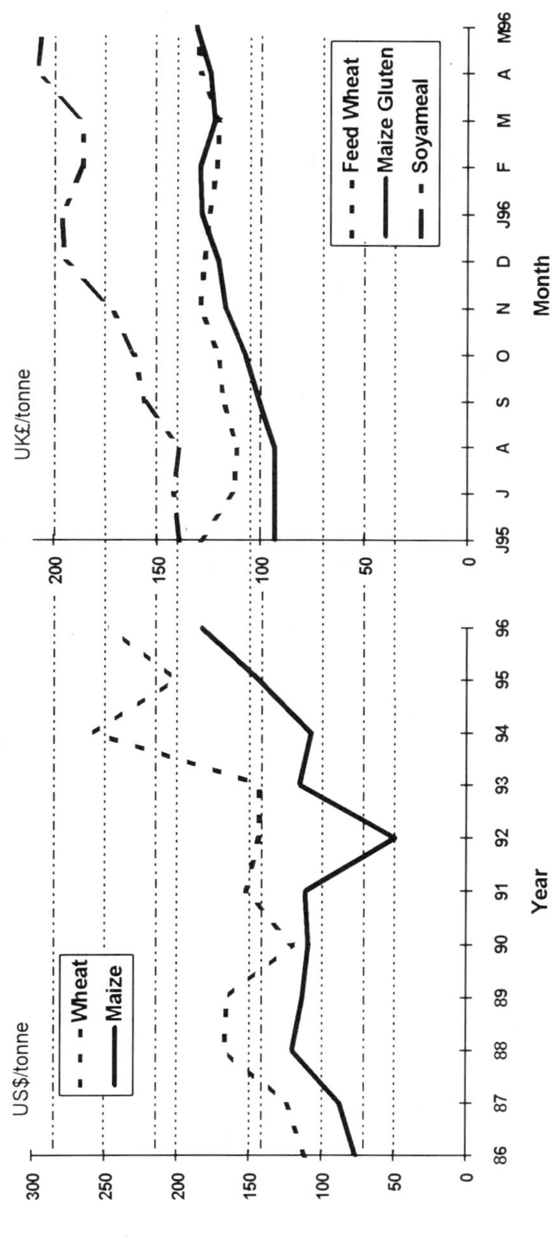

Figure 2. A Comparison of US Export Prices 1986–96 (left) and UK Domestic Cereal Product Prices 1995/96 (right).

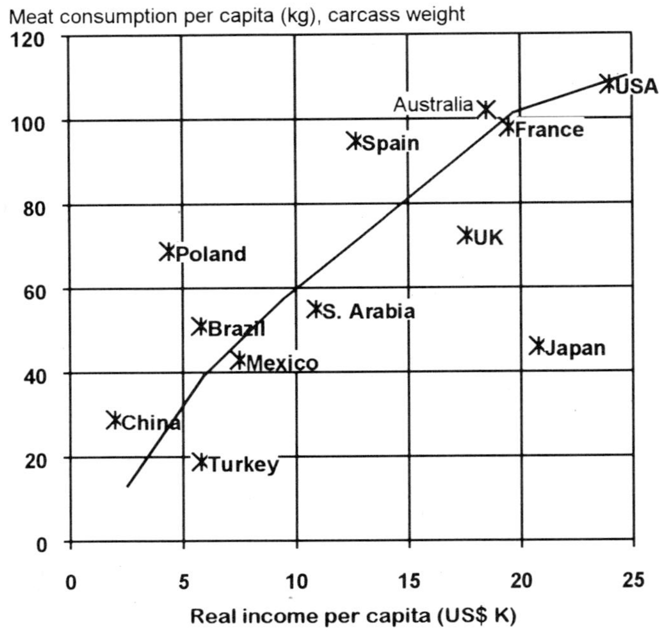

Figure 3. Meat consumption per capita versus real income per capita, 1993.

85%, such that the energy balance is now positive, mitigating some of the arguments used in the EU where only small amounts of bioethanol are being produced in France and the Nordic countries. Hemicellulose and eventually cellulose, which can be derived from maize residues and wheat straws, are seen as future raw materials if the biology can be sorted out.

Polymers

This is the area of greatest current commercial activity, covering products for both food use and fabrication (packaging, in particular). In the food industry, novel starches and starch derivatives are being perfected to meet manufacturers' needs in terms of low temperature stability, shear resistance, pH resistance, etc. These include derivatised, cross-linked, cold water swellable, heat stable, oxidised and bleached products. Starch is also seen as a major ingredient in biodegradable polymers, with many companies entering the market, although market share is still only about 1% of that of petroleum-based products. This is clearly a major opportunity if product quality and price criteria can be met.

5. INNOVATION

Within the EU and US, there are in excess of 100 Mt of cereal residues which could be utilised, and only slightly lower amounts of pulp mill resides. Bioconversion of the hemicellulose and cellulose components of these materials remains one of the key opportunities. In the short term, such technology could be linked to corn wet-milling and the paper pulp industry in order to utilise components of hemicellulose (mainly xylose). At present, this possibility is limited by the ability of yeasts to ferment 5 carbon sugars and the sensitivity of bacteria (which can utilise them) to end product inhibition. Both problems are being tackled by genetic engineering. Strains of *Escherichia coli* and *Saccharomyces cerevisiae* have been engineered to contain enzymes to facilitate this, however improvements in performance, stability, yields and resistance are still required. The second key area concerns the use of enzymes to hydrolyse cellulose in an efficient manner. Current enzymes lose out on stability and rate of catalysis, although attempts continue to improve these.

6. IMPLEMENTATION

Both the US and EU, as well as other countries such as Canada and Japan, seek new uses for agricultural products. The EU is supporting an information system: Non Food Agro-Industrial Research Information Dissemination (NF-AIRID) Network (Mangan et al, 1995). New Uses Councils have been established in the US (Anon, 1995) and Canada; many other national initiatives have also been established. In general, these reflect agricultural push, while market pull is weak. However, consumer pressure and resultant political initiatives remain the key factor in many areas since the normal market forces can be distorted by legislation as discussed above. Even so, raw material prices remain the key, if not the only issue, as far as both traditional and new uses of cereals are concerned. For unsupported markets, raw material price will influence the choice of raw material. Where government (tax) support is required, the extent of such support will influence the extent of commitment, in terms of both the time and net cost that governments, faced with growing budget problems, are prepared to risk.

REFERENCES

Anon (1995) "The 1995 New Uses Briefing Book." New Uses Council Inc, Glenwood Springs Colorado

Campden JR (1995) "Corn's potential continues to soar." in Anon 1995, Part II markets pp 7–8

Mangan C, Kerckow B and Flanagan M (1995) "AIR, Agriculture, Agroindustry and Fisheries, catalogue of Non Food Projects." EUR 16206en, European Commission, Luxembourg

USDA (1995) "Industrial Uses of Agricultural Materials." United States Department of Agriculture Economic Research Services, IUS-5, September, ERS-NASS Herndon VA, US

USDA (1996) "Sugar and sweetener, situation and outlook report." States Department of Agriculture Economic Research Services, SSSV21N1, March, ERS-NASS Herndon VA, US

HGCA (1996) Weekly Digest, various dates April, May, Home Grown Cereals Authority, Market Information, London

2

PRESENT AND FUTURE USES OF CEREAL STARCHES

David Howling

Hollings Faculty
Manchester Metropolitan University
Old Hall Lane, Manchester M14 6HR, United Kingdom

1. INTRODUCTION

Starch is one of the major photosynthetic products and is therefore a constantly renewable resource. It is laid down exclusively by plants to be a source of energy, being converted to sugars by enzymes on the germination of the seed.

As such man has used this for himself since the dawn of time by using it for a food material, a source of energy for life. It is still to the food and beverage industry that we must turn to see the major use of starch today. Figures 1 and 2 for the EU and UK respectively show that the food industry still uses the majority of starch, some 2.9 million tonnes per annum or 48% of the EU market. In the UK the figure is 70% if one takes into account the fermentation sector for potable alcohol.

The food industry has found a number of properties other than energy for the starch molecule. It is now used as a thickener, a binder and a source of sugars - the glucose syrups. This major position, illustrated by the use of starch extracted from maize, wheat or potato, is even more dominant if one considers the vast quantities of flour and cereal that are used in the baking, brewing and breakfast cereal markets.

Thus starch is a major food and animal feed ingredient, yet other non food uses for starch have been devised. Although these constitute a minority they are significant and one of the subjects of this book. Chapters 3 and 10 also consider the range of applications of starch.

2. NON-FOOD USES OF STARCHES

Figure 3 shows the 250,000 tonnes of starch used in the UK market, broken down into the main sectors. It can be seen from this and the equivalent European position, shown in Figure 4, that the paper industry dominates this sector (see also Chapter 3). Starch is used in paper to provide sheet strength, by acting as an adhesive to hold the cel-

Cereals: Novel Uses and Processes, edited by Campbell *et al.*
Plenum Press, New York, 1997

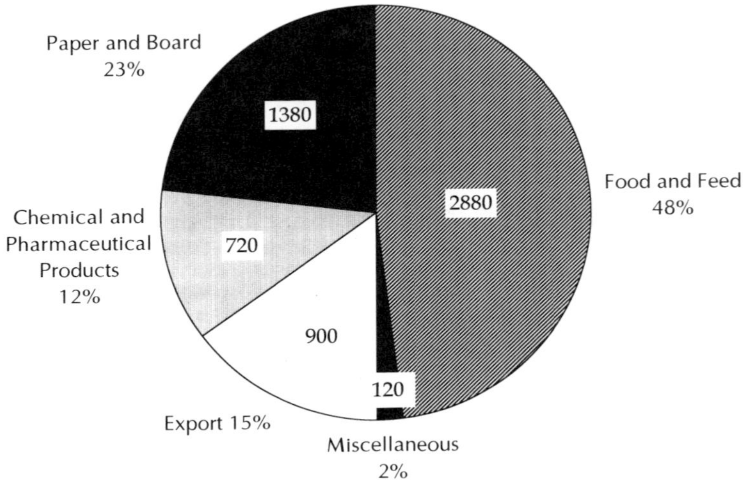

Figure 1. Starch uses in the EU, 1990/91 ('000 tonnes).

lulose fibres together, to provide desirable properties such as sheet and surface strength, sizing, printability and smoothness. It is also used in coating formulations to give the many attractive surface effects found on paper.

Corrugating adhesives make up a significant sector with some 400,000 tonnes in Europe and 54,000 tonnes in the UK being used in this area in 1994. After that binders and the chemical industry use 600,000 tonnes in Europe and 50,000 tonnes in the UK. The type of applications covered by these sectors are shown in Table 1.

Figure 5 illustrates the many ways in which starch can be used. Starch may be used as a powder or a viscous hydrocolloid directly or in blends. It may be modified chemically or physically to impart different properties, whilst remaining a macromolecule; alternatively, it may be considerably modified by hydrolysis with either acid or enzymes to give

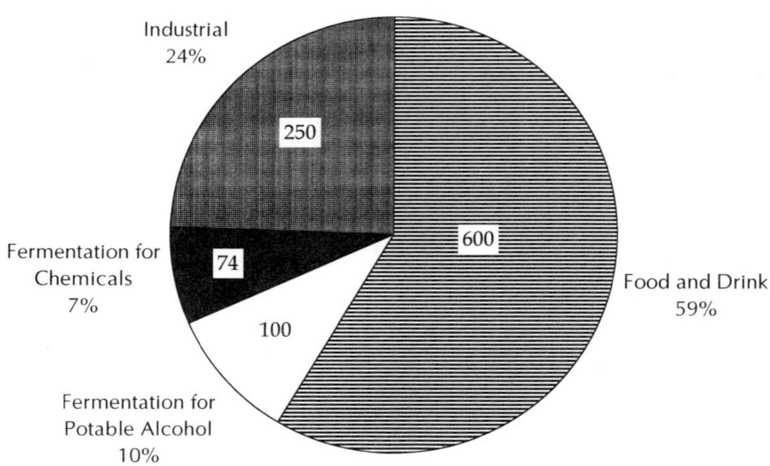

Figure 2. Starch use in the UK, 1994 ('000 tonnes).

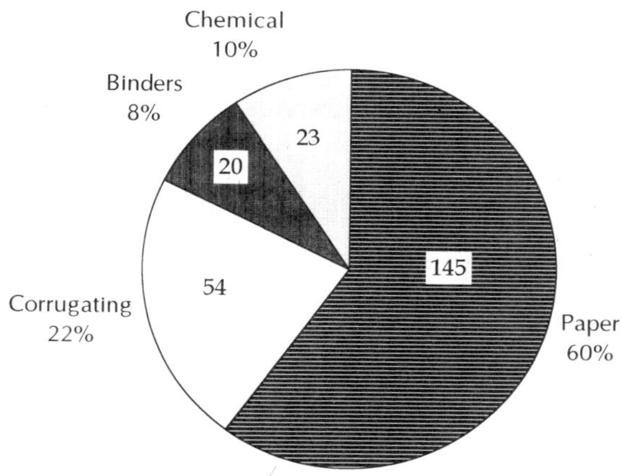

Figure 3. UK industrial markets for starch, 1994 ('000 tonnes).

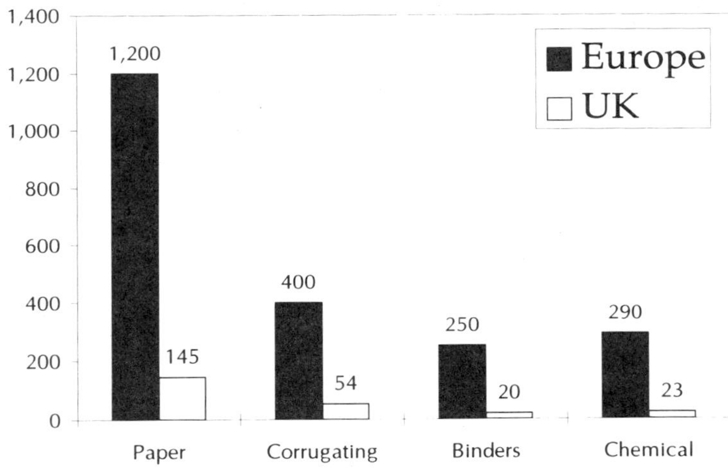

Figure 4. Comparison of UK and European markets for starch, 1994 ('000 tonnes).

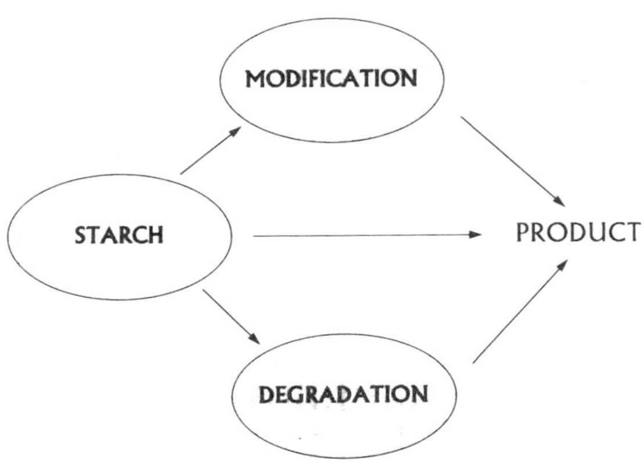

Figure 5. Alternative routes for starch utilization.

Table 1. Applications of starch in the binders and chemical industries

Binders	Chemical and miscellaneous
Pellet binding	Oil well drilling muds
Tableting	Textile sizing
Coal briquetting	Fermentations
Plasterboard	Polymers
Foundry core binding	Plastics
Ceiling tiles	

smaller molecules, reducing sugars or, after hydrogenation, sugar alcohols. These may themselves be used for non food applications.

Several processes may be carried out on starch to give new products. The first and most common is fermentation (see Chapters 8, 9, 25, 26 and 27). Starch hydrolysates are ideal substrates for fermentation, being biologically derived, while the use of cereal based starch for brewing goes back to the dawn of time.

Starch hydrolysates are not as cheap as molasses, which is still the most commonly used fermentable raw material. It constitutes about 70% of the volume used worldwide, several hundred thousand tonnes per annum. However as the cost of downstream purification of pharmaceuticals and fine chemicals increases and the demand on waste treatment grows, then the use of the purer starch hydrolysates as starting materials becomes more attractive. In 1995 some 73,500 tonnes of starch hydrolysates were used in fermentations in the UK; Table 2 shows the range of products. Oxidation of starch hydrolysates to gluconic acid and glucono delta lactone is also a fermentation process and accounts for some 20,000 tonnes in Europe.

Reduction of starch hydrolysates, usually by catalytic hydrogenation over Raney nickel catalyst gives a series of polyol products; the most common is sorbitol, while others include mannitol and maltitol. Again some of the production is used in the food industry, for example in sugar free confectionery. However significant volumes are used in the non food industry, for example as a humectant in toothpaste or as a starting material in the synthesis of vitamin C. The main sectors, with UK volumes for 1995, are shown in Table 3.

Table 2. Volumes of starch used in fermentation processes in the UK (1994 data)

Product	Starch utilised (tonnes)
Biodegradable plastic	1000
Mycoprotein	5000
Yeast	500
Xanthan gum	5000
Sodium benzoate	1000
Citric acid	12000
Clavulinic acid	4000
Antibiotics	35000
Total	**73500**

Table 3. Volumes of starch used in non-fermentation processes in the UK (1994 data)

Product	Starch utilised (tonnes)
Polyols	11500
- Surfactants	(2500)
- Toothpaste	(6000)
- Pharmaceuticals	(3500)
Other pharmaceuticals	5000
Vitamins	20000
Chemicals	23000
Total	**59500**

3. FUTURE PROSPECTS FOR STARCH

The current situation, however, that faces the starch industry in this industrial sector today is a static one with low, even negative growth, though significant quantities are still being used. What are the prospects for the future?

The early seventies saw a quadrupling of the oil price as a result of the Arab-Israeli war; again recently the Gulf War saw oil supplies threatened. Oil supply has survived, and current estimates suggest that there are adequate supplies of fossil fuel to last well into the next millennium. When much of the research was done in industry in the seventies on the alternative chemistry derived from starch, as opposed to mineral hydrocarbons, the cost of crude oil had to exceed \$30 per barrel to be economically viable. It is still not above \$20 today. This is in broad agreement with the observations that the price of oil has to double before *existing technology* becomes viable for a significant move away from fossil fuels towards starch derived processes.

In the light of the above, the need for a renewable resource is not proven on economic grounds, but what of the environmental considerations? Here, sadly, people show little sign of moving to "green" products in large quantities unless both the quality and price match the existing product. For the reasons given above this is seldom the case, and closing the economic and quality gap is the great challenge to science and technology today.

The main hope that people will move significantly towards the use of biologically derived renewable resources lies in the use of legislation or subsidy, as discussed in Chapter 1. People will need either a carrot or a stick to make the move. Starch has already made some progress in this direction, in that since 1986 starch used in non-food applications has been available at competitive prices based on the difference between the EU and the world price.

Examples of the legislative route could include the compulsory inclusion of a proportion of ethanol to replace lead in petrol; the compulsory use of biodiesel in some city centres; the recent German moves on packaging; or the EU regulation that 90% of surfactants must be biodegradable.

The best example of such an approach is the bioethanol story. France, Italy, Brazil and the US have all tried this, and .heir experience has pointed to the obvious technical feasibility of producing and using bioethanol as a liquid fuel. However, in all cases the programmes have relied heavily on government subsidy and legislation for their establishment and maintenance. Chapter 23 presents the current outlook for bioethanol in Europe, concluding that it is likely to remain unviable compared with fossil fuels.

In the face of this less than optimistic picture, where are the major hopes for the starch derived chemicals in the next generation? The following are potential future applications:

- Detergent builders
- Detergent bleaching boosters
- Additives in plastic forming
- Polymer blends
- Thermoplastic starch
- Starch extrusion for insulation
- Starch films
- Graft co-polymers - super absorbents
- Fully biodegradable polymers

Several of the fifteen or so chemical constituents of detergents could be derived from starch (see Chapter 10). Their main advantages are that they are biodegradable and safer in terms of human health. One estimate suggests that 800,000 tonnes of starch could be used in the detergent industry by the turn of the century.

Another potential major area is plastics where biodegradability has obvious attractions (see Chapters 4, 5, 6 and 10). There are two basic approaches here. The first is to integrate starch into existing plastics during formation (see Chapter 4 for an example incorporating starch into polyurethane foam as an extender). The theory here is that in landfill the starch will readily biodegrade, leaving the plastic subdivided and more prone to oxidative processes. The second approach is to exploit the thermoplastic potential of starch by, for example, extruding it with plasticisers to give plastic containers made of as much as 95% starch (see Chapter 6). If starch is extruded without a plasticiser it forms aerated products which can be used for insulation.

A new approach that is developing is the production of monomers by the fermentation of starch (see Chapters 8, 9 and 26). These monomers can then be polymerised into products which are fully biodegradable. An example of this is polyhydroxybutyrate. Production of this biopolymer was 1000 tonne per annum in 1990 and is growing. Its cost was £17.50/kg, so its use remains restricted to special high value areas for the moment.

Starch graft copolymers have been made which have water holding capacities of 1000 times their own weight, hence may be used in incontinence pads, diapers and sanitary products.

4. CONCLUSIONS

Starch is already used widely in non food areas, and its use will continue to grow, particularly as the developments in biotechnology open up the potential for producing specific products by low cost fermentation routes. Its growth will be steady rather than spectacular in the area currently dominated by the petrochemical industry where economics are heavily against it. The challenge is to find more cost effective routes in these areas rather than to rely on wars, subsidies or laws.

REFERENCES

The constraints of length on this paper has necessarily meant that it is only a very abbreviated coverage of a vast subject. I have not attributed or referred to sources in the paper; most of the data comes from the following publications, which I commend for further study.

Carruthers SP and Vaughan CMA (1994) "Sugar and starch as industrial feedstocks." CAS Report 15, Crops for industry and energy. Edited by Carruthers SP, Miller FA and Vaughan CMA. University of Reading

Koch H and Roper H (1988) "New Industrial Products from Starch." Starch/staerke 40,121–131

Leygue JP (1993) "Cereals as Industrial Feedstock." Aspects of Applied Biology, 36

Roper H (1993) "Industrial Products from starch, New Crops for Temperate Regions." edited by Anthony KRM, Meadley J and Robbelen G. Published by Chapman and Hall, London.

Woelk HU (1990) "Carbohydrate feedstocks in Europe-a world perspective." In "Towards a Carbohydrate based Economy" Edited by Ellwood DC, Sageant K, Van Bekkum H and Woelk HU, EUR 12757 EN. Luxembourg: Commission of the European Communities

Descotes G (Ed) (1992) "Carbohydrates as Organic Raw Materials II." VCH New York

3

INDUSTRIAL MARKETS FOR UK-GROWN CEREAL STARCH

S.E. Batchelor,[1] G. Entwistle,[1] K.C. Walker,[1] E.J. Booth,[1] I.M. Morrison,[2] G.R. Mackay,[2] A. Hacking,[3] and T. ap Rees[4]

[1]Scottish Agricultural College
Aberdeen, United Kingdom
[2]Scottish Crop Research Institute
Dundee, United Kingdom
[3]Dextra Laboratories
Reading, United Kingdom
[4]Plant Science Department
University of Cambridge, United Kingdom

1. INTRODUCTION

Starch is an important ingredient in a wide range of foods. It is used as a thickener, to adjust texture, to improve appearance or to act as a filler. The starch industry also supplies a diverse range of non-food markets with starch and starch derivatives. These markets account for approximately 37% of the output of the European starch industry and 24% of the total UK starch supply, but starch crops do not represent a significant proportion of industrial cropping in the UK.

A LINK project was commissioned under the Crops for Industrial Use programme, and funded by the BBSRC, EPSRC, SOAEFD, HGCA and PMB. The aim of the study was to identify and quantify current non-food applications of starch, to assess the potential for growth of established and developing industrial starch-using sectors, and to determine the opportunities for UK agriculture and the UK starch industry. This paper focuses on the paper and surfactants industries which were highlighted by the study as offering the best opportunities for increased industrial utilisation of UK-grown starch.

2. METHODS

Industrial starch-using sectors were identified by reviewing relevant literature and interviewing starch processors. The quantities of starch used in different sectors was obtained from the EU intervention board, and more detailed data on use within particular

Cereals: Novel Uses and Processes, edited by Campbell *et al.*
Plenum Press, New York, 1997

sectors was obtained directly from starch users. Information on current research and development work was obtained from academic researchers, primary starch processors and starch-using manufacturers. Conclusions on the potential for growth of various sectors were drawn from information obtained from starch manufacturers, industrial manufacturers using starch in the production of their products, and end users of products containing or derived from starch. This information, together with economic assessments of starch production carried out on the basis of information obtained from production engineers, was used to determine opportunities for UK industry.

3. RESULTS

3.1. Starch-Using Sectors

Industrial markets for starch identified and studied in detail were paper and board, detergents, flocculation products, textiles, plastics, adhesives, cosmetics and toiletries, pharmaceutical, mineral oil drilling, and agrochemical industries.

3.2. Raw Materials for the Starch Industry and Processing Margins

The major sources of starch world-wide are maize, potato, and wheat. Although forage maize has become popular in recent years for whole crop silage production in the UK, maize is not widely grown for grain production as existing varieties are unsuitable for UK growing conditions. In contrast, potatoes grow well in the UK and high yields are regularly obtained. However, quota restrictions on the allocation of EU support payments currently prevent any development of a UK potato starch industry. In the year 1993/94, 216 221 tonnes of starch was used for non-food markets within the UK. Of this, 30% was imported potato starch, 57% was maize starch processed in the UK from imported maize, and only 13% was wheat starch most of which is both grown and processed in the UK. Figures 1 and 2 shows how industrial starch is utilised within the UK and the EU. The proportions of wheat, maize and potato starch currently used for industrial markets within Europe as a whole match those in the UK, but the pattern of end-use is different.

In general, industrial starch use within the UK is dominated by the paper and board industry (Figure 1). Wheat starch is used in a relatively small number of sectors within the UK: the paper and cardboard, organic chemicals and industrial chemicals industries. Wheat starch represents 17.9% of the starch used in the production of organic chemicals where the main competitor is maize starch (only 5% of the starch used in this sector is potato starch), 12.2 % of the starch used in the production of industrial chemicals, where again the main competitor is maize starch (only 3% of the starch used in this sector is potato starch) and 16.6% of the starch used in the paper and board industry a sector in which potato and maize starch are both strong competitors.

In Europe as a whole, wheat starch is used in a wider range of sectors (Figure 2). In addition to use in the production of organic chemicals, industrial chemicals and paper and board, wheat starch is used for the production of pharmaceuticals, organic surfactants, starch ethers and esters, glues, enzymes, albuminoid substances, plastic products and cotton fabrics, although quantities of wheat starch used in some of these sectors is relatively small.

Typical 1994/95 processing margins achieved by the wheat, potato and maize starch industries were calculated as (per tonne of starch): wheat, £56; maize, £53; potato £18.

3.3. Opportunities Arising from Established Starch-Using Sectors

Industrial use of starch in the UK is dominated by the paper and board industry, and it seems likely that starch use by this sector will increase, as demand for paper is forecast to increase. Starch is used in paper making to improve the strength of paper and as a component of coating formulations. In corrugated board manufacture starch is used as an adhesive, bonding the layers of the board together. Wheat starch accounted for 17 % of the starch used by this industry in the year 1993/94. Nearly all of the wheat starch accounted for in a survey of UK paper manufacturers carried out by the Scottish Agricultural College as part of this study was purchased as native starch, and modified on site by the paper manufacturer for use in surface sizing.

Although potato starch has traditionally been favoured for paper manufacture, in the UK it accounted for just over a third of the starch used by this industry in the year 1993/94, and less than a third in Europe as a whole. Developments in secondary modification have reduced quality differences and the lower cost of maize gives it a competitive advantage. This situation was compounded by the high potato starch prices in 1995 which were due to the poor harvest of 1994. This resulted in paper manufacturers looking for more secure sources of starch. The main opportunity for UK agriculture, taking into account current policy restrictions and patterns of use, may therefore lie in the exploitation of the emerging trend away from the use of potato starch by encouraging increased use of wheat starch in paper manufacturing. This may be aided by the fact that wheat currently has a 5% advantage over maize in terms of processing margins, but there are still concerns over the quality of wheat starch for paper making. Investment in R&D will therefore be required to exploit this opportunity. It is interesting to note that in Europe as a whole, wheat starch accounts for a greater proportion of the starch used in the paper and board industry than in the UK. In Europe 23% of the starch used in this industry is wheat starch, as opposed to 17% in the UK.

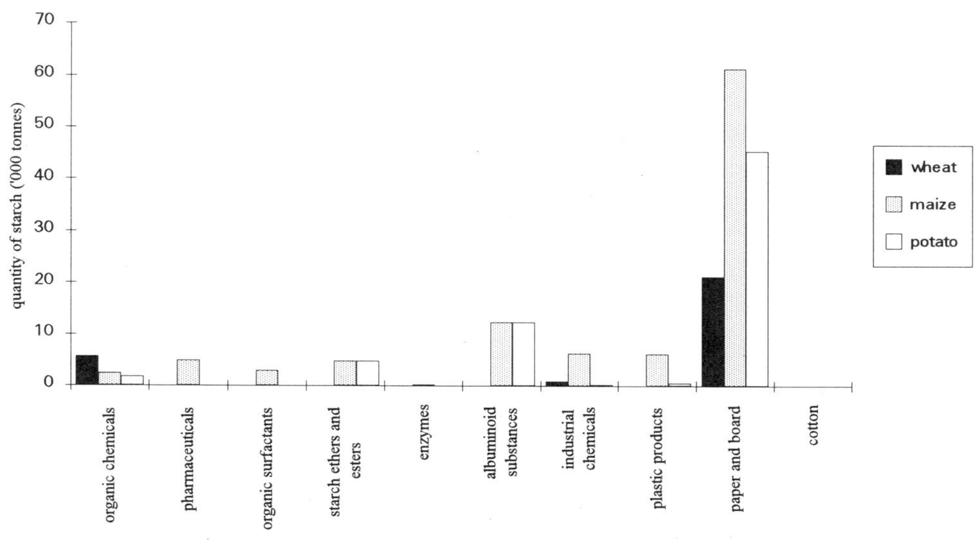

Figure 1. Industrial use of starch in the UK (1993/94).

3.4. Opportunities Arising from Developing Sectors

Of the starch-using sectors studied, those with the greatest opportunity for development appeared to be those based on starch derivatives, rather than markets in which the structure of starch is utilised. The production of surfactants for use in the detergents industry may offer one of the best opportunities. Detergents are complex mixtures which contain, on average, about 15 different compounds. Surfactants are the primary cleansing agents within detergents. Surfactants are low molecular weight amphiphilic molecules consisting of a hydrophilic head group and a hydrophobic hydrocarbon tail. The trend towards natural products in the surfactants industry has two aspects: the use of oleochemical feedstocks for the hydrophobic group and the use of plant-derived carbohydrates to provide the hydrophilic end. Interest in starch-derived products in the detergents industry has arisen from an increasing consumer concern over environmental issues, resulting in a trend towards more "natural" products.

Within the UK in the year 1993/94, organic surfactants accounted for 2962 tonnes of starch, none of which was wheat starch. Within the EU as a whole, however, wheat starch accounted for 11% of the starch used in the production of organic surfactants, indicating that UK grown wheat starch could be used in this sector. No potato starch is used in this sector in the UK and the quantity used in Europe as a whole is negligible (6 tonnes or 0.08%). This is because starch is broken down into its constituent sugar units for the production of surfactants and the high quality of potato starch is of no advantage. The selection of starch source in this sector is very much price driven and consequently development of this sector may open up opportunities for UK-produced wheat starch, assuming it maintains its current price advantage over maize.

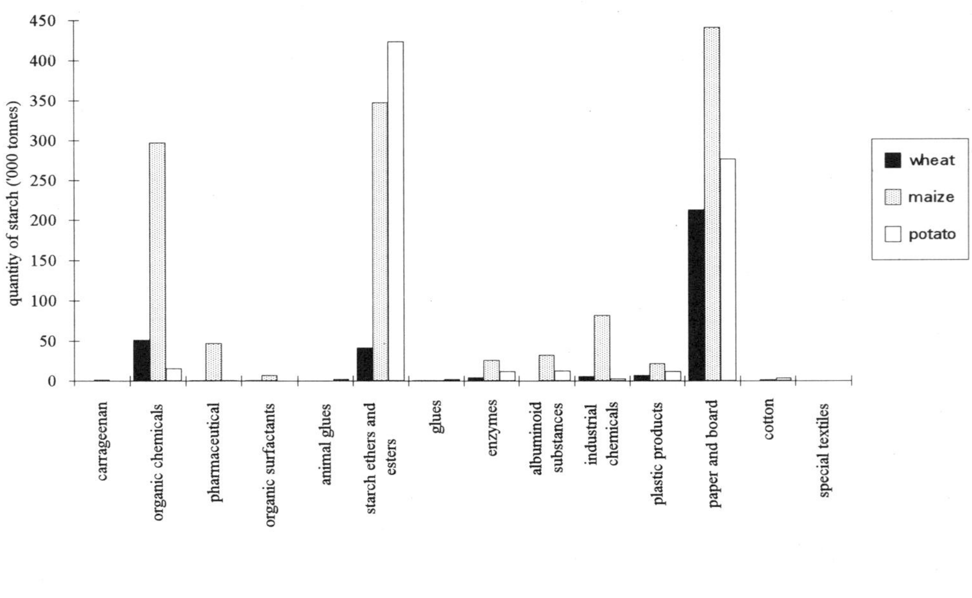

Figure 2. Industrial use of starch in the EU (1993/94).

3.5. Other Opportunities for UK Cereals

Although oat starch is not widely processed for industrial use (it is important within the EU only in Sweden and Finland), it has been suggested that its small starch granules can be technically exploited. Because of their very low granule size (3–10μm), which favours coating applications in paper manufacture, oat starch granules could be particularly suitable for the production of graphics papers, as an improved printability with a less glossy surface could be achieved. Another use for oat starch has been developed recently by a Canadian company, Canamino: when the starch is surface treated it flows and feels like talcum powder.

4. CONCLUSIONS

Increased use of wheat starch appears to offer the best opportunity for the development of a starch industry based on UK-grown starch from the point of view of agronomic suitability, support policy, future market demand and processing margins. Varieties of maize grown for starch production are not suited to UK conditions, and EU support policy currently prevents any development of a UK potato starch industry. However, analysis of the markets for starch indicate that the most promising markets are those based on the use of cereal starch, and wheat starch currently has an advantage in terms of processing margins as compared to maize and potato starch.

4

FLEXIBLE POLYURETHANE FOAM EXTENDED WITH CORN STARCH

Ying-chun Lin, Harold E. Huff, and Fu-hung Hsieh

Department of Biological and Agricultural Engineering
University of Missouri
Columbia, Missouri 65211

1. INTRODUCTION

The use of polyurethane foam is continuing to grow at a rapid pace throughout the world. This growth can be attributed to their light weight, excellent strength/weight ratio, energy absorbing performance (including shock, vibration, and sound), and comfort features of the polyurethane foams (Klempner and Frisch, 1991). Recently, there has been an increased interest in the use of renewable resources in the plastics industry (Bhatnagar et al, 1993; Carraher and Sperling, 1981; Cunningham and Carr, 1990; Cunningham et al, 1991, 1992a, and 1992b; Donnelly et al, 1991; Yoshida et al, 1987 and 1990). In addition, many patents covering processes for utilizing the plant components in the preparation of polyurethane foam have been issued in recent years (Dosmann and Steel, 1961; Hostettler, 1979; Kennedy, 1985; Otey et al, 1968). However, most of these studies focused on rigid polyurethane foam. Less attention has been paid to the flexible polyurethane foam system. Corn starch is a renewable raw material. As a carbohydrate, it contains many active hydrogens and hydroxyl groups. Thus, a great opportunity exists for using corn starch to modify or improve the physical and chemical properties of flexible polyurethane foams.

A blowing agent is usually required for polyurethane foam formation. There are three types of blowing agent: 1) water that reacts with isocyanate and produces carbon dioxide; 2) low boiling liquid chemicals that can be evaporated due to the exothermic reaction of the polyols and isocyanate; and 3) air that blown in or whipped into the polyols and isocyanate mixture. The first reaction which uses water as a blowing agent is preferred for the manufacture of flexible polyurethane foams (Dieterich et al, 1993).

The objectives of this study were to develop flexible polyurethane foams extended with corn starch using water as a blowing agent, to characterize their physical and mechanical properties, and to investigate the effects of biomass concentration on the foam properties.

Table 1. Foam formulations for flexible polyurethane foams

Ingredients	Parts by weight
Component A	
glycerol-propylene oxide polyether triol	100.0
tertiary amine	0.1
corn starch	0, 10, 20, 30, 40
triethanolamine	0.7
surfactant (L-560)	1.0
blowing agent (distilled water)	4.5
Component B	
toluene diisocyanate	(105)*
dibutyltin dilaurate	0.1
stannous octoate	0.3

*The quantity of isocyanate is required to meet an isocyanate index 105, defined as the actual amount of isocyanate used over the theoretical amount of isocyanate required, multiplied by 100.

2. MATERIALS AND METHODS

2.1. Materials

The ingredients used in the preparation of flexible foams were unmodified common corn starch (PF Powdered Starch, American Maize Products, Hammond, IN). Other components used in the flexible polyurethane foams were toluene diisocyanate (OLIN TDI 80, Olin Corp., Stamford, CT), glycerol-propylene oxide polyether triol (ARCOL LHT-42, Arco Chemical Co., Newtown, PA), tertiary amine (DABCO, Aldrich Chemical Co., Milwaukee, WI), triethanolamine, and dibutyltin dilaurate (Aldrich Chemical Co., Milwaukee, WI), stannous octoate (Sigma Chemical Co., St. Louis, MO), surfactant (L-560, Union Carbide Corp., Danbury, CT), and distilled water. The distilled water was used as a blowing agent.

2.2. Experimental Design and Formulations

The effects of corn starch (0, 10, 20, 30 and 40 parts per hundred weight of polyol) in the foam formulation on the properties of water-blown flexible polyurethane foams were studied. Other factors in the foam formulation such as catalyst, surfactant, cross-linking agent, and isocyanate index were fixed. They were determined in a preliminary study to assure that all foam products could be produced in a normal amount of time (10 minutes). The foam formulation for water-blown flexible polyurethane foam is shown in Table 1. The amount of isocyanate added in each formulation was based on the total hydroxyl content of polyether polyol, triethanolamine and water, including water originally present in the corn starch. The amount of water was varied to maintain the same isocyanate index in each formulation (Table 2). Two replicate foams were produced with each foam formulation.

Table 2. Toluene diisocyanate and water added to foam formulation at different levels of corn starch addition

Parts of corn starch per 100 parts of polyol	Added water(g)	Toluene diisocyanate(g)
0	4.5	54
10	3.5	54
20	2.5	54
30	1.5	54
40	0.5	54

2.3. Foam Preparation

Foams were prepared by adding a mixture of toluene diisocyanate, dibutyltin dilaurate and stannous octoate (component B) to a premix of glycerol-propylene oxide polyether triol, tertiary amine, corn starch, triethanolamine, and distilled water (component A). A standard mixing procedure for making foams was used in this study (Bailey and Critchfield, 1981). This procedure involved intensive mixing using a commercial drill press (Colcord-Wright, St. Louis, MO) fitted with a 25.4 cm shaft with a 5 cm impeller arranged to turn at 1845 and 3450 rpm. Component A was sequentially weighed and placed into a disposable paperboard container (0.95 litres) fitted with a steel frame with four baffles, and mixed at 3450 rpm for 30 seconds. The stirring was then stopped, allowing the mix to degas. After 15 seconds, component B was rapidly added and stirring was continued for another 10 seconds at the same speed. The reacting mixtures were then poured immediately into wooden boxes with a dimension 200 x 200 x 100 mm and allowed to rise at ambient conditions. Foams were removed from boxes after 3 hours and allowed to cure at room temperature (23°C) for one week before cutting into test specimens.

2.4. Foam Property Measurements

Foam density, defined as mass per unit volume, was tested according to ASTM D 3574 (Section 9–15). The test specimens (100 x 100 x 50 mm) were calipered and weighed to determine the density in kilograms per cubic metre. Four specimens were tested and the average value was reported. The indentation force deflection value was determined according to ASTM D 3574 (Section 16–22) by the Instron Universal Testing Machine, Model 1132 (Instron Corporation, Canton, MA), fitted with a data acquisition system. The indentation force deflection values at 25, 50, and 65% were calculated by dividing the forces at 25, 50, and 65% deflections, respectively, by the indented area.

The comfort or support factor is defined as the ratio of the 65% indentation force deflection to the 25% indentation force deflection. According to ASTM D 3574, seating foams with low support factors will usually bottom out and give inferior performance. The resilience test is also referred to as the "ball rebound test." For flexible polyurethane foams, the resilience is defined as the rebound height of the ball over the drop height of the ball multiplied by 100. A higher percentage corresponds to a foam having better resilience. The instruments and the methods used conform to the ASTM D 3574 (Section 68–75).

The compression set test under constant deflection was conducted according to ASTM D 3574 (Section 37–44). This instrument consists of two flat plates arranged so that the plates are held parallel to each other and the space between the plates is adjustable to the required deflection thickness by means of calipers. The initial thickness (about 50 mm) of a specimen sample (100 x 100 x 50 mm) was measured. The sample was compressed by 50% of its original thickness between plates and held for 22 hours in an oven at conditions of 70 ± 2°C and 5 ± 1% relative humidity. Thickness was measured 30 min after removal of the plates. The compression set value was calculated as follows:

$$C = \frac{(T_o - T_f)}{T_o} \times 100\%$$

where C=compression set expressed as a percentage of the original thickness, T_o=original thickness of test specimen, and T_f=final thickness of test specimen.

Three samples were tested and the median was reported.

2.5. Data Analyses

A Least Significant Difference rule was applied to compare the means of the foam properties of different treatments and different types of biomass (soybean fibre, isolated soybean protein, and corn starch).

3. RESULTS AND DISCUSSION

3.1. Density

Table 3 shows that the density of corn starch-extended flexible foam rises with increasing weight percentage of biomass added to the foam formulation. This may be explained in terms of formulation and structure difference among these foams. The density of a plastic foam is determined by the density or specific gravity of the material making up the matrix of the foam, the density of the gas in the cells, and the percentage of the material made up of foam network. The plastic phase composition includes polyol, isocyanate and all additives such as surface active agents, stabilizers, cross-linking agents and biomass extenders. The gas phase composition includes gases, either generated by the physical blowing agents which lib-

Table 3. Physical properties of water-blown flexible polyurethane foam extended with corn starch

	Added corn starch, %				
Foam properties	0	10	20	30	40
Density, kg/m^3	27^a	29^b	31^c	33^d	37^e
Resilience, %	22^a	26^b	31^e	27^c	29^d
Indentation force deflection values, kPa					
25% deflection	6.7	7.0	8.7	6.5	9.9
50% deflection	8.9	9.9	12.4	9.1	15.4
65% deflection	13.5	15.9	20.0	15.5	25.8
Comfort factor	2.0^a	2.3^{ab}	2.3^{ab}	2.4^{ab}	2.6^b
Compression set, %	46^d	44^c	42^a	43^b	43^b

Means with the same letter in the same row are not significantly different at 5% level.

erate gases as a result of elevated temperatures (e.g. thermal decomposition sodium bicarbonate) or produced by chemical blowing agents which release gases through chemical reactions (e.g. the chemical reaction between isocyanate and water), and the air which is either introduced into the reaction vessel during the foaming process or diffuses into the cells during the aging process. In this study, with the exception of the percentage of corn starch, each foam formulation has the same amount of water (blowing agent) and other components. As expected, the density increases as the amount of extender increases.

3.2. Resilience

Foams containing corn starch had higher resilience values when compared to that of the control foam (Table 3). The maximum resilience occurred at 20% corn starch addition. This property is particularly important in determining the degree of comfort in a cushion material. Comfort, however, is a subjective property that can vary from one person to another. Hartings and Hagan (1978) demonstrated that the resilience value obtained from the ball-rebound test was correlatable to sitting comfort rated by a panel of judges. As the resilience increases, the comfort rating of the cushion foam also increases. Thus, the incorporation of corn starch into water-blown flexible foam system appears to increase the comfort value of the foam, a desirable trait in cushioning applications.

3.3. Indentation Force Deflection

The major market for flexible polyurethane foam is as a cushioning material in furniture, bedding and automotive seating applications. The load-bearing properties of a flexible foam can be determined by studying the manner in which the structure deflects under a known applied load (Woods, 1982). Figure 1 shows the behavior of the load-deformation, stress-strain relationship under indentation for polyurethane foams extended with corn starch. Foams containing less than 20% corn starch exhibit a plateau stress region. The stress-strain shape for the foam extended with 40% corn starch does not show any sig-

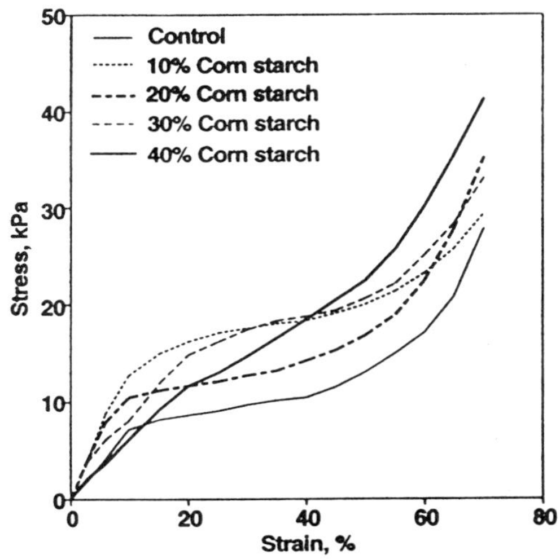

Figure 1. Stress-strain curves for polyurethane foams with or without corn starch.

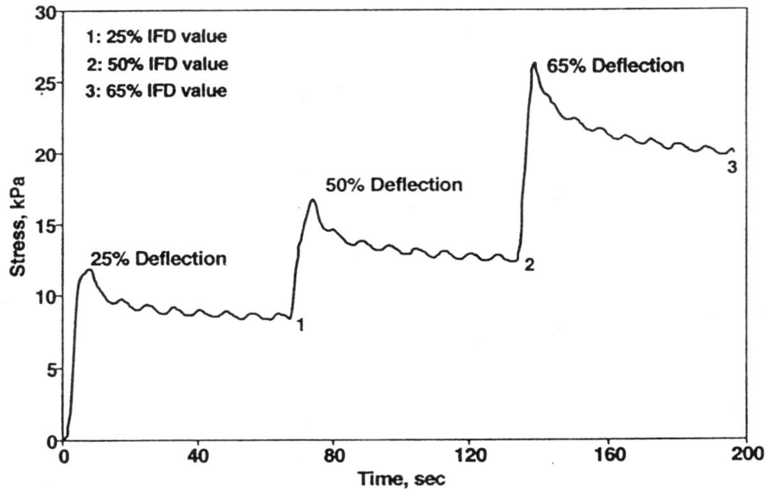

Figure 2. Load-deflection curve for polyurethane foam (with 20% corn starch) in indentation force deflection test.

nificant plateau region and has the highest indentation hardness. Wolfe (1982) suggested that when the stress-strain curve of a foam contains a considerable plateau stress region, it will have a low comfort value. Therefore, the addition of 40% corn starch into the flexible foam system appears to increase the foam comfort value most, based on the stress-strain curves shown in Figure 1.

Another indicator of comfort of the cushion foam is the comfort factor. Figure 2 shows a typical stress-strain curve under indentation test and displays the 25, 50, and 65% indentation force deflection values. The results are shown in Table 3. Foams containing corn starch display a greater comfort factor than the control foam. Only foam containing 40% corn starch exhibits a significant improvement in the comfort factor, however.

3.4. Compression Set Value

Compression set value is a measure of the non-recoverable loss in the thickness of a flexible foam after a static load is removed. This property is important for material-handling applications, such as an interplant container, or where this foam is designed for multiple uses. Table 3 shows the compression set results for polyurethane foams extended with corn starch. All extended foams have smaller compression set values than the control foam. This means that incorporating corn starch into the flexible foam appears to improve the compression set value. The minimum compression set value occurs at 20% corn starch addition. It should be noted, however, that the compression set results obtained in this study are under an accelerated test environment and may not correlate closely with the real end-use situations.

4. CONCLUSIONS

All foams extended with corn starch exhibited significantly higher values in density and resilience than the control foam. An increase in corn starch percentage increased the foam density. The comfort factor increased with increasing the percentage of corn starch in the foam formulation. Foams containing 40% corn starch had a profoundly greater comfort factor than the control foam. Lower compression set values were also observed for foams containing 10–40% corn starch than the control foam.

REFERENCES

Bailey FE and Critchfield FE (1981) "Chemical reaction sequence in the formation of water blown urethane foam." Journal of Cellular Plastics 17, 333–339

Bhatnagar S, Hilton RR and Hanna MA (1993) "Physical mechanical and thermal properties of starch based plastic foams." Paper No 936532 ASAE International Winter Meeting Chicago IL Dec 14–17

Carraher Jr CE and Sperling LH (1981) "Polymer Applications of Renewable Resource Materials", Plenum Press New York

Cunningham RL and Carr ME (1990) "Cornstarch and corn flour as fillers for rigid urethane foams." In "Corn Utilization Conference III Proceedings" National Corn Growers Association and Ciba-Geigy Seed Division, St Louis, MO, pp 1–16

Cunningham RL, Carr ME and Bagley EB (1991) "Polyurethane foams extended with corn flour." Cereal Chemistry 68, 258–261

Cunningham RL, Carr ME and Bagley EB (1992a) "Preparation and properties of rigid polyurethane foams containing modified corn starches." Journal of Applied Polymer Science 44, 1477–1483

Cunningham RL, Carr ME, Bagley EB and Nelsen TC (1992b) "Modifications of urethane-foam formulations using Zea mays carbohydrates." Starch/Starke 44, 141–145

Dieterich D, Grigat E, Hahn W, Hespe H and Schmelzer HG (1993) "Principles of Polyurethane Chemistry and Special Applications." In "Polyurethane Handbook" Ed G Oertel Hanser Publishers, Munich

Donnelly MJ, Stanford JL and Still RH (1991) "The conversion of polysaccharides into polyurethanes: A review." Carbohydrate Polymers 14, 221–240

Dosmann LP and Steel RN (1961) "Flexible shock-absorbing polyurethane foam containing starch and method of preparing same." US Patent 3004934, October 7

Hartings JW and Hagan JH (1978) "Fatigue investigation of urethane seat pads." Journal of Cellular Plastics 14, 81–86, 105

Hostettler F (1979) "Polyurethane foams containing stabilized amylaceous materials." US Patent 4156759, May 29

Kennedy RB (1985) "Pectin and related carbohydrates for the preparation of polyurethane foams." US Patent 4520139, May 28

Klempner D and Frisch KC (1991) "Handbook of Polymeric Foams and Foam Technology." Oxford University Press, New York

Otey FH, Bennett L and Mehltretter CL (1968) "Process for preparing polyether-polyurethane-starch resins." US Patent 3405080, October 8

Wolfe HW (1982) "Cushioning and Fatigue." In "Mechanics of Cellular Plastics" Hilyard NC ed., Applied Science Publishers, Ripple Road, Barking, Essex, England

Woods G (1982) "Flexible Polyurethane Foams: Chemistry and Technology." Applied Science Publishers, London

Woods G (1990) "The ICI Polyurethanes Book", 2nd ed. John Wiley & Sons, New York

Yoshida H, Morck R, Kringstad KP and Hatakeyama H (1987) "Kraft lignin in polyurethanes I. Mechanical properties of polyurethanes from a Kraft lignin-polyether triol-polymeric MDI system." Journal of Applied Polymer Science 34, 1187–1198

Yoshida H, Morck R, Kringstad KP and Hatakeyama H (1990) "Kraft lignin in polyurethanes II. Effects of the molecular weight of Kraft lignin on the properties of polyurethanes from a Kraft lignin-polyether triol-polymeric MDI system." Journal of Applied Polymer Science 40, 1819–1832

5

BIODEGRADABLE COMPOSITES OF STARCH AND POLY(HYDROXYBUTYRATE-CO-VALERATE) COPOLYMERS

J. L. Willett[1] and G. S. O'Brien[2]

[1]National Center for Agricultural Utilization Research
USDA-ARS, Peoria, Illinois
[2]Zeneca Biopolymers
Wilmington, Delaware

1. INTRODUCTION

The use of starch in biodegradable plastics applications has received considerable attention in recent years. Its low cost makes it an attractive filler for high-cost biodegradable polymers which compete with commodity polymers such as polyethylene and polystyrene in disposable, one-use applications such as cutlery, cups, and food trays. The US Department of Agriculture's Agricultural Research Service has conducted research in starch utilization in plastics for many years. Recently, there has been interest in utilizing starch in composites with poly(hydroxybutyrate-valerate) copolymers (PHBV). PHBV copolymers are produced via fermentation of agricultural feedstocks by microorganisms such as *Alcaligenes eutrophus*. These biodegradable, thermoplastic polyesters have been produced and marketed under the trade name Biopol by Zeneca Bioproducts. Under a Cooperative Research and Development Agreement (CRADA) between Zeneca and the USDA's Agricultural Research Service, composites of PHBV with starch and other environmentally benign materials have been developed with a wide range of properties. This paper discusses the effects of composition variables on the mechanical properties and biodegradation of these materials.

2. MECHANICAL PROPERTIES OF STARCH/PHBV COMPOSITES

Composites of PHBV with starch, inorganic fillers, and other additives can be formulated to provide a wide range of properties, from flexible to rigid (Kotnis et al, 1995). The starch in these materials is in its native granular state, and acts as a rigid filler. Statistical design methods were used to formulate a series of materials to provide predictive equations for the various properties as functions of the composition variables. The goal

Table 1. Formulation Design Table (values are wt%)

Formula #	PHBV	Starch	$CaCO_3$	Plasticizer
1	80	15	0	5
2	65	25	0	10
3	70	15	0	15
4	60	25	10	5
5	65	15	10	10
6	50	25	10	15
7	65	15	15	5
8	50	25	15	10
9	55	15	15	15
10	45	25	15	15
Control	100	0	0	0

was to find a systematic method of optimizing formulations for minimum PHBV content under the constraint of consequent loss in the relevant mechanical properties. This would allow for minimum materials cost while maintaining properties within acceptable limits. Formulations were selected to construct a $3^2 2^1$ experimental design plan; the fractional factorial design table is shown in Table 1.

The PHBV resin was grade D400P (8% HV), with 1% by weight BN added as a nucleating agent. The starch was an unmodified corn starch, which was dried to less than 0.5% moisture content before use. The filler was calcium carbonate, grade Omyacarb FT (Omya Corporation)*. The plasticizer was a citrate ester compatible with PHBV. Components were dry blended, and then compounded in a Brabender 19 mm single screw extruder, using a fluted mixing screw with good dispersive mixing action. Test specimens were injection molded. Tensile properties were measured after conditioning for 28 days at 50% relative humidity and 23°C. The results are given in Table 2. (See Chapter 6 for similar tensile strength measurements for coated starch films, and Chapters 14 and 15 for gluten films)

Table 2. Tensile properties of PHBV/starch/filler/plasticizer composites

Formula #	Tensile strength (MPa)	Elongation (%)	Modulus (GPa)
1	22.0	24.2	1.60
2	14.9	28.1	1.21
3	14.1	26.4	0.88
4	15.0	15.2	1.90
5	14.5	17.0	1.23
6	8.9	15.2	0.98
7	16.9	11.6	1.77
8	10.8	11.4	1.38
9	10.6	13.7	0.91
10	8.1	11.4	0.95
Control	31.8	13.2	2.10

* Names are necessary to report factually on available data; however, the USDA neither guarantees nor warrants the standard of the product, and the use of the name by USDA implies no approval of the product to the exclusion of others that may also be suitable.

The tensile property data were analyzed using stepwise regression to provide predictive equations. The results for the tensile strength σ (MPa), elongation ε (%), and modulus E (GPa) are given as follows:

$$\sigma = 31.8 - 0.4\,\omega_S - 0.4\,\omega_{CaCO_3} - 0.9\,\omega_P + 0.01\,\omega_S\omega_P + 0.01\,\omega_{CaCO_3}\omega_P \tag{1}$$

$$\varepsilon = 13.7 + 0.54\,\omega_S + 1.2\,\omega_P - 0.91\,\omega_{CaCO_3} - 0.05\,\omega_S\omega_P \tag{2}$$

$$E = 2.0 - 0.09\,\omega_P + 0.004\,\omega_S + 0.007\,\omega_{CaCO_3} \tag{3}$$

where ω is the weight per cent of the indicated components (S and P are starch and plasticizer, respectively). Correlation coefficients for these equations are 0.99, 0.95, and 0.98, respectively. Note that Equation 1 predicts a decrease in tensile strength regardless of filler type. These equations adequately predicted the properties of other formulations prepared within this composition range (Kotnis et al, 1995).

The addition of starch and $CaCO_3$ to PHBV lowers the tensile strength, and offers only slight stiffening. This suggests poor adhesion at the PHBV/filler interface. The tensile strength data for the composites with the highest plasticizer level agree well with models which predict loss in strength due to reductions in effective surface area as a result of poor filler-matrix adhesion. Scanning electron micrographs (not shown) show that there is little adhesion between the PHBV matrix and the starch or $CaCO_3$ filler particles. Minimization of property loss through better adhesion has therefore been investigated.

3. PHBV COMPOSITES WITH SURFACE MODIFIED STARCH

One method of improving the adhesion between filler and matrix is by coating the filler particles with a polymer which is somewhat compatible with the matrix. A variety of natural and synthetic polymers were examined as coatings for starch granules by Dr. Randal Shogren of NCAUR, including zein, shellac, cellulose esters, polyvinyl alcohol, and polyethylene oxide (PEO) (Shogren, 1995). PEO is known to be partially compatible with PHBV, so it was expected that a PEO coating on the starch granule would improve adhesion with the PHBV matrix.

Starch granules (unmodified corn starch) were coated by slurrying in a solution of the selected polymer; in the case of PEO, the solvent was water. The granules were then separated and dried, and blended with PHBV and plasticizer. Compounding was performed with the same extrusion apparatus described above. Starch levels in the composites were 30% and 50% by weight. Tensile properties of the PEO coated starch are shown in Table 3.

The presence of the PEO on the granule surface clearly enhances the tensile properties. The effect increases with PEO content up to a level of approximately 9%. High molecular weight PEO provides better adhesion than lower molecular weights; when a PEO with a molecular weight of 100,000 was used, the tensile strength was approximately 30% lower than with the high MW PEO. This result suggests the formation of entanglements between the PEO coating and the PHBV matrix may be important.

Other polymer coatings did not improve the properties to the extent observed with PEO. In some cases, no improvement over untreated controls was observed. This result may be due in part to the formation of agglomerates of coated starch granules during the

Table 3. Tensile properties of PEO-coated starch/PHBV composites (Shogren, 1995)

Starch (wt%)*	Tensile strength (MPa)	Elongation (%)	Modulus (MPa)
30 (0)	15	32	250
30 (9)	19	21	220
50 (0)	10	11	300
50 (2)	10	12	280
50 (5)	15	15	210
50 (9)	18	21	170

*Numbers in parentheses are weight % PEO, based on starch. PEO MW = 4×10^6.

drying process. If the coating polymer does not soften sufficiently during extrusion, the agglomerates would not break up, and thereby reduce the mechanical properties.

Another approach to improve adhesion between filler and matrix is covalent bonding. Starch granules were reacted with glycidyl methacrylate via free radical polymerization to produce starch-GMA graft copolymers. The epoxide groups of the GMA graft provide reaction sites for the endgroups of the PHBV to form covalent bonds; stress transfer across the granule-matrix interface would thereby be improved. A series of starch-GMA graft materials were prepared using ceric ammonium nitrate as an initiator, with GMA levels up to 19% by weight. These grafted materials, in which the starch retained its granular structure, were compounded with PHBV and plasticizer, and injection molded. Tensile and flexural properties are shown in Table 4.

The presence of the graft clearly increases the tensile and flexural strength of the composites, although the effect on modulus is not as strong. SEM micrographs of fracture surfaces (not shown) indicate grafting improves adhesion between the granules and the PHBV matrix. While the improvement in properties is significant, grafting increases the cost of the starch filler.

4. BIODEGRADATION OF STARCH/PHBV COMPOSITES

Composites of PHBV with polysaccharides are known to degrade more rapidly than PHBV alone (Ramsay et al, 1993; Yasin et al, 1989). Ramsay and co-workers showed that the starch in these materials degraded faster than the PHBV (Ramsay et al, 1993), while Yasin and co-workers found that hydrolysis was substantially enhanced by the presence of a variety of polysaccharides (Yasin et al, 1989). The effects of starch treatment and other additives were not examined in these studies. The effects of starch treatments and addi-

Table 4. Properties of grafted starch/PHBV composites

Graft content (%)	Flexural modulus (GPa)	Flexural yield strength (MPa)	Tensile modulus (MPa)	Tensile strength (MPa)
0	1.9	31	465	17.1
7.4	1.9	38.2	484	22.2
13.4	1.9	41.8	539	23.6
19.0	1.8	43.5	372	24.3

tives such as plasticizers need to be clarified, since starch/PHBV composites of commercial interest will contain these types of materials.

Imam and co-workers have examined the biodegradation of PEO-coated starch/PHBV composites in municipal activated sludge (Imam et al, 1995). All of the composites showed significant weight loss over a 35 day exposure, up to 78%. Weight loss was accompanied by deterioration of mechanical properties. Degradation of starch was slowed by the presence of the PEO coating. The PHBV with no starch degraded quite rapidly in the sludge environment, and the addition of starch did not enhance the rate of weight loss. Interestingly, the level of starch (30% or 50% by weight) had little effect on the rate of degradation, whether the starch was coated or not.

More recently, we have investigated the effects of various additives and the levels on degradation of starch/PHBV composites during soil exposure. A series of formulations with different levels of starch, plasticizer, and inorganic filler were prepared using a 2^3 factorial design. Starch levels were 10% or 25%, plasticizer levels were 7.5% or 15%, and filler levels were 0% or 20%. Extruded ribbons and injection molded plaques were buried at a depth of 4 inches. Weight loss and mechanical properties were measured as a function of exposure time.

After six weeks of soil exposure, several of the ribbons were fragmented, so that determination of mechanical properties was not possible. When fragmentation occurred, as many of the fragments as possible were recovered for the weight loss determinations. By 11 weeks, most ribbon samples had little mechanical integrity. All specimens were highly discolored after 3 weeks of exposure; formulations with 25% starch and filler were the

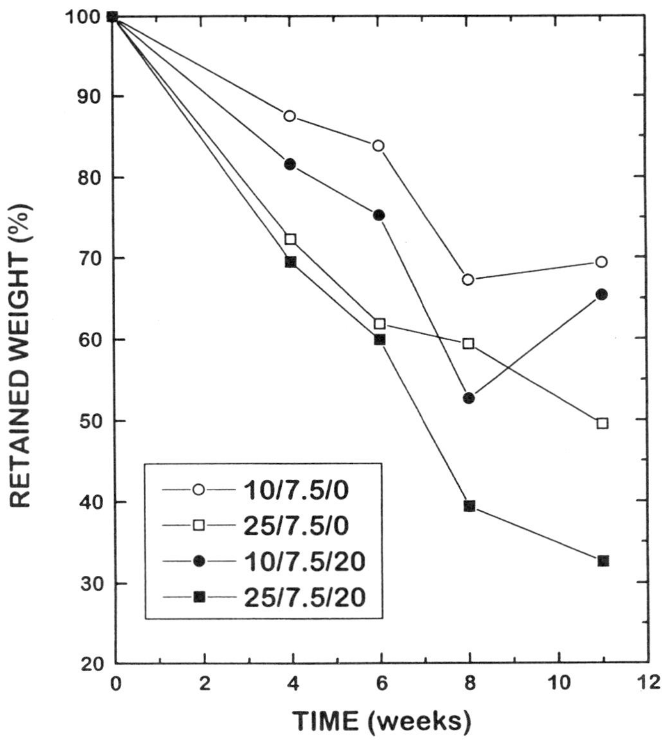

Figure 1. Weight loss of PHBV/starch composite extruded ribbons during soil burial.

most highly discolored. The rate of weight loss was increased by higher starch content; at constant starch content, the inorganic filler substantially increased the rate as well. Weight losses of up to 80% were recorded after 11 weeks of exposure. Representative weight loss data are shown in Figure 1.

Weight loss data for the molded plaques are shown in Figure 2. As seen with the ribbons, the weight loss is more rapid with the higher starch content. The rate of weight loss for the plaques is much slower than the ribbons, which is due to the reduced specific surface area of the thicker plaques. The presence of the filler increases the rate of weight loss at both starch levels. At higher plasticizer levels, the rate of weight loss is slightly reduced.

Scanning electron micrographs of exposed samples show that the starch is rapidly degraded. The voids produced by starch exposure increase the surface area of the plaques, and enhance the accessibility of the PHBV matrix. In addition, the voids act as stress concentrators and further degrade the mechanical properties of the composites. Most of the inorganic filler remains after degradation. Ca analysis data indicate that while the relative Ca content increases during soil exposure, some Ca is lost. This result is based on the fact that the Ca content is less than that calculated assuming only the loss of the organic fractions of the composites. It is not clear at this time whether the Ca loss is due to solubilization or to biological activity.

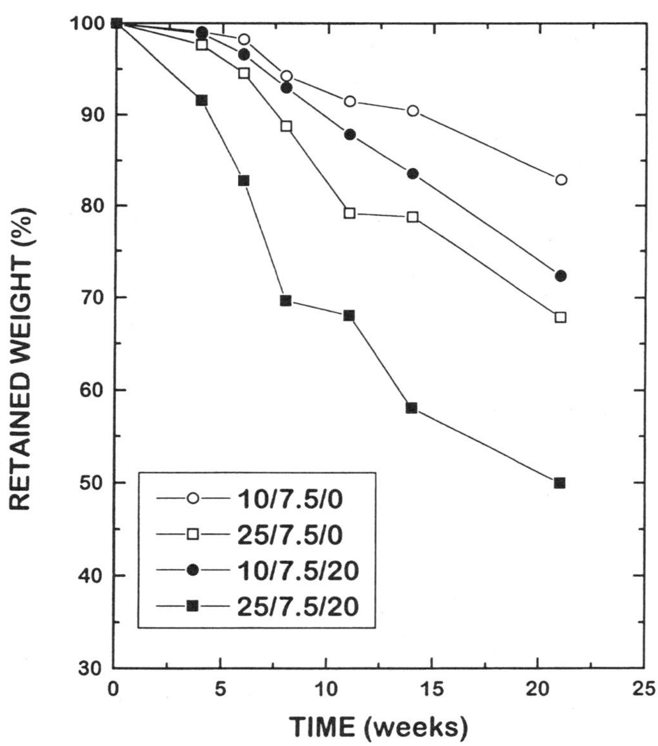

Figure 2. Weight loss of PHBV/starch composite molded plaques during soil burial.

5. CONCLUSION

Methods of incorporating starch and other low-cost fillers into PHBV have been investigated. Using statistical design methods and regression, predictive equations were determined for composites of PHBV with starch, $CaCO_3$ filler, and plasticizer, with correlation coefficients greater than 0.95. Mechanical properties were improved by either coating the starch with PEO or by grafting glycidyl methacrylate onto the starch. Both processes improve the adhesion between the PHBV matrix and the starch granules. Composition effects on biodegradation were studied in activated sludge and soil. PEO-coated starch composites showed a slower rate of weight loss in sludge than either pure PHBV or uncoated starch composites. For samples exposed to soil, degradation was enhanced by increasing starch levels or the presence of inorganic filler.

ACKNOWLEDGMENTS

The authors gratefully acknowledge the excellent efforts of RP Westhoff, RL Haig, GD Grose, and J Fuller in the preparation and testing of the composites used in this study, and A Kelly-Webb for the Ca analysis. Dr GF Fanta prepared the starch-GMA copolymers. This research was performed under CRADA 58–3K95-M-013 between USDA-ARS and Zeneca Biopolymers.

REFERENCES

Imam SH, Gordon SH, Shogren RL and Greene RV (1995) "Biodegradation of Starch-PHBV Composites in Municipal Activated Sludge." J. Environ. Polym. Degrad. 3, 205–213

Kotnis MA, O'Brien GS and Willett JL (1995) "Processing and Mechanical Properties of Biodegradable Poly(hydroxybutyrate-co-valerate)-Starch Compositions." J. Environ. Polym. Degrad. 3, 97–105

Ramsay BA, Langlade V, Carreau PJ and Ramsay JA (1993) "Biodegradability and Mechanical Properties of PHBV/Starch Blends." Appl. Environ. Microbiol. 59, 1242–1246

Shogren RL (1995) "Poly(ethylene oxide)-coated Granular Starch-Poly(hydroxybutyrate valerate) Composite Materials." J. Environ. Polym. Degrad. 3, 75–80

Yasin M, Holland SJ, Jolly AM and Tighe BJ (1989) "Polymers for Biodegradable Medical Devices VI. Hydroxybutyrate -Hydroxyvalerate copolymers: Accelerated degradation of blends with polysaccharides." Biomaterials 10, 400–412

6

BIODEGRADABLE COATINGS FOR THERMOPLASTIC STARCH

John W. Lawton

Plant Polymer Research
National Center for Agricultural Utilization Research
Agricultural Research Service, US Department of Agriculture
1815 North University Street, Peoria, Illinois 61604

1. INTRODUCTION

Over the last few years, there has been renewed interest in biodegradable plastics made from annually renewable, natural polymers such as starch (see Chapters 1, 2, 5 and 10). The fact that starch is receiving considerable attention is understandable, as it is totally biodegradable, is inexpensive compared to other biodegradable polymers, and is available in large quantities. However, starch-based materials and bio-plastics containing starch are only slowly being manufactured and marketed into consumer products, despite the advantages listed above. One reason for this is due to the hygroscopic nature of starch (Whisler and Hillbert, 1944). Starch that comes into contact with water can absorb water, thereby changing the properties of the starch-based material (Swanson et al, 1993). Even starch-based materials that do not come into direct contact with water can be affected by water. Changes in humidity affect the physical properties of starch (Perice, 1928; Lloyd and Kirst, 1963) and starch-based materials (Jasberg et al, 1992). Starch absorbs water under high humidity conditions and loses water under low humidity conditions. Since water is a good plasticizer for starch (Young, 1984; Donovan 1979), any change in the water content of the starch will change the properties of the starch-based article.

One possible way to protect starch from the effects of water is to apply a hydrophobic coating to the starch-based material. This would help in two ways: first, a hydrophobic coating would protect the starch from absorbing water into the starch article; and secondly, such a coating would help in retaining any water added for plasticizing the article. Unfortunately, most hydrophobic coatings do not adhere to starch. The surface of starch needs to be treated with some type of compatibilizing agent before hydrophobic materials will adhere to starch. Otey et al (1974) used toluene diisocyanate as a compatibilizing agent between poly(vinyl chloride) and a starch poly(vinyl alcohol) film. Adhesion between starch (in the granule state) and hydrophobic materials like polyethylene is also a problem in starch/polyethylene composites (Doane et al, 1992).

Cereals: Novel Uses and Processes, edited by Campbell *et al.*
Plenum Press, New York, 1997

2. MATERIALS

Normal commercial grade cornstarch (Buffalo 3401) was purchased from CPC International Inc. (Argo, Illinois). Poly(vinyl alcohol) (PVA) was obtained from Air Products and Chemicals, Inc. (Allentown, Pennsylvania) under the trade name Airvol 325. Airvol 325 is 98–98.8% hydrolyzed with a molecular weight average of 85000–146000. Poly(ethylene-co-acrylic acid) (EAA) was obtained from Dow Chemical Co. (Midland, Michigan) under the trade name Primacor 5981. This EAA had a M_w of about 18000 and M_n of about 7000, and contained about 20% acrylic acid. Reagent grade glycerol was from Fisher Scientific. Poly(lactic acid) (PLA) was a gift from Cargill (Minneapolis, MN). Poly(hydroxybuterate-co-valerate) (PHBV) was a gift of Zeneca Bioproducts (Wilmington, DW). Polycaprolactone (PCL) was purchased from Union Carbide Chemicals and Plastic Co. (Charleston, WV).

3. METHODS

Cast films were prepared by the method previously described by Lawton and Fanta (1994). The film formulation was constant for all the films produced and contained 41% starch, 41% PVA, 3% EAA and 15% glycerol on a dry basis. Starch foamed trays were formed on a baking machine on loan from Frans Haas Machinery of America (Richmond VA). Trays were baked as described by Haas et al (1994) and made with 100% starch. Films and trays were coated with PLA, PHBV and PCL by dissolving the polymers in an appropriate solvent. The dissolved polymer was then applied to the trays by painting on the polymer containing solution and allowing the solvent to evaporate. The films were coated by dipping the films into the polymer containing solution.

Water sensitivity of the films were tested by total immersion of the coated films in water for 15, 30 and 60 minutes. The tensile strength and percent elongation at break were evaluated for each coated film after water immersion using an Instron universal testing machine. Water sensitivity of the trays were tested by putting 20 mL of water into tared coated trays and letting them stand for 30 minutes. The water was poured out of the coated trays and the trays reweighed.

Peel tests were performed on both the coated trays and films using an Instron testing machine. The polymer film was peeled off the tray using a fixture to keep a constant 90° angle during the test. The polymer coating was peeled of the film at 180° angle during testing. Peeling rates were 50.8 mm/minute. Specimen length was 130 mm and the width was 38.1 mm.

4. RESULTS AND DISCUSSION

There was great improvement in the water sensitivity of both the coated films and the coated trays after water immersion. Coated trays absorbed on average only 1.1 g of water whereas uncoated trays absorbed 13.5 g of water. Uncoated trays almost absorbed their weight in water. The same was true for coated films where water absorption was quite high for the uncoated film. PLA coated films absorbed 0.03 g of water after 15 minutes of water emersion while the uncoated films absorbed 0.8 g of water in the same time frame. Coated films of PHBV and PCL could not be tested because these polymers spontaneously delaminated from the films upon drying. Although there was too much water

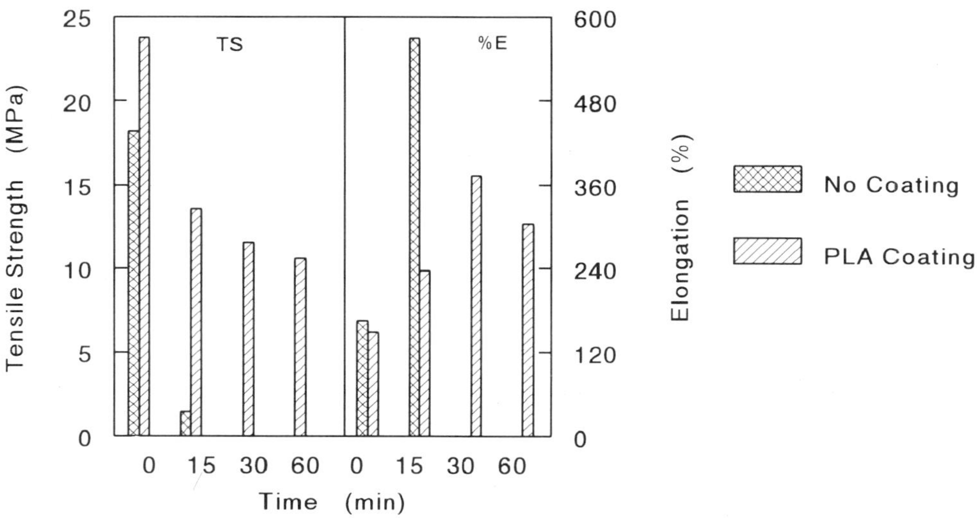

Figure 1. Physical properties of water immersed films.

absorption for this type of coating to be practical, great improvement was shown in the stability of the tensile properties of the PLA coated film (Figure 1; Chapters 5, 14 and 15 give similar measurements for starch and gluten films). Most of the absorbed water probably came from the edges of the film due to the great difficulty of getting a good coating on the edge.

Coating starch-based articles with water resistant coatings shows great promise in solving the water sensitivities of these type of objects. Hydrophobic polymers that would be good to use as water resistant coatings do not adhere well to starch (Lawton, 1995). The peel strength for films of PHBV, PLA and PCL cast onto the starch trays are shown in Table 1. Peel strength of these films increase on the order of PHBV<PCL<PLA. In some cases, the films of PHBV and PCL spontaneously delaminated from the foam prior to testing.

Peel strength for films of PHBV, PLA, and PCL cast onto starch-PVA film are shown in Table 2. The peel strength of these films increased on the order of PCL<PLA<PHBV. PCL could not be tested directly, because all coatings delaminated off the film prior to peel strength testing. The adhesion of these coatings to the starch-based material is too low for practical applications. Great improvements in adhesion of polymer films to the starch-based material can be made if it is first coated with a natural resin such as shellac (Shogren and Lawton, 1996). Peel strength of these coated materials could not be measured directly since the polymer film, rather than the adhesive joint, would break upon testing.

Table 1. Peel strengths of polymer films on starch foam sheets

Film	Peel strength(N/m)
PHBV	39
PCL	103
PLA	230

Table 2. Peel strength of polymer coatings on starch-PVA films

Coating	Peel Strength(N/m)
PHBV	3.8
PCL	*
PLA	3.2

*could not be tested

For good adhesion to take place between adherents, the interfacial energy between the two must be small (Bonnerup and Gatenholm, 1993). Interfacial tension between polymers can be estimated from Wu's harmonic-mean equation (1971). Surface free energies for PHBV, PCL, and PLA were obtained using a Cahn dynamic contact angle analyzer, using water and methylene iodide as the probe liquids. Surface free energies used for the polymers are 41.5 x 10^{-3}, 39.6 x 10^{-3}, and 43.9 x 10^{-3} N/m for PHBV, PCL, and PLA respectively. Surface free energy for starch was estimated by the group contribution method (Carré and Vial, 1993). The surface free energy of starch was 53.7 x 10^{-3} N/m. This value is in the range of other estimates for starch (Lawton, 1995; Odidi et al, 1991; Rowe, 1989). The interfacial tension between starch and the polymers ranged from 4.01 to 8.23 x 10^{-3} N/m. The estimate of the interfacial tension between starch and PCL, PLA, and PHBV were all high which is in agreement with the low values obtained from the peel test.

5. CONCLUSION

Coating or laminating starch-based objects shows great promise in alleviating some of the water sensitivity problems associated with these types of products. Of the polymers tested, there was no significant difference in the amount of water protection shown by these polymers. PLA tended to give better adhesion to the starch-based materials than did the other two polymers. However, none of the polymer tested gave good enough adhesion to the starch to be used commercially as coatings for starch-based materials. This was shown again by the high calculated interfacial tension between starch and PLA, PHBV, and PCL. For starch-based products with high percentages of starch to become more prominent, coatings are needed which have a greater affinity for starch than PCL, PHBV, or PLA. Another possibility is to use a substance which can lower the interfacial tension between starch and these polymers.

REFERENCES

Bonnerup C and Gatenholm P (1993) "The effect of surface energetics and molecular interdiffusion on adhesion in multicomponent polymer systems." J. Adhesion Sci. Technol. 7, 247

Carré A and Vial J (1993) "Simple method for the prediction of surface free energy and its components. Application to polymers." J. Adhesion 42, 265

Doane WM, Swanson CL and Fanta GF (1992) In "Emerging Technologies for Materials and Chemicals from Biomass", Ed. Rowell RM, Schultz TP and Narayan R, American Chemical Society p 197

Donovan JW (1979) "Phase transition of starch-water systems." Biopolymers 18, 263

Haas F, Haas J and Tiefenbacher K (1994) "Process for producing biodegradable thin-walled starch-based mouldings." PCT Int. Appl. WO 94/13734

Jasberg BK, Swanson CL, Shogren RL and Doane WM (1992) "Effect of moisture on injection molded starch-EAA-HDPE composites." J. Polym. Mater. 9, 163

Lawton JW and Fanta GF (1994) "Glycerol-plasticized films prepared from starch-poly(vinyl alcohol) mixtures: Effect of poly(ethylene-co-acrylic acid)." Carbohydr. Polym. 23, 275

Lawton JW (1995) "Surface energy of extruded and jet cooked starch." Starch/Stärke 47, 62

Lloyd NE and Kirst LC (1963) "Some factors affecting the tensile strength of starch films." Cereal Chem. 40, 154

Odidi IO, Newton JM and Buckton G (1991) "The effect of surface treatment on the value of contact angles measured on a compressed powder surface." Int. J. Pharm. 72, 43

Otey FH, Mark AM, Mehitretter CL and Russell CR (1974) "Starch-based film for degradable agricultural mulch." Ind. Eng. Prod. Res. Develop. 13, 90

Perice FT (1928) "Influence of humidity on the elastic properties of starch film." J. Textile Inst. 19, T237

Rowe, R.C.: (1989) "Binder-substrate interactions in granulation: A theoretical approach based on surface free energy and polarity." Int. J. Pharm. 52, 149

Shogren RL and Lawton JW (1996) "Enhanced water resistance of starch-based materials." US Pat. Appl. Serial No. 0057/94

Swanson CL, Shogren RL, Fanta GF and Iman SH (1993) "Starch-plastic materials - preparation, physical properties and biodegradability (A review of recent USDA research)." J. Environ. Polym. Degradation 1, 155

Whistler RL and Hilbert GE (1944) "Mechanical properties of film from amylose amylopectin and whole starch triacetates." Ind. Eng. Chem. 36, 796

Wu S (1971) "Calculation of interfacial tension in polymer systems." J. Polymer Sci. 34, 19

Young AH (1984) "Fraction of starch." In "Starch Chemistry and Technology 2ed.", Ed. Whistler RL, Bemiller JN and Paschally EF, Academic Press, New York

INDUSTRIAL APPLICATIONS FOR LEVULINIC ACID

Viswas Ghorpade and Milford Hanna

Industrial Agricultural Product Center
University of Nebraska
Lincoln, Nebraska 68503-0730

1. INTRODUCTION

Levulinic acid has been produced since 1870. Over the years the basic chemistry and properties have been studied extensively. Though levulinic acid has significant potential as an industrial chemical, it has never reached commercial use in any significant volume. A reason for non-commercialization of this chemical may be that most of the research was done in early 40's, when the raw materials were expensive, yield was low, and equipment for separation and purification was lacking. Today, overproduction of raw materials and developments in science and technology have opened doors to reevaluate industrial potential of a forgotten chemical giant, levulinic acid. Levulinic acid can be produced by high temperature acid hydrolysis of carbohydrates, such as glucose, galactose, sucrose, fructose, chitose and also from biomeric material such as wood, starch and agricultural wastes. Isolation of levulinic acid can be accomplished either by partial neutralization, filtration of humin material and vacuum steam distillation, or by solvent extraction. Levulinic acid is a highly versatile chemical with several industrial uses. Literature shows potential uses as resin, plasticizer, textile, animal feed, coating, and as an antifreeze. At the University of Nebraska-Lincoln, efforts are being made to prepare levulinic acid using an extruder as a continuous reactor and possible use as an antifreeze ingredient. This antifreeze will have definite advantages over ethylene glycol. It will be non-toxic and easily digestable by microorganisms. The antifreeze ingredient will be in solid form, hence it will be marketed more readily than liquid forms. It is less corrosive to the iron parts of internal combustion engines than is tap water and has no detrimental effects on rubber hoses used in engines.

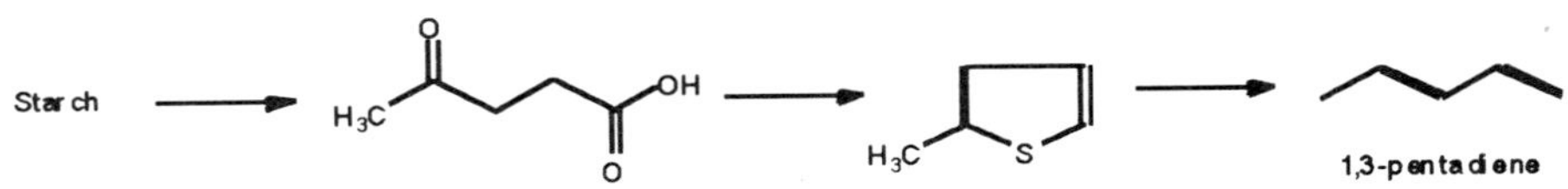

Figure 1. A scheme for synthesis of 1,3-pentadiene (piperylene), with levulinic acid as an intermediary.

2. POLYMERS FROM LEVULINIC ACID

Levulinic acid is a highly reactive and versatile chemical. Valeric-γ-lactone can be obtained in very high yields by hydrogenation of levulinic acid. This compound is a good solvent and may have extensive uses as such. Moreover, it may be hydrogenated to 1,4 pentandiol, which upon dehydration yields 1,3-pentadine (piperylene) (Figure 1). Piperylene is known to polymerize to a rubbery mass, a source of synthetic rubber. Levulinic acid is a solvent for aromatic constituents of crude mineral oil.

A scheme for making rubber from starch was developed in the early 1930's. The Heinemann synthetic rubber process was essentially the production of levulinic acid from starch, and its conversion with phosphorous pentasulfide into methylthiophene which was reduced to 1,3 pentadine.

Esters of levulinic acid (Figure 2) have many industrial applications. Cox and Dodds (1933) described esters of levulinic acid such as the butyl, hexyl and cyclohexyl esters, mentioned as plasticizers in the plastic catalogs of Monsanto Chemical Co. Long chain alkyl esters (Lawson and Salzberg, 1935) of levulinic acid have been described as plasticizers for cellulose esters. Cellulose ester plasticizer (Izard and Salzberg, 1935) has been made by esterifying polyhydric alcohol containing one free hydroxyl group with levulinic acid. Other esters such as methyl isopropyl, isoamyl and 2-pentanol esters of levulinic acid have been described as solvents for synthetic glass like resins (Stampa, 1939). Pseudo esters have been used as paint removers, solvents and lacquers. Other uses of esters are as constituents of hydraulic-brake fluids and as solvent for the extraction of glyceride oils such as linseed oil, soybean and fish oil (Fulton, 1935).

Diphenolic acid, the condensation product of levulinic acid and phenol, is useful in the preparation of the modified phenol formaldehyde resins, polyether resins and as monocarboxilic acid chain stopper in alkyd resin (Bader, 1960). Further, diphenolic acid can be substituted for bis-phenol A, the primary raw material for production of epoxy resin (Figure 3). These phenolic resins can be obtained by heating mixtures of phenol, levulinic acid and concentrated HCl for 6 hours, followed by a separation process. Several other interesting resin and polymers have been obtained from levulinic acid. These or similar polymers may have potential applications in adhesives (Conners, 1989). A heat-setting resin was produced from a fusion of levulinic acid and amine (Hovey and Hodgins, 1940).

Figure 2. Esters of levulinic acid.

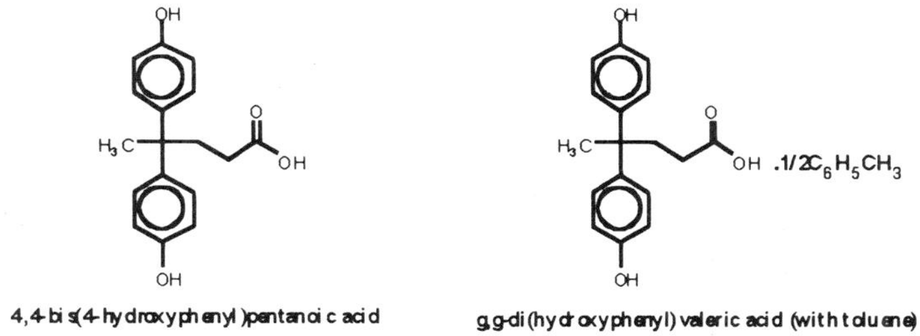

Figure 3. Phenol-formaldehyde resin.

That resin was hard and showed good adhesion properties with glass. A hard flexible resin was formed when levulinic acid reacted with formaldehyde. Hachihama and Hayashi (1954) described a process of making polymides from furfural and levulinic acid. An adhesive formulation with good water resistance was prepared from atactic polypropylene and levulinic acid (Hikotaro, 1973).

3. LIQUID FUEL EXTENDER

Ethanol is a biomass-derived gasoline additive marketed today. However, there are other potential fuel additives that can be derived from biomass. Thomas and Barile (1985) reported use of levulinic acid and its derivatives as fuel extenders.

Alpha-angelicalactone (AL) was synthesized from levulinic acid (Figure 4) and has several industrial uses (Leonard, 1956, 1958). Alpha-angelicalactone was soluble in gasoline and was suggested as a fuel extender. Alpha-methyltetrahydrofuron has been prepared on a commercial scale and used as a solvent and as a chemical intermediate. It is particularly attractive as a fuel additive because it is a cyclic ether with high air-mass/fuel-mass ratio and its structure resembles the anti-knock agent methyl t-butylether. Levulinic acid itself is not very soluble in gasoline but it has a calorific value of 24,000 kJ/litre (86,000 BTU/gallon) and has been described as an antiknock agent. The theoretical weight yield of levulinic acid from hexoses is 64.5%, but the literature shows that only two thirds of the theoretical yield can be obtained. According to Harris (1975) a weight yield of 35–45% can be expected. Conversion of levulinic acid to alpha-angelicalactone occurred at levels of 90% and more. Hydrolysis of starch to glucose can be obtained in high yields in an extruder with mild acid treatment.

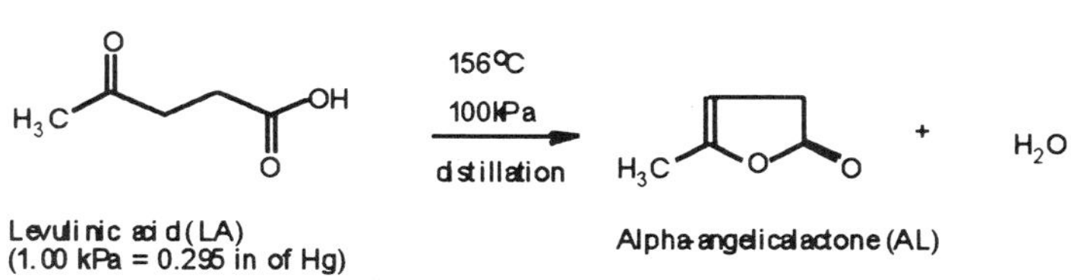

Figure 4. Synthesis of alpha-angelicalactone.

4. LEVULINIC ACID AS AN ANTIFREEZE INGREDIENT

Salts of levulinic acid are water soluble and have several therapeutical uses. These salts were prepared as per the procedures described in work proposed at UNL. Calcium levulinate's solubility in water is temperature dependent. Cox et al (1934) reported higher solubility of calcium levulinate at higher temperature. Wiggins (1949) reported possible use of sodium levulinate as an antifreeze ingredient. This antifreeze has definite advantages over ethylene glycol. It is non-toxic (Tischer et al, 1942) and easily digestible by microorganisms (Harada, 1971). It is a solid and can be more easily marketed than ethylene glycol. It is less corrosive to the iron parts of internal combustion engines than tap water and has no detrimental effects on rubber hoses used in engines (Wiggins, 1949).

5. FOOD AND PHARMACEUTICAL USES

The production of levulinic acid from corn starch at a low cost would result in its consideration for various food applications. Several reports indicate that levulinic acid has been successfully used as an acidulant in carbonated and fruit juice beverages, jams and jellies. On the basis of chemical structure it should act similarly to acetic and propionic acids. Ethyl levulinate is used for flavoring (Leonard, 1956). Alkyl metal halide reacts with levulinate esters to yield a series of g-valerolactones and some of them are used as perfumes and flavors.

The therapeutic use of organic salts of metal depends on the organic part of the molecule to which the metal ion is attached. This part affects the solubility of salt, pH and stability. Salts of levulinic acid, such as calcium, are used therapeutically on various conditions including tuberculosis. Gordon et al (1933) reported intravenous injections of calcium levulinate did not produce irritating effects and it was found to be stable for intravenous administration. Calcium levulinate has advantage over calcium gluconate: it contains 40% more calcium, is soluble in water and 30% solution did not form crystals in ampules for indefinite lengths of time (Proskouriakoff, 1933). Heterocyclic compounds derived from this acid are used as bacteriostatic and analgesic agents (Leonard, 1956).

6. WORK AT UNIVERSITY OF NEBRASKA-LINCOLN

6.1. Synthesis of Levulinic Acid

Levulinic acid has been produced by the action of acids on carbohydrates such as glucose (Sah and Ma, 1930) galactose and sucrose (Wiggins, 1949; Thomas and Schuette, 1931) fructose, glucosamine, chitose, sorbose, deoxypentoses and hexose sugars (Sassaenrath and Shilling, 1966; Thompson, 1940), cane sugar and starch (McKenzie, 1929; Moyer, 1942) and disaccharide and polysaccharide unions (Wiggins, 1949). Table 1 summarizes the biomass feedstock for levulinic acid preparation used by various authors. The effectiveness of dilute acid on the dehydration reaction for levulinic acid preparation is HBr>HCl>H_2So_4>acetic acid. HBr is expensive for manufacture of levulinic acid. Thomas and Barile (1985) reported that better yields can be obtained with H_2So_4 than with HCl.

At the University of Nebraska-Lincoln, starch has been hydrolyzed to glucose with a dilute acid treatment in a twin-screw extruder with mixing screws. Starch amylose content, acid concentration, moisture content and extruder barrel temperature and screw speed

Table 1. Biomass feedstocks for levulinic acid production

Waste plant material: hard wood or beech bark (Kin et al, 1978)
Fiberboard industry waste water (Pajak and Kryczko, 1979)
Bagasse pity, bagasse, molasses (Nee and Yse, 1975)
Post-fermentation liquor (Mel'nikov et al, 1975)
Furfural still residues (Badovskaya et al, 1972)
Aqueous oak wood extracts (Prosinski et al, 1971)
Rice hull (Sumiki, 1948)
Oats residues (Rodriguez, 1973)
Wood sugar slops (Faerber, 1943)
Fir sawdust (Haworth and Shilling, 1966)
Naptha (Kikuchi and Ikematsu, 1974)
Corncob furfural residue (Dunlop and Wells,1957)
Cotton balls rice straw, soybean skin, soybean oil residue, corn husks (Sumiki and Kojima, 1944)
Cotton stems (Minina et al, 1962)
Cottonseed hulls (Akmamedov et al, 1962)
Molasses (Rao et al, 1959)
Starch (Hands and Whitt, 1947)
Potatoes, sweet potatoes, lactose (Takahashi, 1944)
Wastewood pulping residues (Wiley et al, 1955)
Sunflower seed husks (Sil'nikova, 1967)
Tapioca meal (Chapman and McIntosh, 1971)

Adapted from Thomas and Barile (1985)

need to be studied to achieve the desired degree of hydrolysis. Barrel temperature and pressure can be used to optimize conversion of glucose to levulinic acid. The extrusion-processed starch humin should be analyzed for production of levulinic acid by HPLC. Selected high treatments with high yields of levulinic acid will be used for pilot scale production of levulinic acid. Levulinic acid can be separated either by partial neutralization, filtration of humin material and vacuum steam distillation or by solvent extraction.

6.2. Salts of Levulinic Acid

Calcium and sodium salts of levulinic acid were prepared using procedures described by Proskouriakoff (1933). Levulinic acid was diluted with distilled water to make a 20 % solution. The solution was then heated to 90°C on a hot plate and sodium carbonate was added slowly until CO_2 bubbles were released. Heating was continued with constant stirring to evaporate excess water. Solution was removed from hot plate and incubated in 90°C water bath until half the solution was evaporated. As crystals started appearing at the bottom, the solution was removed and cooled to get complete crystallization.

6.3. Freezing and Boiling Point Determination

The freezing point of sodium levulinate is being tested using ASTM D1177 standard method for aqueous engine coolant. Boiling point is being determined by standard ASTM D1120 method. The boiling temperature of the sample will be corrected for barometric pressure and temperature will be noted as boiling point.

ACKNOWLEDGMENT

A special thanks to Nebraska Corn Development, Utilization and Marketing Board and the Agricultural Research Division, University of Nebraska-Lincoln for their financial support.

REFERENCES

Akmamedov K, Minina VS and Usmanov UK (1962) "Cottonseed coverings as a valuable raw material for the hydrolysis industry." Fiz i Khim Prirodn i Sintetich Polimerov Akad Nauk Uz SSR Inst Khim Polimerov 1 78–86, Chem. Abst. 60, 757f

Bader AR (1960) US Patent 2933520

Badovskaya LA, Kul'nevich VG, Firsova IL, Kurzin LM and Chudaev V (1972) "Conversion of still residues from furfural manufacture by their oxidation with hydrogen peroxide to decarboxylic acids and keto acids." Tr. Krasnodar Politekh Inst. 29107–8, Chem. Abst. 76, 115075q

Carison JL and Wash S (1962) "Process for manufacture of levulinic acid." US Patent 3 065 263

Chapman O and McIntosh C (1971) "Photochemical decarboxlation of unsaturated lactones and carbonates." J. , Chem. Soc. D (8) 383–4, Chem. Abst. 75, 35566q

Conners AH (1989) "Carbohydrate in adhesives." In "Adhesives from renewable sources." Hemingway RW, Conners AH and Barnham SJ (eds), ACS Washincgton DC

Cox JG, Dodds LM and Clarence C (1934) "The solubility of calcium levulinate in water." J. Am. Pharm. Asso. 7 662–664

Cox GJ and Dodds MF (1933) "Some alkyl esters of levulinic acid." J. Am. , Chem. Soc. 55, 3391–3394

Dunlop AP and Wells AP (1957) "Levulinic acid." US Patent 2 813 900,, Chem. Abst. 52 9199b

Faerber E (1943) "Recovery of products such as furfural and levulinic acid and its esters from slops from the wood- sugar process or the like." US Patent 2 293 724,, Chem. Abst. 37 10402

Fulton RR (1935) US Patent 1986260

Gordon B, Kough OS and Proskouriakoff A (1933) "Studies on calcium levulinate with special reference to the influence on edema." J Laboratory and Clinical Medicine, 507–511

Hands CHG and Whitt FR (1947) "The preparation of levulinic acid on a semitechnical scale." J. Soc. , Chem. Ind. 66, 415–416

Harris J (1975) "Acid hydrolysis and dehydration reaction for utilizing plant carbohydrates." Appl. Polym. Symp. 28, 131–144

Hachihama Y and Hayashi I (1954) Makromol. Chem. 13, 201

Harada M (1971) "Metabolism of levulinic acid by microorganisms. IV. Removal of levulinic acid hydrolysate of defatted soybeans by levulinic acid utilizing microorganism." Agri. Chem. Soc. Japan. J. 45(2), 55

Haworth CP and Shilling LW (1966) "Levulinic acid from hexose-containing material." US Patent 3 258 481

Hikotaro Y (1973) Japanese Patent 73 43 178

Hovey AG and Hodgins TS (1940) U S Patent 2 195570

Izard EF and Salzberg PL (1935) U S Patent 2004115

Kikuchi Y and Ikematsu K (1974) "Separation and recovery of levulinic acid from naptha liquid - phase oxidation waste liquid." Japan Kokai 74 51200, Chem. Abst. 81, 119997t

Kin Z, Kowalczyk H, Gorski L, Klajenski R, Tonzewski B, Jaworski J and Przybylak E (1978) "Simultaneous preparation of furfural levulinic acid and humic nitrogen fertilizer from waste plant material." Pol. 99 185, Chem. Abst. 88, 3643c

Lawson WE and Salzberg PE (1935) Ibid US Patent 2 008720

Leonard HR (1956) "Levulinic acid as a chemical basic raw material." J. Ind. Eng. , Chem. 48(8), 1331–1341

Leonard HR (1958) "Conversion of levulinic acid into alpha- angelicalactone." US Patent 2809203, Chem. Abst. 52 2819b

McKenzie BF (1929) "Levulinic acid Organic Synthesis." An annual publication of satisfactory methods for the preparation of organic chemicals 50–1

Mel'nikov NP, Levitin BM and Sergeeva LA (1975) "Levulinic acid." USSR 463657, Chem. Abst. 83, 428387

Minina VS, Sarukhanova AE and Usmanov KU (1962) "Preparation of furfural and levulinic acid by hydrolysis of pressed cottton stems." Fixi Khim Prirodn i Sintetich Polimerov Akad Nauz Uz SSR Inst. Khim. Polimerov 1 78–86, Chem. Abst. 60, 757f

Moyer WW (1942) "Preparation of levulinic acid." US Patent 2 270 328

Nee CI and Yse JW (1975) "Furfural and levulinic acid prepared concomitantly from bagasse pith." Taiwan Sugar 22(2), 49–53, Chem. Abst. 83, 117532e

Pajak J and Kryczko M (1979) "Treatment of fiberboard industry wastewater." Pol. 99 879, Chem. Abst. 91, 18099lj

Prosinski S, Adamski Z and Kwasniewski A (1971) "Analysis of chemical components of a hydrolyzate obtained from oak extraction chips after of distilling of furfural." Rocz Wyzsz Szk Roln Poznaniu 52, 89–98, Chem. Abst. 77, 903162

Proskouriakoff A (1933) "Some salts of levulinic acid." J. Am. , Chem. Soc. 55, 2132–34

Rao CK, Reddy GS, Sidhu GS, Kachler IK, and Zaheer SH (1959) "Isolation of levulinic acid from molasses." Indian 70, 171, Chem. Abst. 56, 2623h

Rodriguez ER (1973) "Jointly producing Furfural and levulinic acid from bagasse and other lignocellulostic Materials." US 3701789, Chem. Abst. 78, 16017g

Sah PT and Ma SY (1930) "Levulinic acid and its esters." J. Am. , Chem. Soc. 524880–3

Sassenrath PC and Shilling LW (1966) "Preparation of levulinic acid from hexose-containing material." US Patent 3 258 481

Sil'nikova LL (1967) "Complex processing of plant raw materials with the production of furfural and levulinic acid." Khim Pererab Drev 8, 7–9, Chem. Abst. 67, 118315t

Stampa G (1939) Intern. Sugar J. 41–270

Sumiki Y and Kojima A (1944) "Preparation of levulinic acid and its utilization I. Levulinic acid from agricultural produce waste." J. Agr. , Chem. Soc. Japan 20, 651–2, Chem. Abst. 42 5422

Sumuki Y (1948) "Levulinic acid." Japan 176 438, Chem. Abst. 45, 7589i

Takahashi T (1944) "Studies on decomposition of carbohydrates by strong mineral acids I. Determination of decomposition products." J. Agr. , Chem. Soc. Japan 20553–6, Chem. Abst. 42, 8166f

Thomas RW and Schuette AH (1931) "Studies on levulinic acid I. Its preparation from carbohydrates by digestion with hydrochloric acid under pressure." J. Am. , Chem. Soc. 53, 3485–9

Thomas JJ and Barile GR (1985) Biomass Wastes 8, 1461–94

Tischer RG, Fellers RC and Doyle JB (1942) "The non-toxicity of levulinic acid." J. Amer. Pharm. Assoc. 31, 217–20

Thompson A (1940) "Method of making levulinic acid." US Patent 2 206 311

Wiggins WF (1949) "Utilization of sucrose." Advances in Carbohydrate Chemistry 4, 306–14

Wiley AJ, Harris JF, Salman JF and Locke EK (1955) "Wood industries as a source of carbohydrates." Ind. Eng. , Chem. 47 1397- 1405

8

PRODUCTION OF LACTIC ACID FROM STARCH

Simulation and Optimization

Christina Åkerberg and Guido Zacchi

Department of Chemical Engineering 1
University of Lund
PO Box 124, S-221 00 Lund, Sweden

1. INTRODUCTION

There is an increased interest for the production of lactic acid from renewable resources, such as starch, to be used for the production of biodegradable polylactic acid (see Chapters 1, 6, 14 and 26). The development of a cost-effective and energy efficient process with high yields of lactic acid, using a minimum of resources with a minimum of waste, requires process integration and optimization. The aim of this work is to create a methodology based on mathematical models and simulation tools for the development and optimization of this integrated process.

2. LACTIC ACID PRODUCTION

The fermentative production of lactic acid from starch can be divided into the following main steps: pretreatment, fermentation, separation and purification (Figure 1). In the pretreatment the wheat starch is enzymatically hydrolysed to glucose in two steps: liquefaction and saccharification. In the liquefaction step a thermo-stable α-amylase is used to solubilize the starch. In the saccharification step, the oligosaccharides and maltose are converted to glucose with an α-amylase together with amyloglucosidase. The glucose is used as a substrate for the lactic acid bacteria in the fermentation step, producing lactic acid. The acid is separated from the fermentation broth and further purified.

To avoid product inhibition in the saccharification, this step can be performed at the same time as the fermentation. The kinetics for the two steps are determined separately. The results from this investigation will be used to optimize the integrated process.

Cereals: Novel Uses and Processes, edited by Campbell *et al.*
Plenum Press, New York, 1997

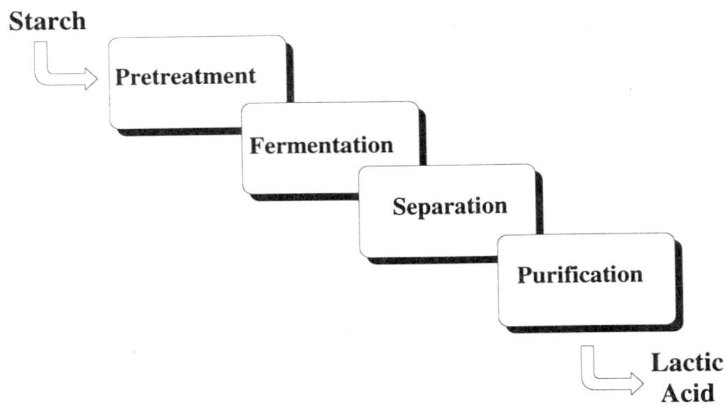

Figure 1. The integrated production of lactic acid from starch.

A project has recently been initiated where the separation of lactic acid from the fermentation broth is investigated. The results from that project will later be integrated with the results from the present study.

3. MODELS

Different mathematical models have been used for the simulation of the saccharification and fermentation steps.

3.1. Saccharification

The following model, which is a modification of the model by Lee et al (1992), is used for modelling of the saccharification of oligosaccharides with sizes of DP1-DP7.

The hydrolysis rate of an oligosaccharide with n glucose units (n = 2 - 7) is calculated by

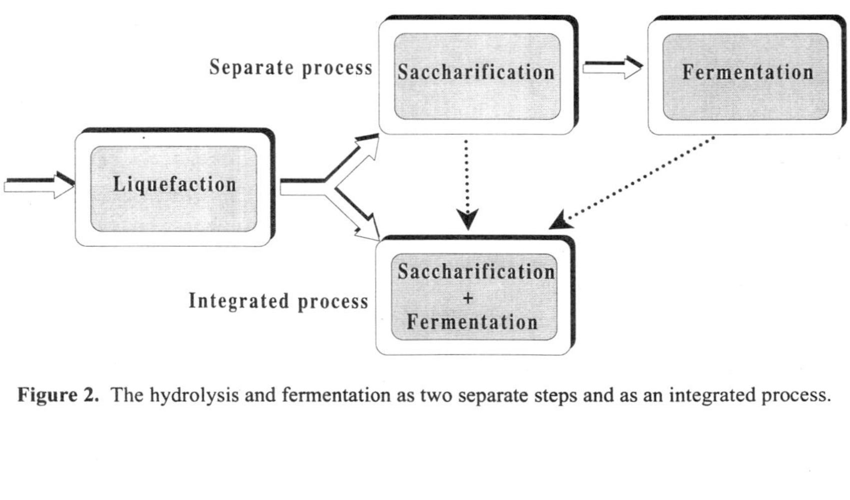

Figure 2. The hydrolysis and fermentation as two separate steps and as an integrated process.

$$\frac{dG_n^*}{dt} = -\frac{V_{\max,n} \cdot G_n}{K_{m,n} + G_n} \cdot \frac{K_g}{K_g + G_1} \tag{1}$$

where G is the oligosaccharide concentration. V_{max} is the maximum reaction rate, K_m is the Michaelis-Menten constant and K_g is the glucose inhibition parameter for glucose production. The index n represents the number of glucose units in the oligosaccharide.

The net rate of consumption of an oligosaccharide with n glucose units (n = 2–6) is

$$\frac{dG_n}{dt} = \frac{dG_n^*}{dt} - \frac{dG_{n+1}^*}{dt} \tag{2}$$

The rate of formation of glucose is

$$\frac{dG_1}{dt} = -\left(\sum_{n=3}^{7} \frac{dG_n^*}{dt}\right) - 2 * \frac{dG_2^*}{dt} \tag{3}$$

The maximum reaction rate $V_{max,n}$ can be expressed as a function of the enzyme concentration E and the substrate concentration S_0

$$V_{\max,n}(E, S_0) = k_{\max,n} \cdot k_{es} \cdot \frac{E}{E + k_e} \cdot \frac{S_0}{S_0 + k_s} \tag{4}$$

where E is the enzyme concentration and S_0 is the initial starch concentraton. Theoretical values for $K_{m,n}$ and $k_{max,n}$ are used while the rest of the parameters, K_g, k_e, k_{es} and k_s, will be determined by non-linear least square fitting to experimental data.

3.2. Fermentation

The rates of cell growth, product formation and substrate consumption are expressed with the following unstructured model including both substrate and product inhibition:

Cell growth (A Monod expression with terms for substrate and product inhibition)

$$r_x = \frac{\mu_{\max} \cdot S}{K_s + S + \frac{S^2}{K_I}} \left(1 - \frac{P}{P_m}\right)^n \cdot X \tag{5}$$

where μ_{max} is the maximum specific cell growth rate, K_s is the saturation constant, K_I is the substrate inhibition constant and P_m and n are constants used for expressing the product inhibition. X, S and P are the cell, substrate and product concentrations respectively.

Product formation (Luedeking and Piret, 1959)

$$r_p = \alpha \cdot r_x + \beta \cdot X \tag{6}$$

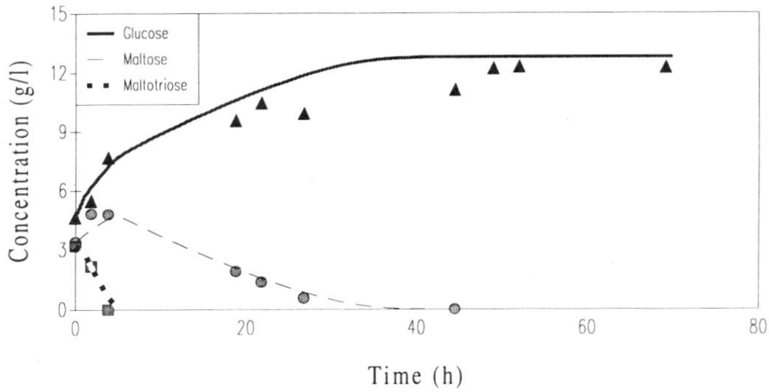

Figure 3. The concentrations of glucose, maltose and maltotriose during hydrolysis with SAN Super at a temperature of 30 °C and a pH of 5.0.

Substrate consumption

$$r_s = -\frac{1}{Y_{P/S}} \cdot r_p \tag{7}$$

where $Y_{P/S}$ is the yield of the substrate conversion to lactic acid.

4. EXPERIMENTAL INVESTIGATION

The kinetics of the liquefaction, saccharification and fermentation steps were investigated. The experimental data from the saccharification and the fermentation steps were used to determine the kinetic parameters in the mathematical models by non-linear least squares fitting.

The thermostable enzyme Termamyl (Novo Nordisk) was used in the liquefaction step and the enzyme SanSuper (Novo Nordisk) was used in the saccharification step. The concentrations of the oligosaccharides were analyzed on a Dionex 500 chromatographic system, with an integrated pulsed electrochemical detector and a post column switching interface; Chapter 9 describes the system. Sample clean up was achieved by microdialysis sampling. The kinetics were investigated for various pH levels (4 - 6), temperatures (30 - 55°C), enzyme concentrations and starch concentrations.

The fermentations were performed at a constant pH with the microorganism *Lactococcus lactis ssp. lactis* ATCC 19435 in a 1 litre fermentor. The glucose and lactic acid were analysed on a HPLC (GILSON, Aminex HPX 87-H from BioRad). The cell concentration was measured as dry weight. The kinetics will be investigated for various temperatures, pH, cell and substrate concentrations for determination of kinetic parameters in the model.

5. RESULTS

The experimental investigations are still in progress. Experimental and simulated data from one hydrolysis and one fermentation are shown in Figures 3 and 4, respectively.

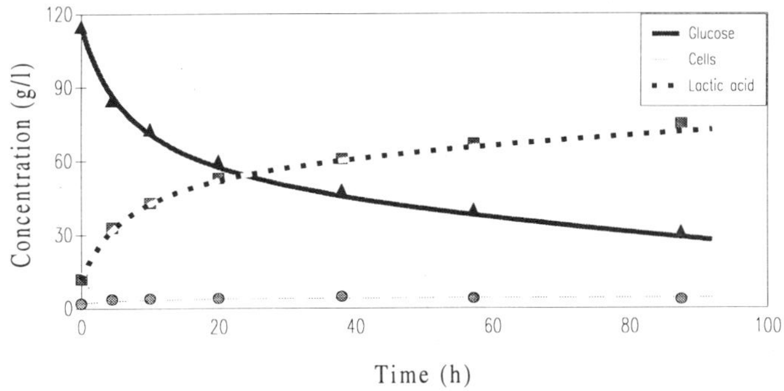

Figure 4. The concentrations of glucose, cells and lactic acid during a fermentation at 30 °C and a pH of 6.0 using glucose as substrate.

6. CONCLUSIONS

The models describe the saccharification and the fermentation steps very well. The experimental work will proceed and the results from these kinetic experiments will be used for the optimization of the integration of the saccharification and the fermentation steps. In the future the separation and purification of the lactic acid will be modelled as well and integrated with the present model.

REFERENCES

Lee C-G, Kim CH and Rhee SK (1992) "A kinetic model and simulation of starch saccharification and simultaneous ethanol fermentation by amyloglucosidase and *Zymomonas mobilis*." Bioprocess Engineering 7, 335–341

Luedeking R and Piret EL (1959) "A kinetic study of the lactic acid fermentation." J. Biochem. Microb. Technol. Eng. 1, 393–412

ON-LINE MONITORING OF ENZYMATIC BIOPROCESSES BY MICRODIALYSIS SAMPLING, ANION EXCHANGE CHROMATOGRAPHY, AND INTEGRATED PULSED ELECTROCHEMICAL DETECTION*

Nelson Torto,[1] Lo Gorton,[2] György Marko-Varga,[2] and Thomas Laurell[3]

[1]On leave from Department of Chemistry
University of Botswana
P/Bag 0022 Gaborone, Botswana
[2]Department of Analytical Chemistry
Center for Chemistry and Chemical Engineering
University of Lund
PO Box 124 S-221 00 Lund, Sweden
[3]Department of Electrical Measurements
University of Lund
Lund, Sweden

1. INTRODUCTION

Microdialysis sampling coupled to column liquid chromatography with integrated pulsed electrochemical detection (IPED) has been shown to be a hyphenation of techniques well suited for the analysis of oligomeric carbohydrates in a continuously changing matrix due to biological activity (Torto et al, 1995). Microdialysis provides a simultaneous sampling and sample clean-up step. Proper choice of a microdialysis membrane with known characteristics, e.g. molecular mass cut-off, porosity and sterilisability ensures enhanced performance of the technique in a crude medium, as it does not perturb the reaction under investigation. Only small amounts of the hydrolysis products (carbohydrates) are removed. Carbohydrates are separated in their enolate form at high pH, eliminating the need for pre- or post-column derivatisation. The chromatographic separation facilitates data evaluation, as carbohydrates are oxidised at the same potential during detection by

* Work carried out in collaboration with Christina Åkerberg and Guido Zacchi, Department of Chemical Engineering, University of Lund/Lund Institute of Technology, Lund, Sweden; see Chapter 8.

Cereals: Novel Uses and Processes, edited by Campbell *et al.*
Plenum Press, New York, 1997

IPED (Johnson et al, 1992). The purpose of this investigation was to develop an analytical system that could be used to study a liquefaction step during the hydrolysis of wheat starch in a fermentation process where glucose is the substrate (see Chapter 8). System development was carried out using soluble starch according to Zulkowsky.

2. EXPERIMENTAL

2.1. Reagents

Glucose, maltose, maltotriose, maltotetraose, maltopentaose, maltohexaose, and maltoheptaose were obtained from Sigma (St. Louis, MO, USA). Zulkowsky starch was obtained from Merck (Darmstadt, Germany) and native wheat starch was supplied by the Swedish Alcohol Industries (Kristianstad, Sweden). 50% w/w NaOH, (JT Baker, Denventer, Holland) was used to prepare 150 mM NaOH mobile phase (eluent A). Eluent B was prepared from 250 mM sodium acetate (Merck) in 150 mM NaOH. Eluents were sparged with helium and continuously kept under a helium atmosphere. Water from a Milli-Q system (Millipore, Bedford, MA, USA) was used as perfusion liquid.

Termamyl 120 L, with an activity of 120 KNU/g of solution, was obtained from Novo Industries A/S (Bagsvaerd, Denmark). 1 KNU (kilo Novo unit) is the amount of enzyme that breaks 5.26 g of starch (Merck, Amylum soluble Erg. B. 6, Batch 9947275) per hour (Product Report, Enzyme Laboratories).

2.2. Equipment

The experimental set-up (see Figure 1) consisted of a Dionex 500 chromatographic system (Dionex, Sunnyvale, CA, USA) with a Carbo Pac PA 1 pre- and analytical column from Dionex. The integrated pulsed electrochemical detector was fitted with a Ag/AgCl reference electrode (Antec, Amsterdam, The Netherlands). A 3-way switch valve (model 225T, NResearch Incorporated, West Calwell, NJ, USA) was connected between the analytical column and the detector to enable post column switching. The wave form employed to the detection unit was: $E_1 = 0.10$ v ($t_d = 0.20$ s, $t_1 = 0.20$ s), $E_2 = 0.70$ v ($t_2 = 0.19$ s) and $E_3 = -0.75$ v ($t_3 = 0.39$ s) (Andrews et al, 1990). Hydrolysis was carried out in reaction vessels housed in a Pierce React-Therm (heating/stirring module no: 18971, Rockford, IL, USA). An in-house designed microdialysis probe (Laurell et al, 1995) fitted with an SPS 4005 or 6005 polysulfone membrane (Freshenius AG, St Wendel, Germany) with a molecular mass cut-off of 5 or 30 kDa, respectively, was used to sample the hydrolysates. The perfusion liquid was delivered using a syringe pump (CMA/100 Microinjection pump, CMA/Microdialysis, Stockholm, Sweden), with an on-line injector controlling unit (CMA/160).

3. RESULTS AND DISCUSSION

Quantitative aspects of coupling microdialysis sampling to anion exchange chromatography with integrated pulsed electrochemical detection depend upon the optimisation of the extraction fraction of the microdialysis membrane to be used. Further, it also depends on the sensitivity of the detector to the analytes under investigation. The extraction fraction is directly proportional to the concentration of the analytes in the bioreactor, as

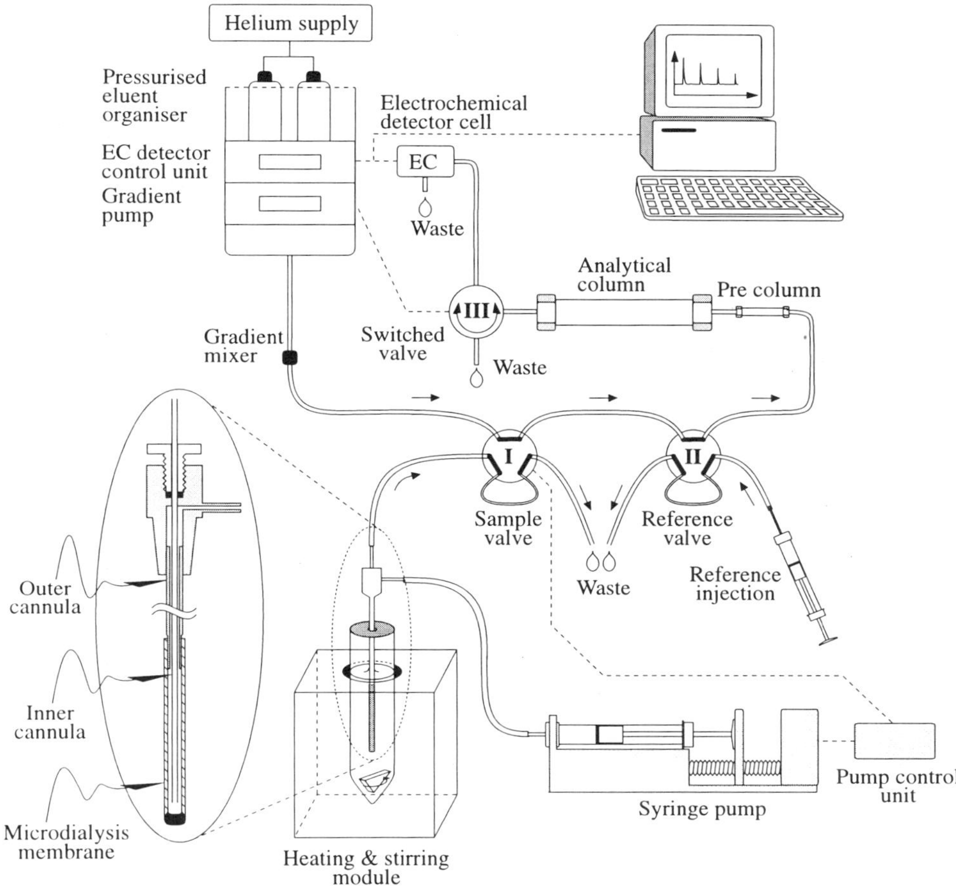

Figure 1. Schematic of combined microdialysis sampling, anion exchange chromatography and integrated pulsed electrochemical detection system for monitoring enzymatic hydrolysis of starch.

shown in equation 1, and it can thus be varied according to the dialysable area of the membrane, molecular mass cut-off, porosity, pore size distribution and complexity of the matrix.

$$E_d = (C_d^{out})/C_b = 1 - \exp(-1/Q_d(R_d+R_m)) \tag{1}$$

where E_d is the extraction fraction, C_d^{out} is the concentration of analyte in the dialysate, C_b is the analyte concentration in the bioreactor, Q_d is the volumetric flow rate, R_d and R_m are the dialysate and membrane resistance, respectively (Bungay et al, 1990).

In this investigation, the extraction fractions of a 5 and 30 kDa SPS membranes with the same dialysable area were compared. The 30 kDa membrane showed higher extraction fractions which resulted in fouling of the electrode, hence further investigations were carried out using the 5 kDa membrane. The low extraction fraction of the 5 kDa membrane combined with the adjustable perfusion rates result in on-line dilution which reduces electrode fouling since only small amounts of analytes reach the electrode. Initial hydrolysis experiments were carried out at room temperature in order to optimise chromatographic

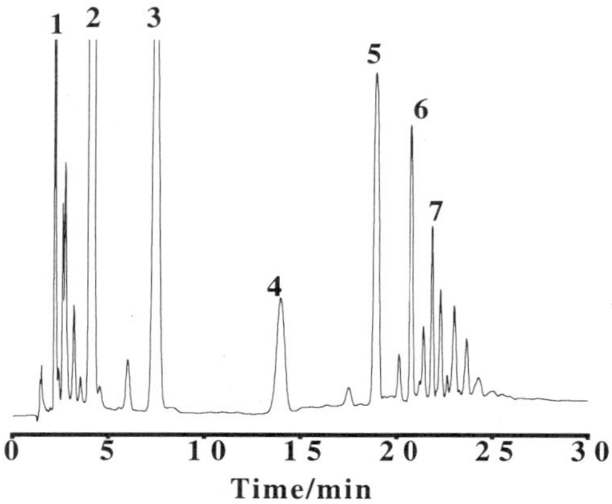

Figure 2. Chromatogram showing degree of polymerisation during starch hydrolysis.

conditions, and samples were taken on-line using continuous flow microdialysis (Torto et al, 1996).

Raising the hydrolysis temperature showed that the reaction reached equilibrium after 1.5 h, although the chromatograms took more than 25 minutes. An off-line procedure was then developed using a 5 mm SPS 4005 microdialysis membrane. Due to the short membrane and low molecular weight cutoff, this offered low extraction fractions and was then an added dilution step to reduce electrode fouling. Figure 2 shows a typical chromatogram obtained during off-line monitoring of the hydrolysis of Zulkowsky starch, where the numbers represent the degree of polymerisation (DP).

4. FURTHER WORK

Work is currently being carried out to make the monitoring more quantitative. Since glucose and maltose are the main products, it is desired that detector response be less than 1000 nC, hence a post column switching interface has been added that would allow detection of higher oligosaccharides without fouling the electrode. It is envisaged that this technique should find wide use not only in the areas of fermentation, brewing and starch industries, but also in other carbohydrate-related fields, especially if more than one hydrolysing enzyme is used.

REFERENCES

Andrews RW and King RM (1990) "Selection of potentials for pulsed amperometric detection of carbohydrates at gold electrodes." Anal. Chem., 62, 2130

Bungay PM, Morrison PF and Dedrick RL (1990) "Steady-state theory for quantitative microdialysis of solutes and water in vivo and in vitro." Life Sci. 46, 105

Johnson DC and LaCourse WR (1992) "Pulsed electrochemical detection at noble metal electrodes in liquid chromatography." Electroanalysis 4, 367

Laurell T and Buttler T (1995) "A microdialysis probe offering arbitrary membrane length and in-situ tunable relative recovery." Analytical Methods and Instrumentation, vol 2, no 4, 197

Novo Enzyme Information, B 552a-GB 1500 September 1990, Novo Industri A/S Bagsvaerd, Denmark

Torto N, Buttler T, Gorton L, Marko-Varga G, Stålbrand H and Tjerneld F (1995) "Monitoring of enzymatic hydrolysis of ivory nut mannan using on-line microdialysis sampling and anion-exchange chromatography with pulsed electrochemical detection." Anal. Chim. Acta. 313, 15

Torto N, Marko-Varga G, Gorton L, Stålbrand H and Tjerneld F (1996) "On-line quantitation of enzymatic mannan hydrolysates in small-volume bioreactors by microdialysis sampling and column liquid chromatography-integrated pulsed electrochemical detection." J. Chromatogr. A 725, 165

10

CEREAL STARCHES

Properties in Relation to Industrial Uses

A. Lynn,[1] R. D. M. Prentice,[2] M. P. Cochrane,[2] A. M. Cooper,[3] F. Dale,[3] C. M. Duffus,[2] R. P. Ellis,[3] I. M. Morrison,[3] L. Paterson,[2] J. S. Swanston,[3] and S. A. Tiller[3]

[1]Food Science and Technology Department
SAC, Auchincruive, Ayr KA6 5HW, United Kingdom
[2]Crop Science and Technology Department
SAC, West Mains Road, Edinburgh EH9 3JG, United Kingdom
[3]Scottish Crop Research Institute
Invergowrie, Dundee DD2 5DA, United Kingdom

1. INTRODUCTION

Starch is the second most abundant biopolymer after cellulose. It is synthesised by plants, stored in organs such as seeds and tubers, and subsequently used as an energy source during germination and growth. Starch is stored in distinct granules, the carbohydrates in these granules comprising two polydisperse polymers, amylose and amylopectin. Both of these polymers are composed of α-D-glucopyranose subunits. Amylose is an essentially linear polymer with the subunits being connected by α-(1→4)-linkages. Amylopectin is a highly branched polymer in which the subunits are connected by α-(1→4)-linkages, and the branches are attached to the linear chains by α-(1→6)-linkages.

2. STARCH GRANULES

2.1. Size and Shape

Starch granules vary considerably in size and shape, depending on their origin and stage of development. Even within the cereals, the diversity of starch granule morphology is large. The starch granules of oats, maize and rice are irregular and polyhedral in shape, those of rice being up to 10 μm in diameter and those of maize being up to 15 μm in diameter. The starches of wheat, barley, rye and triticale exhibit a bimodal size distribution.

Cereals: Novel Uses and Processes, edited by Campbell *et al.*
Plenum Press, New York, 1997

The larger granules are known as "A-type", while the smaller granules are known as "B-type". There has been a report that wheat starch has a trimodal granule size distribution, the third fraction being significant in number rather than in mass (Bechtel et al, 1990). In this trimodal distribution, the smallest granules were designated as "C-type" granules.

In wheat and barley, the A-type granules are lenticular in shape and have an equatorial groove or furrow. The diameter of the A-type granules varies between 10 μm and 40 μm in wheat, and between 10 μm and 30 μm in barley. The B-type granules are spherical or polygonal in shape and in both species their diameter ranges from 1 μm to 10 μm. In barley, the B-type granules make up 80–90% by number of all the granules, but only 10–15% of the starch by weight (MacGregor and Fincher, 1993). In wheat, B-type granules account for about 97% of the starch granules by number and 25–50% of the starch by weight (Evers and Lindley, 1977).

2.2. Composition

The extent of the variation in starch composition is considerable, but, typically, starch contains about 20–30% amylose. Some cultivars of maize, rice and barley produce starch that is almost entirely amylopectin. These so called "waxy" starches may contain as little as 1% amylose (Sargeant, 1982). High amylose starches can be composed of up to 76% amylose (Lineback, 1984). The availability of starches of different compositions has been invaluable in attributing some of the observed properties of starch granules to one or other of their main components.

In addition to amylose and amylopectin, starch may contain a third polysaccharide component called the intermediate fraction. This intermediate fraction is composed of α-glucan that is not readily classified as amylose or amylopectin. Lipids and proteins are also found within the granule. The lipids of wheat and barley starch granules are almost exclusively lysophospholipids. It is believed that this lipid is associated with the amylose fraction, as waxy cereal starches have little or no internal lipid, while high amylose starches have levels of lipid higher than those of normal starches (Morrison, 1995). The protein content of starch varies from 0.05 to 0.5% depending on the origin of the starch. Small amounts of protein are distributed throughout the granule (integral proteins) and larger amounts of protein are found on the starch granule surface. The integral proteins are thought to be starch synthesising enzymes, which have become trapped inside the granule during starch synthesis.

3. PROPERTIES OF STARCH

It is important to realise that the functional characteristics of a starch granule are determined by the properties of the whole granule, and not necessarily only those of its carbohydrate components. Additionally, the functional characteristics of a starch in a given application may not be a result of the chemical composition of the granules at all, but may be determined by the physical properties of the granules. Starches differ in their suitability for use in a given process e.g. potato starch is used for high quality paper production because it contains low amounts of protein. In contrast, while wheat starch can be used for paper production, protein present in the starch takes part in a Maillard reaction causing the paper to discolour, thus lowering its value.

3.1 Gelatinisation

Starch granule gelatinisation is essential for many industrial processes as it alters the rheology and viscosity properties of the system that the starch is in and it also makes the starch more accessible to enzymic action. There are many definitions of gelatinisation, one of these being that gelatinisation is '...the collapse (disruption) of molecular orders within the starch granule manifested in irreversible changes in properties such as granular swelling, native crystallite melting, loss of birefringence and starch solubilisation' (Atwell et al, 1988). An "excess" of water is essential for complete gelatinisation to occur. The disruption of the molecular orders within starch granules requires that energy is supplied to them and this energy is often supplied as heat. On an industrial scale, the energy input necessary for starch gelatinisation is considerable, and may form a large part of the production costs. However, starch granules may gelatinise at a lower temperature as a consequence of the removal of some of the molecular order e.g. by mechanical damage. The energy required to gelatinise a starch sample may be increased if the starch granule structures are allowed to anneal before gelatinisation (Hoover and Vasanthan, 1994). Clearly, this may be undesirable for economic reasons.

The gelatinisation temperatures of wheat and barley starches are similar to those of potato starches, but normal maize starch has a gelatinisation onset temperature that is higher, and a range of gelatinisation temperature that is smaller than those of potato, wheat and barley starches (Inouchi et al, 1991; Shamekh et al, 1994; Svensson and Eliasson, 1995). In both wheat and barley, the gelatinisation temperatures of B-type starch granules are several degrees higher than those of A-type starch granules (Lineback and Rasper, 1988; MacGregor and Fincher, 1993). The gelatinisation temperature of waxy maize starch is somewhat higher than that of normal maize starch (Inouchi et al, 1991) whereas little if any difference was observed between the gelatinisation temperatures of waxy, high amylose and normal barley starches (Lorenz, 1995). The temperature at which the grains of barley and wheat develop has been shown to affect the gelatinisation temperature of the starch they contain (Tester et al, 1991; Tester et al, 1995).

3.2. Pasting

Pasting can be defined as '..the phenomenon following gelatinisation in the dissolution of starch. It involves granular swelling, exudation of molecular components from the granule, and eventually, total disruption of the granules.' (Atwell et al, 1988). The swelling of starch granules in water causes the disruption of some of the covalent bonds that are present within the granules, and also causes some carbohydrate material to be leached from the granules. When the water temperature is low it is short low-molecular-weight amylose that is leached from the granules. As the leaching water temperatures are increased, higher-molecular-weight and branched components are leached from the granules. Oat starches are unusual in that amylose and amylopectin co-leach throughout the swelling and leaching processes (Doublier et al, 1987). Starch granules that have been damaged during extraction or processing also show a different pattern of leaching (Tester and Morrison, 1994). In this case amylopectin is leached into cold water and, as the temperature of the water increases, more linear material is leached from the granule. This leaching of amylose happens at a lower temperature in the damaged starch granules than in undamaged granules because damaged granules are less crystalline and are therefore more easily disrupted.

As the granules swell and material leaches from them into solution, the properties of the starch-containing solution change, and there is a transition from a suspension of granules to a paste. The properties of the paste depend on the source of starch, and whether the granules have been subjected to chemical or physical modification. The pasting properties of a starch are assessed by measuring the viscosity of starch dispersions in a temperature/time profile. The relative proportions of amylose and amylopectin in wheat, potato and maize starches are approximately similar but the viscosity characteristics of these three starches are very different (Galliard and Bowler, 1987). The amylose fractions of these starches have been found to differ considerably not only in intrinsic viscosity, but also in the size and fine structure of the amylose molecules (Shi and Seib, 1989). Starch pastes vary in clarity or turbidity, as well as in their textural properties. Potato starches can be used to produce clear pastes, while pastes produced from quinoa starch rapidly become very turbid. Oat starch gels are more elastic, adhesive and translucent than the corresponding pastes made from unmodified wheat or maize starches (Paton, 1986).

4. INDUSTRIAL USES OF STARCH

Starch is used in a large number of industries and for very varied applications (Table 1; see also Chapters 2 and 3). Long-term users of starch include the paper and textile industries, in which starch is used for processes such as surface sizing and dye binding.

In addition to their natural diversity, starch granules can have their properties modified by either physical and/or chemical processes, thus allowing for the production of specialised starches. The food industries are large consumers of starch. In these industries, the texture, viscosity and colour that starch conveys to food products are of primary interest. Starch is also used as an industrial feedstuff for the production of a wide range of chemicals (Röper, 1993; see Chapters 25 and 26 for alternative systems using whole wheat flour instead of purified starch).

4.1. Viscosity Control

Starches, both modified and native, are used to adjust the viscosity of solutions. Starches are frequently used for this purpose in the food industry. Modified starches have also been used in some oil drilling applications where the starch has been added to drilling mud in order to produce mud with the desired viscosity. In other instances, the viscosity of

Table 1. Industries and how they use starch

Starch application	Industry
Adhesive production	Adhesive
Mulches, pesticide delivery, seed coatings	Agricultural
Absorbent, binder, drug delivery, dusting powder, plasma extender/replacers, excipients, face powders, talcum powders, transplant organ preservation	Cosmetic, medical and pharmaceutical
Viscosity control, glazing agent	Food
Mud viscosity control	Oil drilling
Binding, sizing, coating	Paper and board
Biodegradable component	Plastics
Sizing, finishing & printing, fire resistance	Textile

starch solutions themselves can be very important, especially as incorrect viscosity may cause pumping problems in an industrial plant. Obtaining a consistent viscosity of starch solutions is important to industrial process that involves starch pastes or solutions being handled by mechanical systems, e.g. in the corrugating industry, maintenance of satisfactory viscosity is essential to maintain the desired levels of paper penetration, dewatering and "slinging".

4.2. Production of Biodegradable Materials

The petrochemical industry has spent large amounts of money developing materials that have the necessary mechanical and physical properties for specific applications. These materials are very useful, and have impacted on most aspects of our everyday lives. However, their useful properties have come at a price. These materials are difficult to dispose of in an environmentally friendly manner. Starch is being examined with the aim of developing materials which have the properties of petrochemical plastics, and are also degraded in the natural environment. Starch has the added advantage of being derived from a renewable source, and so its degradation has a zero net effect on the level of carbon dioxide in the atmosphere.

There are different approaches to using starch in plastics (see Chapters 2, 4, 5 and 6). One approach is to incorporate granular starch into the plastic film. The plastic loses structural integrity as the starch granules are degraded, and eventually the film fragments (biofragmentation). Initially, maize starch granules (15 μm) were incorporated in plastic films which were 50 μm in thickness. However, in order to satisfy a demand for starch-filled plastic films as thin as 12 μm, the possibility of using both the A-type and the B-type starch granules of wheat in plastic film production has been explored (Griffin, 1989) and a method of producing small particle starch from maize starch granules has been developed (Jane, 1995). In the production of the starch/plastic blends used in packaging materials, non-granular starch is chemically modified to enable it to interact with the plastic component (Doane, 1989).

Many derivatives of starch are biodegradable. One such derivative is calcium magnesium acetate (CMA). CMA has been used as a road surface de-icer, replacing the sodium chloride and calcium chloride salts normally used. CMA is attractive as it causes less damage to metal and concrete structures, and to the environment generally (Oehr and Barrass, 1992). CMA is also being examined by the users of fossil fuels, as it is able to remove SO_x and NO_x from flue gases (Stechiak et al, 1995).

Another potential use for starch is as a raw material for the production of biodegradable detergents (Röper, 1993). Estimates suggest that approximately 55% of the chemicals in powder formulations, and approximately 70% of those in liquid formulations could be replaced by starch-derived material (Kock et al, 1993).

4.3. Production of Chemicals from Hydrolysed Starch

A major outlet for starch is not for starch in the form of granules, but rather for the products of starch hydrolysis. There are essentially two processes by which starch is hydrolysed. The first is acid hydrolysis, and the second is enzymic hydrolysis. Acid hydrolysis is not a specific process, and so, if a consistent product is required, the reaction must be carefully controlled. Additionally, if the acid hydrolysis of starch is not carefully controlled, unwanted coloured compounds may be produced. The products of enzymic hydrolysis of starch tend to be more consistent in composition than those of acid hydrolysis, this being a result of the inherent specificity of the enzymes used.

It is possible to combine acid hydrolysis and enzymic hydrolysis of starch to produce syrups containing a wide range of sugars. The hydrolysis products of starch can be further processed by microbial fermentations into a very wide range of compounds that include organic acids, alcohols, ketones, polyols, amino acids, nucleotides, biopolymers, lipids, proteins, vitamins, antibiotics, and hormones. The potential for expanding and extending the use of starch in the production of chemicals and pharmaceuticals has been reviewed by Doane (1989) and Röper (1993); Chapters 8, 25, 26 and 27 also consider fermentation technology applied to starch and whole grain cereals.

The source of starch used in an industrial process is determined partly, and indeed sometimes mainly, by economic factors and the traditions of local agriculture, rather than by a scientific assessment of the suitability of the starch for that particular industrial process. Knowledge of the composition and properties of the wide range of starches which can be produced in genotypes of temperate cereals is very incomplete and an appreciation of how environmental conditions during plant growth alters these properties is only beginning to emerge. Our project aims to investigate the composition and properties of starches used in industrial processes, and to compare these results with those obtained from an analysis of starches extracted from a wide range of genotypes of cereals and potatoes grown at different sites, and under glasshouse and controlled environment conditions. Only preliminary results are available so far. It is hoped that by the end of the project we will have accumulated sufficient data for an assessment to be made of the processing potential of UK-grown starches.

Chapter 12 presents a similar systematic study of starch properties from different cultivars of a single genus, *Amaranthus*.

5. MATERIALS AND METHODS

5.1. Sources of Starch

Samples of starch were supplied by starch producers in several countries. Samples of grain of various cultivars of wheat, oats and barley were obtained from the Scottish Agricultural College National/Recommended List Trials at three sites in Scotland in 1995. Grain was also obtained from plants of the same cultivars grown in glasshouse conditions. In addition, high amylose, waxy and normal genotypes of barley were grown at two different temperatures in controlled environment conditions.

5.2. Extraction of Starch

Starch was extracted from the cereal grains by degrading endosperm cell wall material using cellulase in the presence of antibiotics and the α-amylase inhibitor, acarbose, and then removing the protein matrix surrounding the starch granules using proteinase K. The starch granules were washed in water and air dried.

5.3. Analytical Methods

Starch granule size distribution was determined using a Coulter Counter Multisizer II; total polysaccharide was determined by the phenol/sulphuric acid method; total and apparent amylose were determined from the absorbance of the blue complex formed between amylose and iodine; starch damage was determined by assaying the dextrins released after controlled treatment of the granules with fungal α-amylase; values for the initial, peak and final gelatinisation temperatures and for the gelatinisation enthalpy were obtained using a Mettler Differential Scanning Calorimeter (DSC) with a TA10 processor; values for gelatinisation temperatures were also obtained by observing starch granules mounted in a solution of Congo Red using a microscope fitted with a Mettler Hot Stage linked to a Mettler FP90 central processor; starch phosphorus was assayed using inductively coupled plasma atomic emission spectrometry; gel filtration chromatography was carried out on solubilised starch using Sepharose 2BCL to determine the amylose:amylopectin ratio, and on starch debranched by isoamylase using Sephadex G50 Superfine to obtain information on the fine structure of the amylopectin fraction; integral lipids were removed from starch granules using butanol-1-ol:water (84:16) and the lipid content of the starch was determined gravimetrically; integral starch proteins were extracted from starch granules in a buffered solution of sodium dodecylsulphate (SDS) (10% w/v) at 100°C for 10 min and separated using SDS polyacrylamide gel electrophoresis.

6. RESULTS AND DISCUSSION

The laboratory method used to extract starch granules from grains of wheat, oats and barley achieved a quantitative extraction of starch. The granules failed to stain with Congo Red, and gave very low values in the chemical assay for starch damage. It was therefore concluded that the granules had suffered minimal damage during the extraction procedure. Starch extracted in the laboratory was compared with starch which had been extracted commercially from the same sample of wheat. The commercially-extracted starch had a somewhat higher level of starch damage, and a greater difference between the onset and final gelatinisation temperatures in both DSC and hot stage microscopy methods for determining gelatinisation temperatures. SDS-PAGE analyses of starch integral proteins showed that the laboratory-extracted granules lacked a 14 kDa protein which was present in the commercially extracted samples. The protein bands were somewhat less sharp in the electrophoretograms of the proteins extracted from laboratory-extracted starch granules than in those of proteins from commercially extracted starch granules. The greatest difference between laboratory- and commercially-extracted starches was in their granule size distribution. The proportion of B-type granules was very much greater in laboratory-extracted starch than in commercially-extracted starch.

The samples of maize starch obtained from industrial sources included starch from waxy maize, amylomaize and normal maize, and in the laboratory starch was extracted from waxy, high amylose and normal genotypes of barley grown under controlled environment conditions. Gelatinisation data obtained using DSC indicated that the crystallinity of the starch extracted from the waxy genotypes was higher than that of the starch extracted from the normal and high amylose genotypes. In addition, it was found that the crystallinity (but not the gelatinisation temperatures) of the starches from both waxy and high amylose barley, cv. Blenheim, grown in a day/night temperature regime of 20°C/12°C, was different from that of the starches from the same genotypes grown in a constant temperature of 16°C.

Data obtained from determinations of the amylose, lipid and phosphorus content of the starch samples has demonstrated marked differences between waxy, normal and high amylose starches, and has indicated that starches from normal genotypes also vary in composition. The number of samples analysed so far is not sufficient to enable any conclusions to be drawn on whether these small differences in composition are reflected in differences in those properties of starch which determine the suitability for a particular industrial purpose, such as gelatinisation temperatures, gel turbidity, and paste viscosity.

ACKNOWLEDGMENTS

The authors wish to acknowledge financial support from The Scottish Office Agriculture, Environment and Fisheries Department and to thank Bayer plc for supplying acarbose.

REFERENCES

Atwell WA, Hood LF, Lineback DR, Varriano-Marston E and Zobel HF (1988) "The terminology and methodology associated with basic starch phenomena." Cereal Foods World 33, 306–311

Bechtel DB, Zayas I, Kaleikau L and Pomeranz Y (1990) "Size-distribution of wheat starch granules during endosperm development." Cereal Chem. 67, 59–63

Doane WM (1989) "New and potential markets for wheat starch." In "Wheat is Unique." ed. Y Pomeranz, AACC Inc., St Paul, MN USA, 615–631

Doublier J-L, Paton D and Llamas G (1987) "A rheological investigation of oat starch pastes." Cereal Chem. 64, 21–26

Evers AD and Lindley J (1977) "The particle-size distribution in wheat endosperm starch." J. Sci. Food. Agric. 28, 98–102

Galliard T and Bowler P (1987) "Morphology and composition of starch." In "Starch, Properties and Potential." ed.T Galliard Society of Chemical Industry, Great Britain, 55–78

Griffin GJL (1989) "Wheat starch in the formulation of degradable plastics." In "Wheat is Unique." ed. Y Pomeranz, AACC Inc., St Paul, MN USA, 695–706

Hoover R and Vasanthan T (1994) "The effect of annealing on the physicochemical properties of wheat." J. Food Biochemistry. 17(5), 303–325

Inouchi N, Glover DV, Sugimoto Y and Fuwa H (1991) "DSC characteristics of gelatinization of starches of single-, double-, and triple-mutants and their normal counterpart in the inbred Oh43 maize Zea mays L. background." Die Stärke 43, 468–472

Jane J (1995) "Starch properties, modifications, and applications." Journal of Macromolecular Science - Pure and Applied Chemistry A32(4), 751–757

Kock H, Beck R and Röper H (1993) "Starch-derived products for detergents." Die Stärke 45, 2–7

Lineback DR and Rasper VF (1988) "Wheat carbohydrates." In "Wheat, Chemistry and Technology." ed. Y Pomeranz, AACC, St Paul, MN, USA, 277–372

Lineback DR (1984) "The starch granule, organization and properties." Baker's Dig. 58, 16–21

Lorenz K (1995) "Physicochemical characteristics and functional-properties of starch from a high beta-glucan waxy barley." Starch/Starke 47(1), 14–18

MacGregor AW and Fincher GB (1993) "Carbohydrates in the barley grain." In "Barley, Chemistry and Technology." ed. AW MacGregor and RS Bhatty, AACC, St Paul, MN, USA, 73–130

Morrison WR (1995) "Starch lipids and how they relate to starch granule structure and functionality." Cereal Foods World 40, 437–446

Oehr KH and Barrass G (1992) "Biomass derived alkaline carboxylate road deicers." Resources Conservation and Recycling 7, 155–160

Paton D (1986) "Oat starch, physical, chemical, and strucrural properties." In "Oats, Chemistry and Technology." ed. FH Webster, AACC, St Paul, MN, USA, 93–120

Röper H (1993) "Industrial products from starch, recent developments, potential applications and future perspectives." In "New Crops for Temperate Regions." eds Antony KRM, Meadley J and Röbbelen, Chapman and Hall, London, 157–167

Sargeant JG (1982) "Determination of amylose,amylopectin ratio of starches." Die Stärke 34, 89–92

Shamekh S, Forssell P and Poutanen K (1994) "Solubility pattern and recrystallisation behaviour of oat starch." Die Stärke 46, 129–133

Shi Y-C and Seib PA (1989) "Properties of wheat starch compared to normal maize starch." In "Wheat is Unique." ed.Y Pomeranz, AACC Inc., St Paul, MN USA, 215–234

Steciak J, Levendis YA and Wise DL (1995) "Effectiveness of calcium-magnesium acetate as dual SO_2-NO_x emission control agent." AIchE Journal 41, 712–722

Svensson E and Eliasson A-C (1995) "Crystalline changes in native wheat and potato starches at intermediate water levels during gelatinization." Carbohydrate Polymers 26, 171–176

Tester RF, and Morrison WR (1994) "Properties of damaged starch granules V. Composition and swelling fractions of wheat starch in water at various temperatures." J. Cereal Sci. 20, 175–181

Tester RF, South JB, Morrison WR and Ellis RP (1991) "The effects of ambient temperature during the grain filling period on the composition and properties of starch from four barley genotypes." J. Cereal Sci. 13, 113–127

Tester RF, Morrison WR, Ellis RH, Piggot JR, Batts GR, Wheeler TR, Morison JIL, Hadley P, and Ledward DA (1995) "Effects of elevated growth temperature and carbon dioxide levels on some physiochemical properties of wheat starch." J. Cereal Sci. 22, 63–71

11

GRAIN COMPOSITION OF *AMARANTHACEAE* AND *CHENOPODIACEAE* SPECIES

Rolf Carlsson

Department of Natural Sciences
Kalmar University
PO Box 905
S-391 29 Kalmar
Sweden

1. INTRODUCTION

The global demand for more food and industrial raw material produced by agriculture prescribes the optimal utilization of every potential plant resource. Several *Amaranthaceae* species are C4-species and adapted to hot climates. Many *Chenopodiaceae* species are adapted to growth on dry and saline soils, and may be UV-B-tolerant. The plant species are considered as potential new crops for food and industrial raw materials.

Aztecs of Mexico (National Research Council, 1975, 1984) and Incas of the Andes (National Research Council, 1975, 1989; Carlsson, 1994) used the grains of the pseudo-cereals of *Amaranthus* and *Chenopodium*, respectively, as major food staples. These grain crop species have been given an increasing interest for a re-introduction in modern agriculture. For more than a decade a series of conferences in America and Europe have advocated these crops.

Pseudo-cereals are dicotyledonous plants, whose seeds are used for food or feed. However, the same plants are also well-known crops as vegetables and for green crop fractionation for multipurpose industrial uses (Carlsson, 1977; Carlsson, 1994).

The present chapter mainly covers the composition of the grains from *Amaranthaceae* species (*Amaranthus*) and *Chenopodiaceae* species (*Atriplex* and *Chenopodium*) with emphasis on grain proteins. Chapter 12 presents a case study examining grain amaranth as a source of specialty starches, and gives additional detail on the agronomics and use of amaranth.

Most of the data presented have been obtained from plant material cultivated by the author during different growth conditions in Sweden (100 - 400 kg N/ha, 18 - 20 weeks), USA (California; 0 - 200 kg N/ha, 16 weeks), Puerto Rico (105 kg N/ha, 12 weeks), and Brazil (Minas Gerais; 0 kg N/ha, 14 weeks).

Cereals: Novel Uses and Processes, edited by Campbell *et al.*
Plenum Press, New York, 1997

2. CHEMICAL COMPOSITION OF THE GRAINS

The grain yields ranged for *Amaranthus* from about 900–1,800 kg DM/ha and 200 to 300 kg protein/ha for a growth period of 12–20 weeks (cf. below), while the yields of *Atriplex/Chenopodium* ranged from about 2,000 to 4,000 kg DM/ha and 400 to 900 kg protein/ha (Carlsson, 1980). A higher fertilizer level gave higher yields, at comparative cultivations. The chemical composition of the grains are given in Tables 1, 2 and 3. A good review is given by Ruales (1992).

Table 1. Proximate composition of grains of species of *Amaranthaceae*

Species		Protein	Fibre	Fat	Ash (% of DM)
Amaranthaceae:					
Amaranthus:					
Sweden:					
A. bouchonii	(a)	14.1	—	—	—
A. caudatus	(a)	15.8	—	—	—
A. hybridus	(b)	16.2	—	—	—
A. monstrosus	(a)	14.4	—	—	—
A. paniculatus	(b)	17.1	—	—	—
California, USA:					
A. anclancalius		17.9	5.3	5.4	3.4
A. ascendens		16.9	5.4	4.9	3.5
A. caudatus		15.1	4.2	6.9	3.2
A. flavus		17.0	5.0	4.4	3.7
A. gangeticus		16.1	5.4	5.1	3.5
A. hypocondriacus		16.5	5.2	6.2	3.0
A. paniculatus		16.6	5.0	4.9	4.1
A. retroflexus		14.1	6.4	6.4	3.1
A. species —Taiwan		17.8	6.2	6.5	3.6
Puerto Rico:					
A. anclancalius		16.5	—	3.2	—
A. cruentus		16.8	—	5.2	—
A. gangeticus		16.6	—	3.5	—
A. hypocondriacus		15.0	—	4.2	—
A. mantegazzianus		15.1	—	4.2	—
A. species - Taiwan		15.9	—	5.2	—
Brazil:					
A. anclancalius		14.4	2.9	5.1	2.5
A. cruentus		14.4	2.4	4.8	2.8
A. gangeticus		14.4	2.6	6.6	3.5
A. hypocondriacus		14.4	2.6	6.4	2.7

Note: Sweden: a = 155 kg N/ha, b = 400 kg N/ha; California: 200 kg N/ha; Puerto Rico: 105 kg N/ha; Brazil: 0 kg N/ha.

Reference: Carlsson (1980) for Sweden, California, and Puerto Rico, and Correa (1983) for Brazil, if further details are wished.

2.1. Starch

The starch granules of the *Amaranthus* and *Chenopodium* pseudo-cereals investigated are extremely small (1–3 μm) and have a crystalline structure (National Research Council, 1984; Saunders, 1984; Ruales, 1992). This makes the starch commercially interesting. Analyses of the starch have showed that the amylose content was low and that amylopectin dominated (e.g. Saunders, 1984; Carlsson, 1994).

The granular amylopectin starch is apparently not much available as an energy source, without cooking or treatment by hot water of the grains (cf. below). It seems otherwise possible to use the starch as a "low calorie cream substitute". Also, positive effects for industrial baking have been obtained.

The starch content was analysed for grains of California grown *Amaranthus* plants, which contained 62–65% starch of the DM. Plants grown in Sweden, such as, *A. hortensis* contained 50%, and *C. quinoa* 60%. When not analysed the starch content can be estimated as the residual amount of the dry matter not specified in Tables 1, 2 and 3.

Table 2. Proximate composition of grains of species of *Chenopodiaceae*

Species	Protein	Fibre	Fat	Ash (% of DM)
Chenopodiaceae:				
Atriplex:				
Sweden:				
A. hastata	19.2	—	6.6	—
A. hortensis:		—		—
brown seed	25.2		7.0	
black seed	19.0		5.8	
A. nitens	27.7	—	4.3	—
Chenopodium:				
Sweden:				
C. album	18.3	—	12.0	—
C. ambrosoides	12.4	—	7.3	—
C. bonus henricus	18.3	—	8.2	—
C. foliosum	13.6	—	6.0	—
C. giganteum	19.3	—	5.7	—
C. gigantospermum	16.0	—	7.4	—
C. glauca	14.8	—	9.3	—
C. hydridum	14.0	—	6.8	—
C. opulifolium	17.4	—	5.0	—
C. pumilio	16.8	—	7.6	—
C. quinoa BP—183	17.3	2.0	6.3	3.6
C. rubrum	16.9	—	14.8	—
C. schraderianum	15.2	—	5.4	—
C. urbicum	17.0	—	5.1	—
C. viride	16.0	—	6.8	—
C. vulvaria	19.9	—	7.7	—
Puerto Rico:				
C. quinoa BP—83	17.9	—	5.0	—

Note: Sweden: 155 kg N/ha. Puerto Rico: 105 kg N/ha.
Reference: Carlsson (1980) for further details.

Due to the high protein content of the pseudocereal grains, relative to normal cereal grains, the starch content is lower in pseudo-cereals than in cereal grains.

2.2. Protein

In general the grains are rich in protein that is well balanced from a nutritional point of view (Carlsson, 1980; Carlsson, 1994).

Cultivation of the plants under different growth conditions showed that grains of *Amaranthus* species contained about 14 to 18% protein (Table 1; Cheeke et al, 1980; Carlsson, 1980; Correa, 1983) and *Atriplex* and *Chenopodium* species grains contained from 14 to 28% (Table 2; Carlsson 1980; Carlsson, 1994). For other *Chenopodiaceae* species (e.g. *Kochia* and *Salsola* species) similar values were noted (Carlsson, 1994). An increased level of nitrogen fertilizers increased the grains protein contents with a few %-units (*Amaranthus* spp.: Table 1; *Chenopodium quinoa*: Table 3).

2.3. Amino Acids in the Whole Grain Protein

The protein amino acid composition of the *Amaranthus* and *Chenopodium* grains is most favourable for human demands as well as for animal feeding (Tables 4 and 5). The lysine and the sulphur amino acids (methionine, cysteine/cystine) contents are high and not limiting for growth, compared to the contents of cereal proteins (Saunders, 1984; Carlsson, 1994): The leucine content may be limiting for *Amaranthus*/*Chenopodium* grains as a sole protein source, but not in feed mixtures.

An increased nitrogen fertilizer level from 0 to 200 kg N/ha increased the level of e.g. lysine of the whole grain protein. An increase in maturation time before harvest (11 and 16 weeks) did not affect the amino acid composition (Carlsson, 1980).

There is an interesting variation of the protein amino acid composition in *Amaranthus* grains (Table 5: see different essential amino acids), which might indicate a possibility for selection for an even better protein amino acid composition of the grains. Part of

Table 3. *Chenopodium quinoa* grain protein content. Effects of cultivation year and nitrogen fertilizer levels (kg/ha)

	C. quinoa BP 183			*C. quinoa BP 183*	
Year	% of DM	kg N/ha	Year	% of DM	kg N/ha
1968	16.7	400	1982	12.3	100
1969	14.5	200	1984	16.3	200
1970	14.8	310	1988	14.7	100
1971	14.9	260	1989	15.9	200
1972	14.8	300	1990	16.6	200
1973	17.3	450	1991	14.3	100
1976	15.7	190	1992	18.3	200
1977	14.5	200	1993	15.7	150
1978	15.7	265			
1979	13.6	110			
1980	14.8	200			
1981	15.8	200			

Note: Nitrogen is given as NPK (14-4-17). Cultivated areas from 100m to 1 ha.
Averages:14.2 (100 kg N/ha), 15.4 (200 kg N/ha), 15.1 (300 kg N/ha), 17.0 (400 kg N/ha).

Table 4. Amino acid composition of grain protein of *Atriplex hortensis*, *Chenopodium album*, *Chenopodium pallidicaule*, *Chenopodium quinoa*, *Kochia scoparia*, *Portulaca oleracea* (g amino acid per 100 g protein or 16 g nitrogen)

Latin Name:	Cys	Met	Asp	Thr	Ser	Glu	Pro	Gly	Ala	Val	Ile	Leu	Tyr	Phe	Lys	His	Arg	Try
A. hortensis(author)	1.7	2.1	—	3.7	—	—	—	—	—	4.8	4.1	6.0	3.2	4.0	5.2	2.6	—	—
C. album(author)																		
Sweden	2.6	2.1	9.2	4.8	4.9	15.8	4.5	6.2	4.6	5.3	4.0	8.2	3.9	4.6	5.3	2.8	9.9	—
India	2.2	2.5	9.5	5.3	5.2	15.9	5.2	5.9	4.9	5.9	3.3	7.9	4.3	4.3	6.7	2.8	8.2	—
C. pallidicaule	—	—	—	4.8	—	—	—	—	—	4.6	6.8	5.8	—	3.6	6.0	2.5	7.9	0.8
C. quinoa(author)	2.0	2.3	9.0	4.1	4.8	17.8	3.2	6.4	4.8	5.2	4.9	7.3	3.5	4.5	7.0	3.4	11.9	—
K. scoparia	1.1	—	8.0	3.3	3.7	12.8	3.5	5.2	4.0	4.3	3.0	5.9	2.9	4.6	4.8	3.3	7.2	—
P. oleracea	—	1.9	6.5	2.6	2.9	13.0	3.3	6.4	2.8	3.3	2.7	4.6	3.8	3.3	2.8	2.1	8.1	—
FAO-human	3.5	—	4.0	—	—	—	—	—	—	5.0	4.0	7.0	5.6	5.5	—	—	—	

Note: Data from Carlsson (1994)

Table 5. Amino acid composition of grain protein of *Amaranthus caudatus*, *A. cruentus*, *A. hypocondriacus*, *A. gangeticus* and *A. mantegazzianus* (g amino acid / 16 g N)

Latin Name:	Cys	Met	Asp	Thr	Ser	Gly	Pro	Glu	Ala	Val	Ile	Leu	Tyr	Phe	Lys	His
A. caudatus (200)[1]	2.3	2.4	—	3.5	5.9	6.9	—	—	—	4.1	3.6	5.3	2.8	3.4	5.3	2.5
A.cruentus (0)	2.3	2.2	—	3.5	5.9	6.9	—	—	—	4.3	3.8	5.5	3.7	3.7	5.1	2.4
A. cruentus (200)	2.2	2.3	—	3.6	5.9	7.2	—	—	—	4.6	4.0	5.7	4.0	4.0	5.5	2.6
A. hypocon. (0)	2.3	2.5	—	3.0	6.3	6.6	—	—	—	4.0	3.5	5.2	3.0	3.5	5.1	2.3
A. hypocon. (200)	2.5	2.6	—	3.6	6.3	7.4	—	—	—	4.5	3.9	5.7	3.3	4.0	5.5	2.5
A. gangeti (200)	2.2	2.6	—	3.5	6.7	7.8	—	—	—	4.5	3.9	5.6	2.8	3.9	5.2	2.5
A. mantega (200)	2.0	2.3	—	3.5	5.3	6.5	—	—	—	4.6	3.9	5.6	3.3	3.9	5.2	2.5
Human ref.	1.3	2.2	—	4.0	—	—	—	—	—	5.0	4.0	7.0	2.8	2.8	5.5	—

Note: 1: (kg N/ha) ; Human reference protein according to FAO 1965 and 1973

Table 6. Proportions between the grain protein fractions albumin, globulin, prolamin, and glutelin (%)

Latin name	Albumin	Globulin	Prolamin	Glutelin
Amaranthus sp (Correa, 1983)	65	17	11	7
Chenopodium quinoa (author)	44	37	10	10

the variation could be due to a difference in the proportions between the different major grain protein fractions.

2.4. Lipids

Both *Amaranthaceae* and *Chenopodiaceae* grains have higher lipid contents (oil and fat) than traditional cereal grains (Bressani and Elias, 1984; Saunders, 1984; Carlsson, 1994). Thus these grains are more energy-rich than traditional cereal grains. A few *Chenopodiaceae* genera, such as *Salicornia*, are even looked upon as possible commercial oil crops.

The oil of the grains of *Amaranthus* and *Chenopodium* is rich in oleic acid and linoleic acid. In general, for *Amaranthus* species 20 to 35% of the fatty acids are oleic acid, 40 to 60% linoleic acid (Carlsson, 1980; Becker et al, 1981), while for *Chenopodium* species, 20 to 25% of the total fatty acids are oleic acid and 40 to 60% are linoleic acid (Carlsson, 1994). The content of palmitic acid is about 20% for both types of species. For *Chenopodium* species a few per cent of C-22 and C-22 fatty acids are present. *Amaranthus* grains have been reported to have relatively high squalene values (Saunders, 1984).

Table 7. Amino acid composition of extracted protein fractions of *Amaranthus anclancaliu* L. and *A. hypocondriacus* (mg amino acid / 100 mg protein)

	Albumin		Globulin		Prolamin		Glutelin	
Fraction	*A.a.*	*A.h.*	*A.a.*	*A.h.*	*A.a.*	*A.h.*	*A.a.*	*A.h.*
Amino acid:								
Aspartic acid	6.6	8.3	9.0	9.1	7.1	8.2	8.8	11
Threonine	2.8	3.5	3.1	3.0	4.6	5.4	4.0	4.5
Serine	3.9	4.7	4.6	4.4	3.6	4.6	4.8	5.1
Glutamic acid	28	24	21	22	21	16	25	20
Proline	3.6	3.7	4.2	4.2	6.0	6.1	4.7	4.6
Glycine	7.2	7.8	4.5	4.4	7.5	7.8	7.5	8.5
Alanine	2.6	3.5	3.4	3.0	6.8	7.7	5.0	6.1
Valine	3.9	4.2	4.7	4.7	7.2	8.0	5.0	5.9
Isoleucine	4.0	3.8	4.1	3.9	5.0	4.8	4.0	4.0
Leucine	4.6	5.1	6.4	6.2	6.9	6.8	6.2	6.6
Tyrosine	3.8	3.8	4.4	4.2	3.9	4.9	2.9	3.2
Phenylalanine	3.6	3.6	6.1	6.0	4.7	3.9	4.7	4.4
Lysine	7.3	8.2	6.1	6.0	4.3	5.7	4.8	6.3
Methionine	2.2	1.9	4.1	5.3	2.1	2.2	1.0	0.6
Histidine	2.1	2.2	2.9	2.8	1.8	1.9	1.9	1.7
Arginine	14	11	11	11	7.7	5.8	10	7.3

Note: Data from Correa (1983)

2.5. Vitamins and Minerals

The vitamin and mineral composition of *Amaranthus* grains (Saunders, 1984) and of *Chenopodium* grains (Risi and Galwey, 1984; Ruales, 1992) have been well documented. Compared to cereal grain the contents of both vitamins and minerals are high, probably due to occurrence of dicotyledon leaves in the grains.

The levels of provitamin A (beta-carotene), vitamin B (as folic acid, riboflavin, niacin and thiamine) and tocopherols, as well as vitamin C are fairly high.

The content of minerals, such as Ca, P, Mg, Fe, and K, is also high. The content of iron is several times higher than literature values for whole wheat grains. The high calcium values are connected to high contents of insoluble oxalates.

3. NUTRITIVE VALUE OF GRAIN PROTEINS

3.1. Amaranthaceae

The excellent amino acid compositions of the protein of the raw *Amaranthus* grain would indicate a high *in vivo* value of the protein (Carlsson, 1980; Cheeke et al, 1980; Correa, 1983). However, this was not the case (cf. below). Rat assays were mainly used for nutritive tests, unless stated otherwise (cf. Saunders, 1984: rats, chicks and pigs). The nutritive effects on small children have been investigated (Morales, 1984). Cooking, also when including drum drying of the cooked grains, increased the nutritive value (Cheeke et al, 1980; Bressani and Elias, 1984; Saunders, 1984). Heating the grain (+60°C) as such and popping them did not increase the nutritive value.

The reason for an unexpected low nutritive value could be due to unavailable energy of the starch, having a low digestibility. Cooking the grains to open up the starch molecule structure apparently increased the nutritive value of the grain (Carlsson, 1980; Bressani and Elias, 1984; Saunders, 1984). On the other hand, occurrence of antinutritive substances could lower the availability of the protein amino acids. The contents of trypsin inhibitors, phytohaemagglutinins, oxalate, nitrate, and saponins are relatively low (Carlsson, 1980; Correa, 1983; Saunders, 1984), while the content of phenolics, especially in dark-seeded types could be high (Carlsson, 1980).

There was a strong negative correlation between the phenolic content of the whole grain flour and its pepsin/pancreatin *in vitro* value. Oxidized phenolics, quinones, makes lysine in proteins unavailable. By grinding and sieving the flour of *Amaranthus* grains, it was noted that the starch-enriched fraction, with less phenolics, had as low a nutritive value as whole grain flour, while a finely ground dark-grey high-content protein fraction with a 3 times higher phenolic content had a fairly good nutritive value (Carlsson, 1980). Saunders (1984) also showed that the starchy perisperm fraction had a low nutritive value, while a seed coat-embryo meal fraction (comparable to the dark-grey fraction) had a relatively high value.

In all cases, the seed coat-embryo meal fraction had a high protein content, up to about 40% of the dry matter (Carlsson, 1980; Sanchez-Marroquin, 1984; Saunders, 1984).

The "high" nutritive value of the seed coat-embryo meal, in spite of the high phenolic content, might be explained by a better composed protein, probably based on albumins with a high lysine content from the dicotyledons and the other outer parts of the grain.

By supplementing both raw and cooked grain flours by leucine, Bressani and Elias (1984) showed that the protein's nutritive value increased considerably. Thus, raw starch, some phenolics, and the limiting amino acid leucine lower the putative high nutritive value of the whole grain protein of *Amaranthus*.

3.2. Chenopodiaceae

Grain proteins of cultivated *Chenopodium* species seem to have a relatively high *in vivo* nutritive value, while the value of *Atriplex* species grains are lower, although both types of grains have well-balanced protein amino acids (Carlsson, 1980). The nutritive tests were mainly based on rats; other cases have investigated poultry and pigs (Gandarillas et al, 1968), as well as undernourished young boys (Lopez de Romana et al, 1981). Weanling food from *C. quinoa* can be useful (Ruales, 1992).

In contrast to *Amaranthus* grains the content of saponin as a antinutritive factor can be high for both *Atriplex* and *Chenopodium* grains. The content of antinutritive phenolics can also be high in dark-coated seeds. Otherwise, as for *Amaranthus* grains above the content of other antinutritive substances are low.

The nutritive value for *C. quinoa* and *C. pallidicaule* grain proteins estimated by rat assays indicates values similar to or better than milk protein (White et al, 1955; Quiros-Perez and Elvehjem, 1957; Mahoney et al, 1975; Telleria Rios et al, 1978).

However, as for *Amaranthus* grains, cooking or water extraction of saponins at a high temperature (70 - 87°C) could give higher nutritive protein values than for non-treated grains (Mahoney et al, 1975; Telleria Rios et al, 1978). Thus, as for *Amaranthus* grains, the starch structure should be changed before grain consumption.

On the other hand, bitter, high-saponin content types of *C. quinoa* and *A. hortensis* grains have low nutritive values. Water or ethanol extraction can in such cases increase the nutritive value of the protein (Carlsson, 1980). Water, often at pH 8, removes mainly saponins, but ethanol extracts also phenolics (cf. above).

Sweet grain types of *C. quinoa* generally have a higher nutritive value of their grains than bitter types (Telleria Rios et al, 1978). A low value for saponin-containing grains could be due to a low feed intake or an effect on the protein digestibility (Cheeke, 1976). Fractionation of *Chenopodium* grains also gives a starchy perisperm and an embryo/seed coat meal. The latter may contain several times more saponin than the whole grain, but its protein content is around 40% of the dry matter.

The high saponin content in *Chenopodiaceae* species can be interesting as a pharmaceutical tool for anticholesterol effects, as the saponins from *Medicago sativa* (lucerne, alfalfa) (Ueda et al, 1987).

The starch of the present pseudo-cereal grain has to be "cooked", and antinutritive substances as phenolics or saponins have to be removed before the full nutritive value of the grains can be used. These antinutritive substances, biological pesticides, can thus not fully be removed from the crop without negative consequences for its cultivation.

4. GRAIN PROTEIN FRACTIONS

A simple way to extract and fractionate proteins into albumin, globulin, prolamin and glutelin has been used to determine which fraction could be used for selection work or production of protein health drinks (Carlsson, 1980; Correa, 1983; Rosenlund, 1989; Carlsson, 1994). The whole grain protein of various species of *Amaranthus*, *Atriplex*, and *Chenopodium* have been fractionated in this way. The proportions of the fractions are given in Table 6.

The amino acid compositions of the protein fractions are given in Tables 7, 8, 9 and 10, for *Amaranthus anclancalius* and *A. hypocondriacus* (Correa, 1983), *Atriplex hortensis*, *Chenopodium album*, and *C. quinoa*, respectively (Carlsson, 1994). The content of ly-

Table 8. Amino acid composition of extracted protein fractions of *Atriplex hortensis* L. cv BP 150 (mg amino acid / 100 mg amino acids)

	Protein fraction			
Amino acid	Albumin	Globulin	Prolamin	Glutelin
Cystine	1.6	1.7	5.3	1.9
Methionine	3.1	2.2	1.4	2.5
Aspartic acid	10.1	11.4	7.6	10.3
Threonine	5.9	5.3	3.7	3.4
Serine	5.4	5.8	4.3	6.0
Glutamic acid	14.6	15.2	13.4	20.0
Proline	3.0	2.1	15.5	4.4
Glycine	5.8	5.8	8.6	4.5
Alanine	5.1	5.1	5.6	3.6
Valine	6.0	6.3	3.7	5.0
Isoleucine	5.0	4.8	3.6	4.8
Leucine	7.6	7.9	5.2	5.8
Tyrosine	3.8	4.1	5.2	3.6
Phenylalanine	5.2	5.5	3.6	4.4
Lysine	7.3	7.0	3.1	4.2
Histidine	2.7	2.4	2.6	3.5
Arginine	8.1	7.4	4.4	12.6

Note: Data from Carlsson (1994)

Table 9. Amino acid composition of extracted protein fractions of *Chenopodium album* L. cv C4-India (mg amino acid / 100 mg amino acids)

	Protein fraction			
Amino acid	Albumin	Globulin	Prolamin	Glutelin
Cystine	2.0	1.1	6.0	1.6
Methionine	3.3	3.3	2.2	3.3
Aspartic acid	10.8	9.9	7.9	9.6
Threonine	5.9	4.1	4.7	3.7
Serine	5.7	5.6	4.9	5.2
Glutamic acid	15.9	17.0	17.4	18.5
Proline	2.5	3.7	6.0	4.7
Glycine	6.1	5.5	9.4	5.6
Alanine	5.6	4.3	6.8	4.0
Valine	5.8	5.6	4.4	5.0
Isoleucine	4.8	5.5	3.4	5.1
Leucine	7.6	7.3	6.2	6.4
Tyrosine	3.8	4.2	5.8	4.0
Phenylalanine	4.9	4.8	4.6	4.8
Lysine	6.6	5.6	3.4	4.5
Histidine	2.2	3.1	1.3	3.6
Arginine	6.3	9.2	5.6	10.6

Note: Data from Carlsson (1994)

Table 10. Amino acid composition of extracted protein fractions of *Chenopodium quinoa Willd.* cv Sajama-Bolivia (mg amino acid / 100 mg amino acids)

	Protein fraction			
Amino acid	Albumin	Globulin	Prolamin	Glutelin
Cystine	1.4	1.3	3.9	0.9
Methionine	4.4	1.7	1.7	2.2
Aspartic acid	9.9	10.8	7.7	11.0
Threonine	5.2	4.0	4.5	4.5
Serine	5.5	6.3	5.0	5.0
Glutamic acid	16.1	18.8	14.0	18.0
Proline	2.2	3.0	10.1	5.3
Glycine	5.5	5.5	9.4	5.8
Alanine	5.3	5.4	4.6	5.6
Valine	5.8	5.0	3.8	6.4
Isoleucine	4.2	4.8	2.6	5.9
Leucine	8.0	8.0	6.4	8.3
Tyrosine	3.8	3.8	5.8	4.1
Phenylalanine	4.8	5.2	3.5	5.8
Lysine	6.9	4.4	6.0	3.7
Histidine	2.9	3.4	3.6	2.5
Arginine	8.1	9.6	7.3	8.2

Note: Data from Carlsson (1994).

sine was highest in the albumin fractions (6.6 - 8.1% of the protein amino acids.) For *Amaranthus* protein fractions, the methionine content was highest in the globulin fractions (4.1 - 5.3%; cysteine was not analysed), while the *Chenopodiaceae* species indicated that the methionine content was highest in the albumin fractions (3.1- 4.4%). For the latter species the cysteine content was highest in the prolamin fractions (3.9 -6.0%). The differences in the amino acid composition of the protein fractions could be of great value for the food product industry and for genetic breeding work, including bio-engineering to produce transgenic plants such as transgenic high-lysine rice.

5. CONCLUSIONS

The ancient pseudo-cereals from *Amaranthus* and *Chenopodium* species of the Aztec and Inca empires, as well as from India, are being re-introduced at a global scale due to their excellent grain compositions. The grains are rich in a well-balanced protein and contain, relative to traditional cereals, a high fat content, a high content of most vitamins, and essential minerals such as iron. The nutritive value of well-processed grain flour is high. The flour is excellent for food formulas, especially for young children. The grains can be used as feed. The amylopectin type of the starch can have interesting industrial applications.

REFERENCES

Bressani R and Elias LG (1984) "Development of 100% amaranth food." Proceedings of 3rd Amaranth Conf., Rodale Press Inc., Emmaus, PA 18049, USA, pp 8–19

Carlsson R (1977) "*Amaranthus* species and related species for leaf protein concentrate production." Proceedings of 1st Amaranth Conf., Rodale Press Inc., Emmaus, PA 18049 USA, pp 83–99

Carlsson R (1980) "Quantity and quality of *Amaranthus* grains from plants in temperate cold and hot and subtropical climates - A review." Proceedings of 2nd Amaranth Conf., Rodale Press Inc., Emmaus, PA 18049 USA, pp 48–58

Carlsson R (1994) "*Chenopodiaceae* species: Salt-tolerant plants for green biomass and grain production." In "Handbook of Plant and Crop Stress." Ed. M Pessarakli, Marcel Dekker Inc., New York USA, pp 543–558

Cheeke PR (1976) "Nutritional and physiological properties of saponins." Nutr. Reports Int. 133, 315–324

Cheeke PR, Bronson J and Carlsson R (1980) "Feeding trials with *Amaranthus* grain forage and leaf protein concentrates." Proceedings of 2nd Amaranth Conf., Rodale Press Inc., Emmaus PA 18049 USA, pp 5–30

Correa AD (1983) "Estuda da proteina e de outros constituentes da semente de algumas especies de amaranto." MSc thesis in Biochemistry, Department of Biochemistry and Immunology, Federal Univ. Minas Gerais Belo Horizonte MG Brazil, 78 p

Gandarillas HSC, Cardozo A and Alandia SB (1968) "La alimentation con quinoa en el crecimiento de pollos cerdos." Boletin Experimental No 33, DivInvestAgric Estacion, Experimental Ganadera de Patacamaya, Ministerio de Agricultura Bolivia, 12 p

Lopez de Romana G, Graham GG, Rojas M and MacLean WC Jr (1981) "Digestibilidad y calidad proteinica de la quinoa: Estuda comparativo en ninos entre semilla y harina de quinoa." ArchLatinoamericanos de Nutricion 31(3), 485–498

Mahoney AW, Lopez JG and Hendricks DG (1975) "An evaluation of the protein quality of quinoa." J. Agr. Food Chem. 23(2), 190–193

Morales E (1984) "Digestibility and utilization of grain amaranth protein and energy by small children." Proceedings of 3rd Amaranth Conf. Rodale Press Inc., Emmaus PA 18049 USA, pp 157–166

National Research Council (1975) "Underexploited tropical plants with promising economical value." National Academic Press Washington DC USA

National Research Council (1984) "Amaranth - Modern prospects for an ancient crop." National Academy Press, Washington DC, USA

National Research Council (1989) "Lost crops of the Incas: Little-known crops of the Andes with promise for worldwide cultivation." National Academic Press, Washington DC, USA

Quiros-Perez F and Elvehjem CA (1957) "Nutritive value of quinoa proteins." J. Agr. Food Chem. 5:7 538–541

Risi JC and Galwey NW (1984) "The *Chenopodium* grains of the Andes: Inca crops for modern agriculture." Adv. Appl. Bio. 10, 145–216

Rosenlund S (1989) "Identification of seed proteins in *Chenopodium* album L cv 2." BSc Thesis, Dept of Natural Science, Kalmar University, Kalmar, Sweden

Ruales Najera J (1992) "Development of an infant food from quinoa (*Chenopodium* quinoa Willd) - Technological aspects and nutritional consequences." PhD Thesis, Department of Applied Nutrition and Food Chemistry, Lund University, Lund, Sweden

Sanchez-Marroquin A (1984) "Amaranth as an enriching product in staple foods." Proceedings of 3rd Amaranth Conf., Emmaus, PA 18049, USA, pp 20–45

Saunders RM (1984) "Nutritional and starch composition studies with grain amaranth." Proceedings of 3rd Amaranth Conf., Rodale Press Inc., Emmaus, PA 18049, USA, pp 46–62

Telleria Rios ML, Sgarbieri VC and Amaya-F J (1978) "Evaluacion quimica y biologica de la quinoa (*Chenopodium* quinoa Willd): Influencia de la extraccion de las saponinas por tratamiento termico." Arch Lationamericaos de Nutricion 28, 253–263

Ueda H, Ohshima M and Akimoto I (1987) "Nutritive value and hypocholesterolemic effect of alfalfa leaf protein concentrates prepared from two different varieties in chicks." Jpn. J. Zootech. Sci. 58(4), 347–355

White PL, Alvistur E, Dias C, Vinas E, White HS and Collazos C (1955) "Nutrient content and protein quality of quinoa and canihua - Edible seed products of the Andes mountains." J. Agr. Food Chem. 3(6), 531–534

12

DEVELOPING SPECIALTY STARCHES FROM NEW CROPS

A Case Study Using Grain Amaranth

Harold Corke,[1] Huaixiang Wu,[1] Shaoxian Yue,[2] and Hongliang Sun[2]

[1]Department of Botany
University of Hong Kong
Pokfulam Road, Hong Kong
[2]Institute of Crop Breeding and Cultivation
Chinese Academy of Agricultural Sciences
Beijing 100081, China

A wide range of variation was found in the properties tested among *Amaranthus* species and among genotypes within the same species. It was generally found that the amylose content of cultivated genotypes of *Amaranthus* was lower than that of non-cultivated genotypes; starch of cultivated genotypes had more stable pasting properties (i.e. higher peak viscosity, lower viscosity drop during shear thinning and lower retrogradation) than noncultivated genotypes; starch of cultivated genotypes had lower T_p and higher ΔH than non-cultivated genotypes; the starch pastes of cultivated genotypes were stable during cold storage, i.e. hardness, cohesiveness and modulus of cultivated starch pastes were lower, and adhesiveness was higher, compared to non-cultivated genotypes. The values for pasting, functional, and thermal properties of *Amaranthus* starch were highly correlated, especially the pasting and functional properties. Amylose content was closely related to the physical and functional properties of *Amaranthus* starch. The environmental effect on the properties of *Amaranthus* starch was different for different species. Compared to the reference corn, rice, potato and wheat starches, *Amaranthus* starch tended to have more stable paste, i.e. lower shear thinning and lower retrogradation, and higher T_p and ΔH; *Amaranthus* starch paste was more resistant to cold storage. Generally, many *Amaranthus* starches would be good thickeners and stabilizers in food processing. The wide genetic diversity necessitates specific choices for specific uses.

1. INTRODUCTION

Grain amaranth, an annual food and feed crop, is a dicotyledonous C_4 plant belonging to the *Amaranthaceae,* genus *Amaranthus,* and consisting of several species. It is a

Cereals: Novel Uses and Processes, edited by Campbell *et al.*
Plenum Press, New York, 1997

fast-growing plant, resistant to stress conditions, and of high nutritional value. It was a staple food in ancient Aztec culture and is still cultivated as a minor food crop in Central and South America and some areas of Asia and Africa. This crop is now attracting world-wide attention because of its superior agronomic traits and its high potential for food and feed uses. Grain amaranth has been recognized as a high-potential new food crop for the 21st century, and in China particularly we think it has equal potential as a feed crop.

2. CHARACTERISTICS OF GRAIN AMARANTH IN CHINA

There are two types of amaranth, vegetable amaranth and grain amaranth. On top of the stem is an indefinite inflorescence, the panicle. Amaranth grain is very small (about half the size of millet), and may be light yellow, brown-yellow or brown-black in color. The 1000 seed weight is 0.6 – 0.9 g, and 60,000 – 100,000 seeds can be produced by a single plant. Grain amaranth has four notable characteristics:

- High grain protein content (around 16%) and quality (high lysine content); leaf protein comparable to alfalfa (lucerne).
- High yield potential. It typically gives a grain yield of 2,250 – 4,500 kg ha^{-1} and a fresh weight of leaf and stem of 30,000 – 60,000 kg ha^{-1}.
- High stress tolerance, to drought, salinity, alkalinity, or acidic soil conditions.
- Very low seeding rate and high germination rate, making it suitable for reclamation of barren land using aerial sowing.

Chapter 11 presents more detail on the composition and nutritional quality of Amaranth species.

3. INTRODUCTION OF GRAIN AMARANTH TO CHINA FROM ABROAD

Since 1982 we (Professors SX Yue and HL Sun) have introduced many varieties of grain amaranth from the United States and planted them in more than twenty regions in China with good results (Yue et al., 1993). Through 13 years of screening and breeding, five varieties have been approved by the Government and released throughout the country. These varieties are *Amaranthus cruentus* R104, *A. cruentus* K112, *A. hypochondriacus* 1023, *A. hypochondriacus* 1024, and *A. hybridus* 1004. The area planted to commercial amaranth (grain and forage) annually in China has now reached 86,000 ha, mainly distributed in Sichuan and Yunnan Provinces and areas of north and northeast China, and in coastal shoally land, etc.

4. THE PRESENT STATUS OF AMARANTH RESEARCH AND DEVELOPMENT IN CHINA

4.1. Amaranth as a Feed Source

Using amaranth grain meal or dry leaf meal in compound feed for chicken, pig, and cattle can raise the quality and yield of the animal products. For raising fish in coastal fishponds a complete feed has been made from a combination of grain amaranth and Sudan grass. See Chapter 11 for other examples of Amaranth used for animal feed and human food.

4.2. Food Applications

Many food products have been made and sold on the retail market, e.g. amaranth instant flour, dried noodles, cakes, biscuits, popped amaranth, and soysauce made from soybean and brown-black seeded amaranth.

4.3. Cultivation Techniques

Recommendations for amaranth cultivation for high yield production in different regions and systems of China have been proposed. The highest grain yield achieved to date has been 5,340 kg ha^{-1}, and the highest yield of silage was 172,500 kg ha^{-1}.

5. DEVELOPMENT PROSPECTS

5.1. Developing Barren Land with Grain Amaranth

There is a vast area of barren land, saline – alkaline soil, coastal shoally land, acid soil, sandy soil, etc., which is wasted or underutilized. Grain amaranth could be planted as a pioneer plant to exploit these lands to improve the soil and obtain feed for developing animal production.

5.2. Planting Amaranth and Raising Livestock and Poultry in Peasant Courtyards

Chinese peasant households often have a courtyard for themselves of 70 – 200 m^2, used to plant flowers and trees and raise livestock and poultry. The yield of 70 m^2 of planting amaranth can provide the silage for 3 – 5 pigs, a major boost to livestock production in peasant households.

5.3. Aerial Sowing of Perennial Pasture and Annual Grain Amaranth for High Quality Forage

Aerial sowing of perennial pastures is popular in China, especially in plateaux of mountain areas where soil and water loss is serious in exposed soil. A mixed sowing of perennial pasture and an annual feed crop like grain amaranth may be used.

5.4. Combined Utilization of Grain Amaranth for Feed and Food

The economic benefits of planting grain amaranth in China are clear. The ratio of investment input to production value for most crops is 3 – 5, but for grain amaranth it is as high as 6 – 10.

Table 1. *Amaranthus* genotypes screened for adaptability in China

Variety	Source	Score	Comments
1. *A. cruentus* K112	USA	5	High yield, disease free, regular
2. *A. cruentus* R104	USA	5	High yield, early maturity, slight disease
3. *A. cruentus* V67	China/Shennongjia	5	High yield, disease free, regular
4. *A. cruentus* V69	China/Shennongjia	5	High yield, disease free, regular
5. *A. cruentus* CrO72	India/Kerala	5	High yield, disease free, regular
6. *A. hybridus* HrO27	Zambia	4	High yield, regular, slight disease
7. *A. pumilus* Au002	US/Florida	4	Medium-high yield, luxuriant leaves, slight disease
8. *A. deflexus* De002	France	4	Medium-high yield, disease free, seed too small
9. *A. cannabinus*	US/Virginia	4	Medium-high yield, deciduous leaves in later stage of growth
10. *A. paniculatus*	China/Tibet	4	High yield, regular, slight disease

5.5. Screening Genotypes for Adaptability in China

In the past three years cooperative research has been conducted between the University of Hong Kong and the Chinese Academy of Agricultural Sciences, to test the adaptability and quality of 250 genotypes in Beijing and Wuhan. Ten useful genotypes have been screened out, as shown in Table 1.

In short, grain amaranth has great development potential in China. In the feed industry it can produce leaf meal feed; for the food industry a high quality protein, a functionally interesting starch, or a tasty flour supplement. In a history of several thousand years of human agricultural development, over many thousands of types and varieties of crop have been selected, but today some 90% of the world's food is derived from less than ten major crops such as wheat, rice, maize, sorghum, barley, soybean, potato, sweet potato, etc. It is essential to supplement and enrich this list with additional crops. In 50 years soybean came from near-zero production to a dominant role in US crop agriculture. In 50 years maize came from near-zero production to a dominant role in Chinese crop agriculture. Grain amaranth is a crop of the future with a long history of use and development.

6. *AMARANTHUS* STARCH: BACKGROUND

Amaranthus starch has been studied since the 1970's and some interesting findings have been reported, e.g. a wide range of viscosity, resistance to shear thinning, stable paste properties, and small starch granule size (Bahnassey and Breene, 1994; Konishi et al, 1985; Lorenz, 1981; Mistry and Eckhoff, 1992; Myers and Fox, 1994; Kazutoshi and Sakaguchi, 1981; Paredes-López et al, 1988, 1994; Paredes-López and Hernandez-López, 1991; Stone and Lorenz, 1984; Sugimoto et al, 1981; Wu et al, 1995; Yanez et al, 1986; Zhao and Whistler, 1994). The focus of such research was restricted to very few *Amaranthus* species or genotypes, and sometimes gave contradictory results due to the genetic variation of the properties of *Amaranthus* starch.

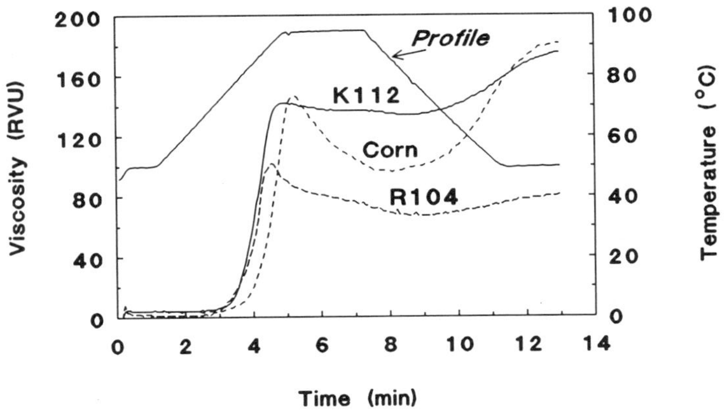

Figure 1. Viscoamylographs of two Amaranthus genotypes (K112 and R104) compared with corn.

One of the key methodologies in starch evaluation is the use of viscoamylography. This is illustrated in Figure 1, drawn from our earlier work (Wu et al, 1995). This shows that even in the simple case of evaluating for food processing applications the two major genotypes (K112 and R104) presently grown in China, the starch properties exhibit major differences. This emphasizes the need for quality assessment with end-uses in mind during breeding and selection. An interesting attribute is also illustrated in Figure 1, i.e. the resistance to shear-thinning of many *Amaranthus* genotypes. In fact, this property resembles that of a lightly cross-linked modified corn starch. This sharp difference between even these two genotypes led us to investigate further a wider range of genetic variation in starch properties among *Amaranthus* species and among the genotypes within a species. The extent of variation was previously investigated in depth, limiting any general understanding for utilization of *Amaranthus* starches. Large-scale studies on genetic variation in starch properties of other crops are also surprisingly limited, but a few have been reported, e.g. for variation of thermal properties of maize (Li et al, 1994). In order to investigate the genetic variation in starch properties among the *Amaranthus* species, a comprehensive survey of the properties of *Amaranthus* starch was conducted. Below we present the results of physical and functional properties of starch in some *Amaranthus* species, giving a general idea of genetic diversity and the effect of growing environment. The study below contrasts with that presented in Chapter 10; that study is investigating starches from different grain types, while our study looked at starches from within the single *Amaranthus* genus.

7. *AMARANTHUS* STARCH: A GENETIC RESOURCE SURVEY

7.1. Materials and Methods

7.1.1. Genetic Materials. Seed of 243 genotypes representing 26 *Amaranthus* species was generously provided by Mr David Brenner from the USDA Plant Introduction Station collection held at Iowa State University, Ames, Iowa. They were grown in field experiments and evaluated for agronomic traits in Beijing and Wuhan, China, in 1994. Of these, only 93 genotypes of 9 species produced enough seed to isolate sufficient starch to complete all testing. All these genotypes were defined as "non-cultivated" genotypes in discussion herein. A further 31 cultivated Chinese were grown and tested under the same conditions. These were grouped as "cultivated" genotypes. Corn, rice, potato, and wheat starch (Sigma Chemical Co) were used as reference standards.

7.1.2. Starch Isolation. The starches were isolated according to Wu et al (1995) with some modifications. *Amaranthus* grains were steeped in 2 volumes of 0.25% NaOH for 24 hours at 4°C, then the steeped seeds underwent a repeated blending-sieving procedure with centrifugation and removal of the brown layer.

7.1.3. Starch Content. An approximate starch content or "extractable starch content" was calculated from the weight of extracted starch relative to the starting weight of seed.

7.1.4. Amylose Determination. A combination of the methods of Williams et al (1970) and Morrison and Laignelet (1983) (basically the IRRI rice apparent amylose method provided by Dr BO Juliano), with minor modifications, was used for amylose determination. This is an iodine-binding spectrophotometric method using 0.2% I_2 in 2.0% KI, and reading absorbance at 620 nm.

7.1.5. Rapid Visco-Analyzer (RVA). A Rapid Visco-Analyzer (RVA) (Newport Scientific Pty. Ltd., Narrabeen, Australia) was used for testing pasting properties, following Wu et al (1995). The time-temperature profile was as follows: starting at 50ºC and holding for 1 minute, heating to 95ºC in 3.7 minutes, and holding at 95ºC for 2.5 minutes, cooling to 50ºC in 3.8 minutes, and holding at 50ºC for 2 minutes. 3.0 g starch (dry basis) and 25 g distilled water (adjusted by the moisture content of the starch) were mixed to make a starch suspension in the aluminum RVA canister. The peak viscosity (PV), hot paste viscosity (HPV), temperature at which the PV was attained (P_{temp}); time to peak viscosity (P_{time}), and cold paste viscosity (CPV) were recorded. From those parameters, the differ-

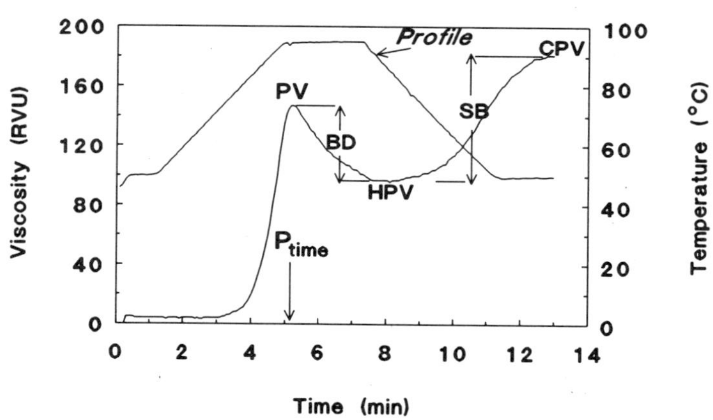

Figure 2. Parameters calculated from viscoamylographs: peak viscosity (PV); hot paste viscosity (HPV); time to peak viscosity (P_{time}); cold paste viscosity (CPV); breakdown (BD); and setback (SB).

ence between PV and HPV was calculated as breakdown (BD), and between HPV and CPV as setback (SB) (Figure 2).

7.1.6. Differential Scanning Calorimetry (DSC). Differential Scanning Calorimetry (DSC) (with a Mettler DSC20 instrument plus a Mettler TC11 data analysis station, Mettler, Naenikon-Uster, Switzerland) was used for the thermal analysis following Wu et al (1995). Only peak gelatinization temperature (T_p) is mentioned in this report.

7.1.7. Texture Analysis. A QTS-25 texture analyzer (Stevens Advanced Weighing Systems, Leonard Farnell and Co. Ltd., England) was used to test properties of starch paste from the RVA after gel formation, as described by Wu et al (1995). Each sample after the RVA testing was stored at 4ºC for 24 hours and 7 days before testing. Hardness, cohesiveness, modulus and adhesiveness were recorded after 24 hours and the further changes after 7 days.

8. RESULTS AND DISCUSSION

8.1. Starch Content

There was variation in extractable starch content among different *Amaranthus* species and within the same species in different growth locations. The mean extractable starch content of *Amaranthus* species was fairly high (20.1%), but higher in cultivated genotypes (36.4%) than in non-cultivated genotypes (14.5%). This is reasonable because *Amaranthus* grain is mainly used for food and a high starch content would tend to be selected by the cultivators. *A. hypochondriacus* seeds contained the highest starch content (37.6%), and *A. cruentus* had about 20% starch content. The remaining species, *A. dubius, A. hybridus, A. pumilus, A. retroflexus, A. spinosus, A. tricolor* and *A. viridis* contained very little starch. The growing environment affected the extractable starch content to some extent. *A. cruentus,* and *A. hypochondriacus* had higher extractable starch content when grown in the north (Beijing) than when grown in the south (Wuhan).

8.2. Amylose Content

The average amylose content of all genotypes tested was 19.2%, with a mean of 10.7% for cultivated and 23.2% for non-cultivated genotypes respectively. The lower amylose content (more waxy characteristic) in many of the Chinese cultivated genotypes may be partially due to consumer preference. *Amaranthus* grains have been widely used in China as a substitute for sesame seeds for cake coating, for which the waxy types are strongly favored (Yue et al, 1993). The wild species *A. retroflexus*, a weed in the field, had the highest average amylose content (34.3%), compared to 7.8% in *A. hypochondriacus* which was the lowest. *A. tricolor* also showed high amylose content (29.0%). The environmental effect on the amylose content was marked, especially for *A. cruentus* which had higher amylose content when grown in Wuhan (25.0%) than in Beijing (19.2%).

8.3. Pasting Properties

8.3.1. Peak viscosity (PV). In general, PV of cultivated genotypes (mean 296) was higher than that of the non-cultivated genotypes (mean 229). The standard deviation of the cultivated genotypes was higher (227) than that of the non-cultivated genotypes (99), showing a wide variation in the cultivated genotypes. *A. cruentus* (mean 288) had the highest PV of all the species tested, followed by *A. retroflexus* (mean 222), *A. hybridus* (mean 213) and *A. spinosus* (207) while *A. pumilus* was the lowest (mean 104). Since PV is related to the swelling power of the starch, the wide variation of PV indicated similarly great differences in swelling properties. The PV of *Amaranthus* starch was in the range of those of rice, wheat and corn starch, but lower than potato starch.

8.3.2. Temperature at Peak Viscosity (P_{temp}). P_{temp} for cultivated genotypes (mean 86.1ºC) was lower than that for non-cultivated genotypes (mean 90.9ºC), indicating that lower temperature was needed to gelatinize the starch of cultivated genotypes. *A. hypochondriacus* P_{temp} was 82.2ºC, much lower than most other species. The environmental effect on P_{temp} was different for different *Amaranthus* species, e.g. P_{temp} was higher for *A. cruentus* when grown in Wuhan (91.7ºC) than in Beijing (89.2ºC), while for *A. hybridus*, it was in the opposite, the P_{temp} was lower for Wuhan (82.0ºC) than for Beijing (94.8ºC). The P_{temp} of *Amaranthus* starch was lower than those of rice starch and wheat starch, but much higher than those of corn starch and potato starch.

8.3.3. Time to Peak Viscosity (P_{time}). The P_{time} of the cultivated genotypes (mean 7.9 minutes) was much lower than that of non-cultivated genotypes (mean 10.0 minutes). *A. cruentus* (mean 7.7 minutes) most rapidly reached the PV, followed by *A. retroflexus* (7.9 minutes), *A. hypochondriacus* (8.3 minutes) and *A. hybridus* (8.6 minutes) while *A. viridis* was the slowest (10.9 min). Environmental effects on the P_{time} seemed inconsistent because the P_{time} of *A. cruentus* was longer for seeds produced in Wuhan than in Beijing, while it was opposite for *A. hybridus*. Mean P_{time} of starch from cultivated *Amaranthus* genotypes was shorter than that of wheat starch, but longer than corn, rice, and potato starches.

8.3.4. Hot paste viscosity (HPV). During RVA viscoamylography, after PV is attained, viscosity decreases with continued shearing at constant temperature, i.e. the starch undergoes shear thinning. This property of starch is one of the key factors affecting the ease of handling of many food systems during processing. Subramanian et al (1994) reported the shear thinning properties of sorghum and corn starches, but there is no systematic report about the shear thinning properties of *Amaranthus* starches. The HPV of cultivated genotypes (mean 151) was less than that of non-cultivated genotypes (mean 176). *A. retroflexus* had the highest HPV (223) followed by *A. spinosus* (217) and *A. cruentus* (202) while *A. pumilus* had the lowest HPV (102). Environmental effects on HPV was different for different *Amaranthus* species, e.g. the HPV of *A. cruentus* was higher for the seeds produced in Wuhan than in Beijing, while the opposite held for *A. hybridus*. *Amaranthus* starches tended to have higher HPV than corn starch, similar to those of rice starch and potato starch, but lower than wheat starch, although the differences were not large.

8.3.5. Cold paste viscosity (CPV). The CPV of the starch paste is very important in food processing (e.g. canning) and for its contribution to textural and sensory properties of the food. The CPV of the cultivated genotypes (mean 185) was lower than that of non-cultivated genotypes (mean 233). *A. retroflexus* had the highest CPV (mean 289) followed by *A. cruentus* (mean 288) *A. spinosus* (mean 253) and *A. hybridus* (mean 244), compared to the lowest in *A. viridis* (mean 90). Generally, starch from cultivated *Amaranthus* genotypes had lower CPV than corn, rice, potato and wheat starches, while starch from non-cultivated *Amaranthus* genotypes had similar CPV to those of corn and rice starches and lower CPV than those of potato and wheat starches.

8.3.6. Breakdown (BD). Breakdown is one of the parameters indicating paste stability. The lower the drop from the peak to the lowest point in shear thinning, the higher the shear resistance. BD of cultivated genotypes (mean 144) was much higher that of the starch from non-cultivated genotypes (mean 52), showing that the starch paste of non-cultivated genotypes was more resistant to shearing. This implied non-cultivated genotypes might be used to improve the cultivated genotypes for this trait. *A. cruentus* was the most sensitive to shear thinning (mean 86) followed by *A. viridis* (mean 76), *A. hypochondriacus* (mean 45) and *A. hybridus* (mean 39).These results contrasted with other species, *A. retroflexus*, *A. spinosus* and *A. tricolor*, which had negative BD values, suggesting that the starch paste of those species were very stable; this might be partially due some further starch swelling. Starch paste of *A. pumilus* was also very stable since the BD was only 2. The BD of cultivated *Amaranthus* genotypes was similar to rice and wheat, but lower than potato and corn starches. The RVA viscoamylograph profile of certain *Amaranthus* starches strongly resembles a typical modified cross-linked corn starch.

8.3.7. Setback (SB). The viscosity changes during the cooling of the starch paste were mainly due to amylose molecular reassociation. The correlation between amylose content and CPV was significant ($r = 0.71$). The SB of the starch from the cultivated genotypes (mean 34) was much lower than that of non-cultivated genotypes (mean 57), further supporting the association between the amylose content and CPV. Generally *Amaranthus* starch had very low setback compared to the four reference starch samples, showing that it has promise as a food thickener and stabilizer.

8.4. Thermal Properties

8.4.1. Gelatinization Peak Temperature (T_p). Mean T_p of starch from cultivated genotypes (75.6ºC) was lower than that of non-cultivated genotypes (77.9ºC). *A. tricolor* and *A. dubius* had the highest mean T_p (82ºC), and *A. hypochondriacus* the lowest (73.9ºC). The environmental effect on T_p differed among *Amaranthus* species, e.g. *A. cruentus* had almost the same T_p when grown in Beijing and Wuhan, but *A. hybridus* had higher T_p in Wuhan than Beijing. *Amaranthus* starch T_p was higher than rice, corn, and wheat starches, but similar to potato starch. Data on T_o, T_c, and ΔH will be reported elsewhere.

8.5. Textural Properties

Textural (functional) properties relate to starch gel utilization in food. Many parameters can be used to describe texture, but hardness, cohesiveness, modulus, and adhesiveness were chosen for this study. Hardness is related to the firmness of the starch gel, cohesiveness is the ability to maintain integrity under mechanical action; modulus is related to resistance to deformation; and adhesiveness is related to the ability of the gel to stick to other objects. Due to space limitations in this report, textural properties are discussed only in relation to correlations with other traits.

8.6. Correlations between the Pasting and Textural Properties of *Amaranthus* Starch

PV and P_{temp} were significantly correlated to hardness and modulus but not to cohesiveness and adhesiveness at 24 hours and 7 days. HPV was highly significantly correlated to hardness (0.72 and 0.65 respectively), modulus (0.70 and 0.64 respectively), and to cohesiveness (0.24 and 0.27 respectively) at 24 hours and 7 days. P_{time} was not correlated to any functional properties. The correlations between CPV and textural properties were similar to those between HPV and textural properties. In general, thermal properties in *Amaranthus* starch had low correlations to both pasting and textural properties, while pasting properties and textural properties were significantly correlated. Screening for pasting properties, which is technically simple, would be useful to identify useful textural variants.

8.7. Dimensions of the Individual Variation

Selection of individual genotypes suited to particular functional needs can be done based on our results. Indicative values for the range in values of major traits are indicated in Table 2.

9. CONCLUSIONS

A wide range of variation was found in the various properties tested both among *Amaranthus* species and among genotypes within the same species. Amylose content of cultivated genotypes of *Amaranthus* was generally lower than in non-cultivated genotypes; starch of cultivated genotypes had more stable pasting properties than noncultivated genotypes; the starch pastes of cultivated genotypes was very stable under cold storage. Correlation analysis showed that pasting properties, textural properties, and thermal properties of *Amaranthus* starch were closely inter-related, especially the pasting and textural properties. Amylose content was essentially related to most physical and textural properties of *Amaranthus* starch. Compared to the reference corn, rice, potato and wheat starches, *Amaranthus* starch paste tended to be more stable; to have higher gelatinization temperatures and higher energy of enthalpy for gelatinization. Also, *Amaranthus* starch pastes was more resistant in cold storage, with lower changes of hardness, cohesiveness, modulus, and adhesiveness.

Table 2. Comparison of the value range of major parameters of *Amaranthus* starch with corn and potato starch

			Amaranthus	
	Corn	Potato	Two highest	Two lowest
Peak viscosity	353	797	441, 434	40, 36
Gelatinization temperature (°C)	72	65	83, 83	71, 71
Gel hardness	481	385	433, 399	11, 10

Amaranthus starch is a good thickener and stabilizer for use in food processing. The environmental effect on the properties of *Amaranthus* starch is being investigated further in field experiments conducted in 1995. There are two key aspects to the utilization of biological variation in *Amaranthus* starch properties: 1) identification of genotypes with useful starch traits among existing cultivars or other agronomically productive lines; and 2) identification of genotypes with useful starch traits that are in themselves unadapted to production agriculture but serve as a source of useful genes for breeding improved quality lines. The research reported here is intended to help guide selection for both these uses.

ACKNOWLEDGMENTS

The authors would like to thank Mr David Brenner (USDA Plant Introduction Station, Iowa State University) for his advice and generous provision of the *Amaranthus* seeds, and Ms Xiaofang Chen for technical assistance in starch isolation. This research project was funded by the Hong Kong Research Grants Council and the University of Hong Kong Committee on Research and Conference Grants. Versions of this material are to be published in *Cereal Chemistry* and *Cereal Foods World.*

REFERENCES

Bahnassey YA and Breene WM (1994) "Rapid Visco-Analyzer (RVA) pasting profiles of wheat, corn, waxy corn, tapioca and amaranth starches (*A. hypochondriacus* and *A. cruentus*) in the presence of konjac flour, gellan, guar, xanthan and locust bean gums." Starch/Stärke 46, 134–141

Konishi Y, Nojima H, Okuno K, Asaoka M and Fuwa H (1985) "Characterization of starch granules from waxy, nonwaxy, and hybrid seeds of *Amaranthus hypochondriacus* L." Agric. Biol. Chem. 49, 1965–1971

Li J, Berke TG and Glover DV (1994) "Variation for thermal properties of starch in tropical maize germ plasm." Cereal Chem. 71, 87–90

Lorenz K (1981) "Amaranthus hypochondriacus - characteristics of the starch and baking potential of the flour." Starch/ Stärke 33, 149–153

Mistry AH and Eckhoff SR (1992) "Characteristics of alkali-extracted starch obtained from corn flour." Cereal Chem. 69, 296–303

Morrison WR and Laignelet B (1983) "An improved colorimetric procedure for determining apparent and total amylose in cereal and other starches." J. Cereal Sci. 1, 9–20

Myers DJ and Fox SR (1994) "Alkali wet-milling characteristics of pearled and unpearled amaranth seed." Cereal Chem. 71, 96–99

Kazutoshi O and Sakaguchi S (1981) "Glutinous and non-glutinous starches in perisperm of grain Amaranths." Cereal Res. Comm. 9, 305–310

Paredes-López O, Bello-Perez LA and Lopez MG (1994) "Amylopectin, structural, gelatinisation and retrogradation studies." Food Chemistry 50, 411–417

Paredes-López O and Hernandez-López D (1991) "Application of differential scanning calorimetry to amaranth starch gelatinization - influence of water, solutes and annealing." Starch/Stärke 43, 57–61

Paredes-López O, Carabez-Trejo A, Perez-Herrera S and Gonzalez-Castaneda J (1988) "Influence of germination on physico-chemical properties of amaranth flour and starch microscopic structure." Starch/Stärke 40, 290–294

Stone LA and Lorenz K (1984) "The starch of Amaranthus - physico-chemical properties and functional characteristics." Starch/Stärke 36, 232–237

Subramanian V, Hoseney RC and Bramel-Cox P (1994) "Shear thinning properties of sorghum and corn starches." Cereal Chem. 71, 272–275

Sugimoto Y, Yamada K, Sakamoto S and Fuwa H (1981) "Some properties of normal- and waxy-type starches of *Amaranthus hypochondriacus* L." Starch/Stärke 33, 112–116

Williams PC, Kuzina FD, and Hlynka I (1970) "A rapid colorimetric procedure for estimating the amylose content of starches and flours." Cereal Chem. 47, 411–421

Wu H, Yue S, Sun H, and Corke H (1995) "Physical properties of starch from two genotypes of *Amaranthus cruentus* of agricultural significance in China." Starch/Stärke 47, 295–297

Yanez GA, Messinger JK, Walker CE and Rupnow JH (1986) "*Amaranthus hypochondriacus*, starch isolation and partial characterization." Cereal Chem. 63, 273–276

Yue SX, Sun HL, and Tang FD (1993) "The Research and Development of Grain Amaranth in China.". Chinese Agricultural Science and Technology Publishing House, Beijing, pp 466, in Chinese

Zhao J and Whistler RL (1994) "Isolation and characterization of starch from amaranth flour." Cereal Chem. 71, 392–393

13

REMOVAL CHARACTERISTICS OF BAKED WHEAT STARCH DEPOSITS TREATED WITH AQUEOUS CLEANING AGENTS

R. A. Din and M. R. Bird

School of Chemical Engineering
Bath University
Claverton Down, BA2 7AY, United Kingdom

1. INTRODUCTION

The fouling and cleaning of surfaces in contact with foods remains one of the major processing problems in the food industry. Baking processes must continuously guard against contamination of their products and reduction in quality due to lack of hygiene. Continuous operation of equipment has led to the introduction of 'Cleaning In-Place' (CIP) methods. The development and criteria affecting cleaning of processing and storage equipment is of increasing concern. Considerable time, detergent and energy may be saved if a clear understanding of the principles involved in cleaning starches and a knowledge of the effect of certain variables upon starch removal were determined. Cost optimisation of dairy CIP cycles has been studied by Bird and Espig (1994). Their study analyses a typical multi-stage acid/alkali dairy CIP cycle to examine the effect of detergent temperature, flow-rate and concentration upon the cost of cleaning. The results show that the cost of cleaning agent concentration and temperature influence costs most, whereas the flow-rate selection requires prior knowledge of specific down-time cost.

Starch is the major component of cereal (40–90% dry matter); its major nutritional property is to provide energy (4.4 kcal/g) (Hoseney et al, 1971). The two glucose polymers in starch, amylose and amylopectin, play important roles in the interaction of starch molecules and other food components (protein, fibres, lipids, etc.) and hence the functional and physicochemical properties of starch (Kulp and Lorenz, 1981). The arrangement of starch components changes continuously under the influence of hydrothermic parameters, during both food processing and storage (Dennet and Sterling, 1979). In particular, the baking of wheat starch can lead to the formation of tenacious deposits on the heat exchange surface.

Previous research (Linderer and Wildbrett, 1994) shows that the behaviour of a starch film in the cleaning process depends significantly on the type of soiling starch. The specific properties of the applied starch, such as gelatinization temperature, swelling

Cereals: Novel Uses and Processes, edited by Campbell *et al.*
Plenum Press, New York, 1997

power, solubility and chemical composition, are important factors influencing the success of the cleaning process: cereal starches of wheat were found to give much higher residues on hard surfaces than potato starch or chemically modified starch types.

An experimental study has been performed to determine the effectiveness of a number of aqueous cleaning agents on the removal of wheat starch. In order to make a feasible comparison between different cleaners, a reproducible deposit must first be generated. This required both the raw material used and the temperatures in the system to be kept constant. The fouling conditions chosen were similar to those occurring in the industrial process of baking. An experimental apparatus has been designed and constructed for cleaning the fouled wheat samples *in-situ* under controlled thermo-hydraulic conditions.

This paper describes the initial cleaning studies using RO water, sodium hydroxide, nitric acid, a formulated alkaline cleaner (Micro-90) and a non-ionic surfactant (alcohol ethoxylate). A gravimetric measurement technique was employed to monitor the cleaning process by studying the starch remaining on the surface. The results can thus be interpreted directly in terms of cleaning mechanisms.

2. EXPERIMENTAL APPARATUS

A detailed description of the apparatus and methods used are described in a previous paper by Din and Bird (1995).

2.1. Fouling Apparatus

Hard breadmaking wheat flour, supplied by Spillers Milling Ltd., was used to produce a baked wheat starch deposit. The flour contained around 70% starch and was also rich in gluten (11%), a protein matrix comprising almost three quarters of the total protein content of wheat.

A wheat starch paste was fouled onto 316 stainless steel test plates, 23mm wide, 150mm long, 3.5mm high with a channel milled through the centre line, 6mm wide and 2mm deep. Each plate was transferred to a fan-assisted oven maintained at 180°C for 10 minutes, to simulate industrial baking conditions. The deposit was allowed to air cool under ambient conditions before weighing the plate to ±0.001g.

2.2. Cleaning Rig

As with the fouling rig, the design and construction of the cleaning apparatus rig reflected the requirements of uniform and reproducible thermo-hydraulic conditions. It was equipped with a data-logging facility to enable continuous monitoring of the parameters affecting cleaning. After the cleaning process, each test plate was oven dried to constant mass.

In all experiments, cleaning solution was first prepared in a 200 litre storage tank which was heated to the required process temperature using a thermostatically controlled oil heater (Conair & Churchill Ltd.). The solution was pumped using a magnetically coupled centrifugal pump (Little Giant) and the flow-rate monitored using an electromagnetic flowmeter (Endress & Hauser).

The test-plate was mounted into the main process stream in a rectangular cross-sectioned glass cell, to allow direct visual observation of the starch removal. The use of this apparatus allowed cleaning to be observed directly without the optical distortion associated with cylindrical systems.

3. RESULTS AND DISCUSSION

The process parameters of temperature and flow-rate were maintained constant at 70°C and 4 litres/minute respectively during this initial investigation of cleaning parameter effects. A range of cleaning concentrations was chosen to cover all but the most extreme industrial cases. Results are given in Figure 1 for the removal of deposit after 25 minutes of cleaning of a 2 mm deposit.

It can be observed from these results that certain cleaning concentration optima exist for the conditions investigated. Cleaning with Micro-90 at 70°C for 25 minutes displayed an optimum concentration at 5wt%. This value was also apparent for the cleaning using nitric acid under similar conditions, although the deposit removal was significantly lower. As for sodium hydroxide cleaning, an optimum concentration is seen at 0.5wt% yielding a 16% deposit removal over the 25 minute duration. The use of higher cleaning concentrations in all these cases resulted in a decrease in overall cleaning efficiency. By far the most effective cleaner proved to be the non-ionic surfactant (alcohol ethoxylate), yielding 99% removal at 3.5wt% under the conditions specified.

Visual observation of the behaviour of the starch deposit during cleaning provides clues to the mechanisms involved in the process. When cleaning was carried out with Reverse Osmosis (RO) water swelling was apparent. At low sodium hydroxide concentrations, stresses set up in the deposit due to swelling caused the removal of deposit, implying a similar removal model to that for proteinaceous deposits (Bird, 1993). Swelling was most significant when surfactant was used for cleaning, resulting in the removal of large amounts of wheat deposit in an aggregated mass. In contrast, the action of nitric acid appeared to react chemically with the starch, causing a browning of the deposit with minimal swelling, and removal of material in small fibres.

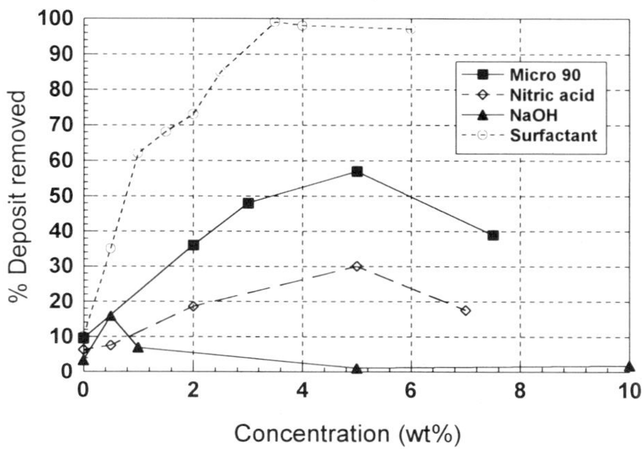

Figure 1. Concentration effect of Micro-90, nitric acid, sodium hydroxide and non- ionic surfactant upon removal of 2 mm wheat deposit cleaned for 25 minutes at 70°C.

4. CONCLUSIONS AND FUTURE WORK

This investigation has shown the existence of concentration optima when wheat starch is cleaned using Micro-90, nitric acid and sodium hydroxide for the time, flow and temperature conditions investigated. The most effective cleaner observed was the alcohol ethoxylate. This result indicates that a ready-made cleaner designed optimally by the chemical industry for a cleaning application would be the most favourable. It also challenges the view of a growing number of commercial organisations which advocate cleaning using sodium hydroxide or nitric acid. An apparatus and protocol has been developed which can determine time-accurate analysis of chemical cleaner performance on baked wheat starch removal. Now that an effective cleaning agent has been isolated and its capability tested, a full kinetic investigation to determine the time course of removal is under way.

REFERENCES

Bird MR (1993) "Cleaning of food process plant." PhD thesis, University of Cambridge

Bird MR and Espig SWP (1994) "Cost optimisation of dairy cleaning in place (CIP) cycles." Trans. IChemE 72,17

Dennet, K and Sterling C (1979) "Role of starch in bread formation." Starch/Staerke 31, 209

Din RA and Bird MR (1995) "The effect of water on removing starch deposits formed during baking." IChemE Research Event pp 187–189

Hoseney RC, Finney KF and Pomeranz Y (1971) "Functional (breadmaking) and biochemical properties of wheat flour components VIII. Starch." Cereal Chemistry 48, 91

Kulp K and Lorenz K (1981) "Starch functionality in white pan breads: new developments." Baker's Digest 55 (5) 24

Linderer M and Wildbrett (1994) "Starch residues in the cleaning process." Proceedings of Fouling in Food Processing, Cambridge University 129–136

14

APPLICATION OF CEREAL PROTEINS IN TECHNICAL APPLICATIONS

Peter Kolster, Leontine A de Graaf, and Johan M Vereijken

Industrial Proteins Division
Agrotechnological Research Institute (ATO-DLO)
PO Box 17, NL-6700 AA Wageningen, The Netherlands

1. INTRODUCTION

In recent years, research and development on biodegradable polymers and materials has increased considerably. These efforts are to a large extent driven by the increased awareness of environmental concerns related to the use of synthetic, non-biodegradable polymers. These environmental concerns have their origin especially in the persistence in the environment of these polymers and in their negative effect on the recycling of materials. Research on biodegradable polymers is also stimulated by the fact that there is an overproduction in Western agriculture, which results in a demand for new applications of agricultural produce.

For the replacement of synthetic materials, biodegradability, although important, is just one of the industrial requirements that should be met by products based on biodegradable polymers. Other properties that are of crucial importance are those related to:

- the *processing* of the polymers. In order to substitute synthetic polymers by biodegradable polymers, it is of prime importance that the same processing equipment (such as extrusion and injection molding equipment) that is now being used for synthetic polymers can also be used for biodegradable polymers.
- the *performance* of the products. The biodegradable products should satisfy industrial specifications with respect to, for instance, mechanical and barrier properties. Particularly important in this context is the water sensitivity (i.e. the deterioration in properties of biopolymer based products after contact with water). In many cases, products should be water stable.

Furthermore, it is worthwhile to note that a fast biodegradation is not always appreciated by industry. An example is the use of temporary protective coatings, which should have a relative long (> weeks) out-door durability. This implies that biodegradability as such is important, but to enhance their range of applications it is of major importance to be able to adjust the *rate of biodegradation*.

Cereals: Novel Uses and Processes, edited by Campbell *et al.*
Plenum Press, New York, 1997

A group of biodegradable polymers that can be used in technical applications are the 'industrial proteins'. Industrial proteins can be defined as proteins produced at such a scale that they can be used in commodity applications. Examples of industrial proteins are collagen, gelatin, casein and proteins isolated from crops as soy, peas and cereals. Wheat gluten in particular is a very promising raw material for technical applications, due to the unique intrinsic properties of this biopolymer and its relatively low price. On the other hand, because of the increase in production volume (estimated to be more than 400,000 tons worldwide in 1995) and the fact that the use of gluten in the main application (bakery sector) is not increasing, the development of new applications is also of major importance for manufacturers of wheat gluten.

In this chapter, a short historical overview of technical applications of proteins is presented, followed by an overview of properties and prices of other biodegradable polymers that can be used in technical applications. With these polymers, industrial proteins should compete on the market for biodegradable polymers. Finally, market segments in which wheat gluten can be used successfully will be discussed based on examples.

1.1. Historical Overview

There is a long history in the development of non-food, non-feed applications of industrial proteins. Proteins such as casein, collagen and blood proteins have been used in adhesives over many centuries (Bye, 1989). Casein has been used, and is being used, in paper coatings, paints, plastics and leather finishes (Lakshminarayana et al, 1985; Detlefsen, 1989; Anonymous, 1991). A well known, large scale technical application of proteins is the use of gelatin in photographic emulsions. In the thirties of this century, the development of technical applications of plant proteins, especially soy proteins, was studied in the framework of the 'chemurgic movement' (Myers, 1993). At that time, products such as fibres, plywood adhesives and paper coatings were developed. As a result of the rise of the petrochemicals, proteins and other agricultural feedstocks were replaced by synthetic polymers. In some applications proteins are still being used, such as gelatin and casein in adhesives and soy proteins in paper coatings. It is estimated that in the USA about 25,000–30,000 tons of soy proteins are used in paper coatings (Myers 1993).

The substitution of proteins by synthetic polymers is caused by the lower price, but also by the better performance of the synthetic polymers. Since World War II, there has been an enormous increase in knowledge of the adjustment of the properties of synthetic materials. As a result, the chemical industry is able to produce tailor-made products that can meet high industrial standards.

Because of the increased knowledge of protein technology and chemistry and the increased demand for biodegradable polymers, research on technical applications of proteins has resulted in the last decades in new protein-based products. For instance, numerous technical applications of wheat gluten, or derivatives thereof, have been described in the (patent) literature, such as plasticizers for synthetic materials, detergents, cigarette filters and inks (Bietz and Lookhart, 1996).

1.2. Comparison with Other Biodegradable Polymers

The development of biodegradable materials that can substitute for synthetic materials has been an important research topic in recent years. There are now a number of biodegradable materials available on the market and others are being developed. An important question is the (potential) market position of industrial proteins in comparison to other biodegradable polymers. Important aspects in this comparison are the properties, water sensitivity, price and availability.

Mayer and Kaplan (1994) have written an excellent review article in which they compare the costs, availability and performance of various biodegradable materials. Table 1 summarizes the results of their study. At this point, it is worthwhile to note that Mayer and Kaplan did not include proteins in their review.

It should be realized that the market for biodegradable polymers is very heterogenous with respect to specific demands. Each biodegradable polymer, analogous to synthetic polymers, has its specific application area. Table 1 shows that for technical applications, starch (or derivatives) is a very attractive biodegradable polymer because of its low price and availability. Furthermore, it biodegrades rapidly. A negative attribute of starch is its hydrophilicity, causing starch based products to be very sensitive to water (Chapters 6 and 15 present work on coatings to reduce the water sensitivity of starch- and protein-based products, respectively). This water sensitivity limits the applicability of this biopolymer. Mayer and Kaplan give as examples of potential applications mulch films, compost bags and packing foams. Poly(hydroxybutyrate-co-valerate) and polycaprolactone are examples of biodegradable materials that are water stable and can therefore be used in products that are, or may come, in contact with water (see Chapter 5). The price of these materials is however clearly higher than that of starch. Cellulose acetate and poly(lactic acid) have good mechanical properties and can be used as a substitute for materials that are produced by injection molding. Again, the price is higher than that of starch. The availability of the biodegradable polymers is, for most applications, sufficient or can be increased easily (such as polylactic acid).

To summarize, based on their costs two classes of biodegradable polymers can be distinguished:

Table 1. Biodegradable polymers; costs, availability and applications (after Mayer and Kaplan, 1994)

Polymer	Costs ($/kg)	Production level (kg/year)	Attributes: Positive	Attributes: Negative	Potential applications
Starch	0.3–1.6	>100 billion	Low cost, rapid biodegradation	Hydrophilicity	Mulch film, compost bags, packing foams
Cellulose acetate	3.5	1 billion	Tensile strength	Reduced biodegradation	Injection molding
Poly(hydroxybuty rate-co-valerate)	12-15 (expected 5)	300,000	Water stable, rapid biodegradation	Costs	Paper laminates
Poly(vinyl alcohol)	3-5	70-100 million	Oxygen barrier	Solubility in water	Pesticide, fertilizer, detergent dispersal
Polycaprolactone	6	<5 million	Water stable, toughness	Low melting point	Compost bags
Poly(lactic acid)	2-6	10 million	Tensile strength	Brittle, hydrolytically unstable	Injection molding, paper coating

- a price of about 1 US$/kg. This group comprises only starch and starch derivatives. Because of the properties of these polymers, only specific commodity products can be developed.
- a price of 3 - 6 US$/kg. This group comprises the other biodegradable polymers in Table 1. Because of the specific properties (e.g. mechanical and barrier properties, water stability), these polymers can be used in higher added value products.

In their review, Mayer and Kaplan did not discuss the use of proteins. In the authors' opinion, industrial proteins fit well within their selection of biodegradable materials. First, concerning the costs, there are large differences in price of industrial proteins, from about US$1- 1,5/kg for wheat gluten, US$4/kg for soy isolates and more than US$10/kg for specific gelatins. The price of wheat gluten but also of soy proteins is competitive to that of most polymers mentioned in Table 1. Also the availabilty is not limiting the industrial application of the industrial proteins. Estimated annual production volumes (worldwide) of wheat gluten are more than 400,000 tonnes/year. The production of soy proteins is clearly higher. For instance, in the United States 28 million tonnes of high-protein meal were produced in 1990.

It can be concluded therefore that, based on the price and the availability, industrial proteins are an attractive raw material for technical applications. Another consideration for technical applications is the performance of protein-based products. The next part of this chapter describes the (potential) markets in which wheat gluten can be used, and gives examples of the properties of the products.

2. GLUTEN BASED PRODUCTS

When developing technical applications of proteins, a first step is to define the application area for which a specific protein is best suited. This selection is based on its intrinsic properties. For technical applications, wheat gluten as produced by wheat starch manufacturers has some unique properties. Most important are:

- insolubility in water
- adhesive properties
- film forming properties
- viscoelastic behaviour

Based on these intrinsic properties, wheat gluten is very well suited for application in coatings/films, adhesives and in products produced by injection molding. Research of, amongst others, ATO-DLO has shown that thermoplastic processing (extrusion, injection molding) of wheat gluten is possible. In the following, examples of the research of ATO-DLO on adhesives and coatings/films will be presented.

2.1. Adhesives

It is well known that proteins have excellent adhesive properties and protein-based adhesives have been, and still are, on the market. Proteins that have been used in adhesives include blood proteins, casein, gelatin and soy proteins.

Technically demanding types of adhesives are labelling adhesives. In cooperation with an industrial partner, ATO-DLO is developing labelling adhesives based on plant proteins, including wheat gluten. Requirements for these type of adhesives include a good

Table 2. Wet-tack of labelling adhesives based on plant proteins

Product	Wet-tack
Commercial adhesive	10–20
Experimental adhesive	40–50

wet-tack (needed for the transfer of the label to, for example, the bottle) and good rheological behaviour. It should be realized that labelling of bottles is a high speed application. In some cases, rates up to 60,000 bottles per hour are used in industry. Therefore, adjustment of the rheological behaviour is of paramount importance.

Table 2 shows the wet-tack of a commercially used and an experimental adhesive, the latter based on plant proteins. By a combination of protein modifications and the use of additives, it is possible to introduce wet-tack in plant protein-based formulations that exceeds that of commercial adhesives. The research furthermore has shown that it is possible to adjust the rheological behaviour of these plant protein-based adhesives within the narrow range required for labelling adhesives.

There are therefore very good opportunities for the use of plant proteins, including wheat gluten, in labelling adhesives. This is also true for other types of adhesives, such as packaging adhesives. For manufacturers of proteins, it is important to realize that adhesives are a large scale (ktonne) application.

2.2. Wheat Gluten Films and Coatings

Because of the excellent film forming properties, proteins offer very good perspectives for use in coatings and films. These coatings and films can be used, for instance, in the food sector. In the technical sector, the coatings can be used in the packaging industry to adjust the barrier properties (for gases, flavours, water vapour) of paperboard packaging materials. Table 3 shows that proteins provide a better barrier for oxygen than synthetic polymers. This is of importance for packaging of fresh products such as tomatoes. In this application, proteins can substitute for synthetic materials such as polyethylene and waxes.

Among the various industrial proteins, wheat gluten is probably the most interesting for producing coatings and films. This is due to the relatively low price of this biopolymer, but also the intrinsic properties of wheat gluten make this a preferred protein for

Table 3. Oxygen permeability of films based on proteins or synthetic polymers (Torres 1994)

Polymer	Oxygen permeability in amol.m/m^{2}sPa
Gluten	3.8
Collagen	2.3
Whey	4.2
Corn zein	7.7
LDPE	1870
HDPE	427
Ethylene vinyl alcohol	12

coatings. Particularly important in this respect are the viscoelasticity of gluten - of importance for coatings that are applied on flexible surfaces such as paper (board) - and its insolubility in water. Accordingly, several research groups have studied coatings and films based on wheat gluten (Gennadios et al, 1993; Ghorpade et al in Chapter 15).

Anker et al (1972) described a method for producing wheat gluten films in which the gluten is first dispersed in a mixture of ethanol, glycerol, NaOH and water. After drying, a closed film is formed. Later, Gontard et al (1992) developed a film forming 'solution' in which wheat gluten is dispersed in a mixture of ethanol, acetic acid, glycerol and water. In these two, and other methods described, the use of ethanol and acidic or alkaline pH was essential. This hampers the industrial application of wheat gluten-based coatings/films.

In co-operation with Latenstein Zetmeel BV, a Dutch wheat starch and gluten manufacturer, ATO-DLO started a few year ago a research programme that aimed at the development of a new method for preparing gluten films and coatings. The method should enable large scale industrial applications of wheat gluten-based coatings and films. Hence, the use of organic solvents should be minimal and extreme pH values should be avoided. This research succeeded in the development of a film forming 'dispersion' of wheat gluten without organic solvents or extreme pH values. In the method developed, which is now the subject of a patent application, gluten is dispersed in water. Plasticizers are added in order to obtain non-brittle films or coatings. The dispersions show excellent film forming properties. In Table 4, the mechanical properties of gluten coatings produced using the Anker method and the developed ATO/Latenstein method are compared. Chapters 5, 6 and 15 present similar measurements for other starch- and gluten-based films.

Compared to the films produced according to the Anker method, the ATO/Latenstein films are somewhat weaker. The extensibility (strain) of the ATO/Latenstein films is however clearly higher at all three plasticizer concentrations tested.

The new method, free of organic solvents, developed for preparing gluten films/coatings enables in our opinion the development of new and large scale applications of these materials. In these applications, the unique properties of wheat gluten can be fully exploited.

2.3. Adjustment of the Properties

For industrial applications of proteins, it is essential to be able to adjust the properties (those related to the processing and to the products) towards the requirements of specific applications. By using additives or blending with other (biodegradable) polymers, it is possible to improve these properties to some extent. A powerful tool is the use of

Table 4. Mechanical properties of gluten films (60% RH, 20°C) produced according the Anker or the ATO/Latenstein method

	Anker		ATO/Latenstein	
% plasticizer (glycerol)	Stress (MPa)	Strain (%)	Stress (MPa)	Strain (%)
20	4.4	46	2.5	320
45	0.7	272	0.5	505
60	0.3	217	0.5	543

chemical and/or enzymatic protein modifications. Examples of properties that can be adjusted are:

- solubility
- rheological behaviour
- adhesion to various substrates
- mechanical properties
- barrier properties
- water sensitivity

The use of chemical modifications will be illustrated by the reduction of the water sensitivity of gluten-based films.

One of the most important shortcomings of proteins, and other biopolymers, is the fact that contact with water (soluble or vapour) has detrimental effects on the properties of protein-based products. This is also true for wheat gluten, although this biopolymer is insoluble in water. Table 5 shows the effect of differences in water content, resulting from incubation at 60 or 75% relative humidity (RH), on the mechanical properties of gluten films.

This table clearly shows that the strength of the film is significantly reduced at increased RH. At the same time, the extensibility (strain) is strongly increased. For protective coatings the sorption of water would result in a very low resistance against mechanical damage, which is unacceptable for technical applications.

To reduce the water sensitivity of protein based products, chemical modifications and cross-linking is being studied. As shown in Figure 1, proteins contain different reactive groups (COOH, SH, NH_2) that can be used in chemical modifications. For instance, hydrophobic groups can be introduced in proteins by reaction of aldehydes with lysine residues. Furthermore, the water sensitivity of proteins can be reduced by esterification of the carboxylic groups of glutamate and aspartate with methanol, thereby reducing the hydrophilicity of the carboxylic groups.

A very powerful method to reduce the water sensitivity of proteins is cross-linking, for which different functional groups of the proteins can be used (NH_2, COOH, SH). If needed, a two step approach can be followed. In the first step, additional reactive groups can be introduced in the protein (e.g. extra NH_2 groups by reaction of -COOH groups with diamine/carbodiimide). These groups can be used in the subsequent cross-linking reaction. The preferred procedure to reduce the water sensitivity of proteins depends very much on the amino acid composition of the protein. For instance, wheat gluten contains only minor amounts of lysine; therefore, cross-linking agents active on the NH_2 groups are not the preferred choice.

Table 6 shows the effect of modification alone, and modification with subsequent cross-linking, on the water sensitivity of gluten films. Water sensitivity was assessed after

Table 5. Effect of water sorption on the mechanical properties of wheat gluten based films. The films contain 10% glycerol

Relative humidity (%)	Stress (MPa)	Strain (%)
60	8.9	4.8
75	1.7	324.7

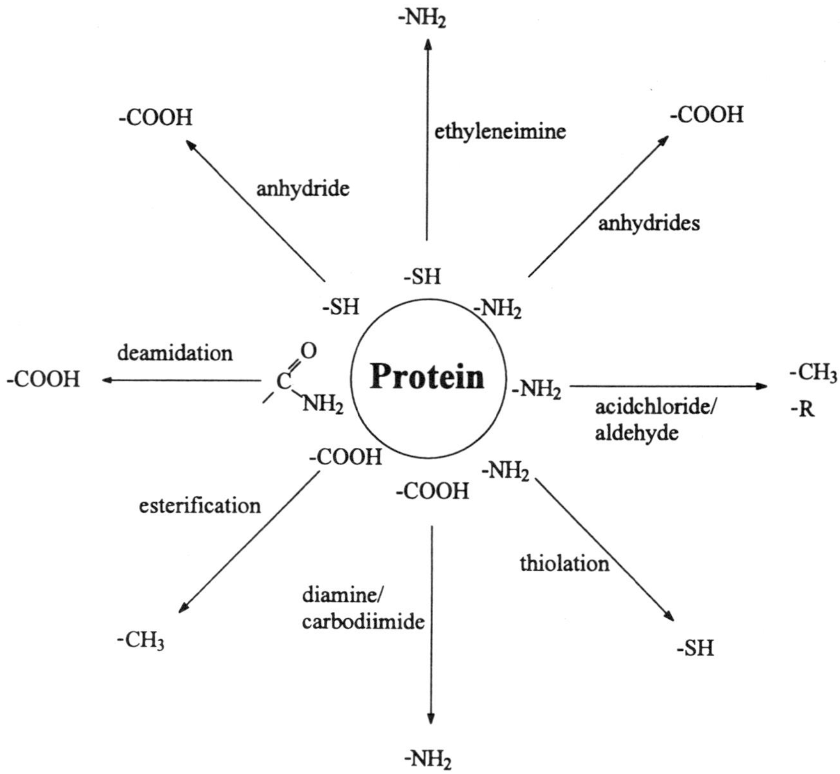

Figure 1. Chemical modifications of proteins.

submerging a disk of the gluten film for more than 1 hour in water. The increase in radius was determined and the strength of the material was assessed manually. Without modification, gluten films swell strongly (> 100%) and are very weak. Modification (hydrophobization) reduces the swelling by a factor of 3, but the mechanical strength is still poor. Modification followed by cross-linking results in a reduction of swelling to about 10%, which is comparable to the swelling of some synthetic polymers (e.g. acrylates, polyurethanes) and result in films with very high 'wet strength'. The combination of a low degree of swelling and high wet strength indicates efficient cross-linking.

Chemical modifications provide a flexible and powerful tool to tailor the properties of protein-based products towards the specific requirements. By using the proper modifi-

Table 6. Effects of modifications on swelling and wet strength of gluten based films

Sample	Swelling (%)	Wet strength
Control	100	low
Modified	30	low
Modified + crosslinked	10	high

cation strategy, both water stable or water soluble products can be made from wheat gluten.

3. CONCLUSIONS

For technical applications, there are at the moment several biodegradable polymers available or close to market introduction, such as starch-based products and poly(lactic acid). Industrial proteins are a group of biopolymers that has received only limited attention as a starting material for biodegradable materials, compared to the other biodegradable polymers. However, there are very good opportunities for this group of polymers on the market for biodegradable materials for the following reasons:

- the competitive prices of proteins, particularly of wheat gluten. The price of wheat gluten is clearly lower than that of many other biodegradable polymers;
- the large scale availability;
- the intrinsic properties of proteins such as wheat gluten: film forming, adhesion, barrier and mechanical properties;
- the wide range of modifications that can be used to tailor the properties of the proteins towards the specific industrial requirements.

In the market for biodegradable polymers, proteins such as wheat gluten will most likely find applications complementary to those of starch and the other, more expensive, biodegradable polymers. An industrial protein such as wheat gluten is somewhat more expensive than starch. However because of their specific properties, proteins are better suited for specific markets (adhesives, coatings, films) than starch. Particularly important in the comparison with starch is the wide range of chemical modifications that is available. These modifications can be used to tailor the properties of protein-based products, including the water sensitivity. On the other hand, it may require extensive modifications to implement in proteins properties comparable to that of the more expensive biodegradable polymers such as poly(hydroxybutyrate-co-valerate) and poly(lactic acid). In applications requiring a high water stability, or a high strength, other biodegradable polymers could well be more appropriate. In conclusion, there are very good opportunities for the development of novel, technical, applications of industrial proteins.

REFERENCES

Anker CA, Foster GA and Loader MA (1972) "Method for preparing gluten containing films and coatings." United States Patent 3,653,925

Anonymous (1993) "Industrial uses of agricultural materials. Situation and outlook report." United States Department of Agriculture, Economic Research Service

Bye CN (1989) "Casein and mixed protein adhesives." In "Handbook of adhesives." Ed. I Skeist, Van Nostrand Reingold. 135

Bietz JA and Lookhart GL (1996) "Properties and Non-Food Potential of Gluten." Cereal Foods World 41, 376–382

Detlefsen WD (1989) "Blood and casein adhesives for bonding wood." ACS Symp. Ser. 385, 445

Gennadios A, Weller CL and Testin RF (1993) "Modification of physical and barrier properties of edible wheat gluten-based films." Cereal Chem. 70, 426

Gontard N, Guilbert S and JL Cuq (1992) "Edible wheat gluten films: influence of the main process variables on film properties using response surface methodology." J. Food Sci. 57, 190

Lakshminarayana Y, Radhakrishnan N, Parthasarathy K, Srinivasan KSV and KT Joseph (1985) "Modified Protein binder with improved wet rub fastness." Leather Science 32, 134

Mayer JM and DL Kaplan (1994) "Biodegradable Materials: Balancing Degradability and Performance." Trends in Polymer Science 2(7), 227–235

Myers DJ (1993) "Industrial applications for soy protein and potential for increased utilization." Cereal Foods World, 335

Torres JA (1994) "Edible films and coatings from proteins. Protein functionality in food systems." IFT Basic symposium series. Ed. Hettiarachchy, Marcel Dekker, 467–507

15

MECHANICAL AND BARRIER PROPERTIES OF WHEAT GLUTEN FILMS COATED WITH POLYLACTIC ACID

Viswas Ghorpade, Curtis Weller, and Milford Hanna

Industrial Agricultural Products Center
University of Nebraska
Lincoln, Nebraska 68583-0730

1. INTRODUCTION

Development of biopolymer films and coatings from protein, polysaccharide, and lipid materials has received increased interest in recent years. In the midst of rising concerns over solid packaging waste and dwindling petroleum reserves, the renewable and degradable nature of biopolymer film ingredients make such films particularly appealing for innovative uses in the field of packaging. However, unlike some proteins, few nonfood applications for wheat gluten have been developed. In 1990 world wheat production was 589 million metric tonnes, of which 12.6% was produced in the US. *Industrial Uses of Agricultural Materials* (June, 1993) reported consolidated sales for low density polyethylene (LDPE) at about 3.1 million metric tonnes for various food and nonfood packaging uses in the US in 1992. The total LDPE sales, agricultural uses and trash bags sales accounted for 770,000 metric tonnes in 1992. A 30% market penetration of wheat gluten polymer in agricultural mulches and trash bags would use around 230,000 metric tonnes of gluten in the US alone.

Protein films are fragile due to the hydrophilic nature of their amino acids. Furthermore, the shape of the polymeric molecules encourage interlocking molecular segments and determine the physical properties of the films. Extent of crosslinking and additives alter the rigidity, toughness, permeability, flexibility and brittleness of the films. Several efforts have been made in recent years to incorporate biopolymers, such as starches and proteins, into plastic materials to enhance properties. Starch-based polyethylene films were pioneered by the National Center for Agricultural Research, Peoria, IL (Doane, 1988; see also Chapters 5 and 6). Archer Danials Midland Corp. (ADM) acquired the Griffin (1977) patent on production of starch-substituted films for bags and commercially marketed a "master blend" to film manufacturers. In that process, only 6% starch was used in the product. Otey and Westhoff (1984) and Otey et al (1977, 1980, 1987) prepared starch-based composite films with polyethylene (ethylene-co-acrylic acid) for agricultural

Cereals: Novel Uses and Processes, edited by Campbell *et al.*
Plenum Press, New York, 1997

mulches. Other films, such as starch-polyvinyl alcohol coated with a water resistant polymer, were also studied (Otey et al, 1974). Dennenberg et al (1978) demonstrated biodegradability by *Aspergillus niger* of starch graft-poly methylacrylate copolymer which exhibited excellent tensile strength and elongation at break properties.

Plant proteins were investigated for applications in edible and non-edible films (Kester and Fennema, 1986; Guilbert, 1986; Gontard et al, 1993; Krochta, 1992; Gennadios et al, 1994a; Kolster et al in Chapter 14). Though protein films have been extensively studied, they have not been commercialized because of mechanical and barrier property limitations. Films from proteins in combination with synthetic plastic monomers are potential candidates for environmentally friendly, compostable resins. Research is under way at several universities to incorporate protein into polyethylene films. Park et al (1993) reported characteristics of zein-filled polyethylene compostable films. Zein was added to low density polyethylene (640 I, Dow Chemicals, USA) at 0, 2, 4, and 6% by weight. Ghorpade et al (1994) studied property evaluations of cast soy protein films by substituting modified soy protein isolate with various amounts of polyethylene oxide. Other efforts have used extruders to make protein-polymer complexes in a continuous process. Thus, this project was designed to explore the utilization of wheat protein partially to replace petroleum-based plastic. As such, the overall objective of this project was to improve mechanical and barrier properties of films by coating them with polylactic acid (PLA) and to estimate the material costs of using PLA-coated films as an agricultural mulch.

2. MATERIALS AND METHODS

2.1. Materials

Wheatpro-80™, a vital wheat gluten, was obtained gratis from Ogilve Mills Ltd. (Quebec, Canada). Other reagents such as ammonium hydroxide (5N), glycerine (USP grade), and ethanol (95%) were of reagent grade from Baxter (McGaw Park, IL). Polylactic acid polymer (Ecopla™) was purchased from Cargill, Inc. (Minneapolis, MN).

2.2. Film Formation

Films were prepared by mixing 3.36 g of glycerine with 48 mL of 95% ethanol, followed by dispersion of wheat gluten (10 g) in the solution with constant stirring and heating for 10 minutes on a magnetic stirrer-hot plate. Distilled water (32 mL) and 5N ammonium hydroxide (8 mL) were added slowly. Heating rate was adjusted so that the temperature of the solution was about 72–75°C at the end of the preparation time.

2.3. Casting Wheat Gluten Films

After removal from a hot plate, film-forming solutions were filtered through cheese cloth to cease bubbling before casting on a glass plate. A thin-layer chromatographic spreader bar (Brinkman Co., New York, NY) was used to spread the film forming solutions onto a glass plate. Each glass plate was taped on either side to restrain movement during casting. Plates with cast wheat gluten solutions were kept at ambient temperature (23°C) for 24 hours. Dried films were peeled from the glass plates and stored in an environmental chamber at 50% RH and 25°C for 48 hours. Films were cut into 7 x 7 cm sample pieces for water vapor permeability tests and into 10 x 2.54 cm pieces for mechanical properties testing.

2.4. Polylactic Acid (PLA) Coating

PLA solutions were prepared by dissolving 0.5, 1.0, 2.0, 4.0, and 8.0 g of PLA (w/v) in 100 mL of chloroform. Chloroform was used for control samples. Film samples were dipped into the prescribed solution for a minute and allowed to dry inside a vented hood at ambient conditions. The dipping process was repeated once to ensure an even coating of PLA on the film samples.

2.5. Thickness Measurements

A hand-held micrometer (BC Ames Co., Waltham, MA) was used for measuring film thickness to the nearest 2.54 µm. For each of these types of film, two samples were taken. From each of these samples, one measurement was made at the center and four on the perimeter. Measurements were taken before and after dipping the film samples in the PLA solutions.

The amount of wheat gluten required to cover 1 ha of land was calculated by assuming 10 g of wheat gluten made a 23 x 33 cm piece of film with an average thickness of 0.135±0.006 mm. Then the PLA required to cover 1 ha was calculated as area x thickness x density. PLA density was 1.25 kg/m^3.

2.6. Tensile Strength and Elongation at Break

Films were conditioned at 50% RH and 25^0C for 3 days before testing. An Instron Universal Testing Instrument (Model 5566, Instron Engineering Corp., Canton, MA) was used to determine Tensile Strength and Elongation according to ASTM Method D 882–88 (ASTM, 1989a). Film specimens 2.54 cm wide and 10 cm long were cut. Mean thickness measurements, as described in the previous section, were used in Tensile Strength calculations. The initial grip separation and cross-head speed were set at 5 cm and 50 cm/min, respectively. Tensile strength was calculated by dividing the maximum (peak) load necessary to pull the specimen apart by the original cross-sectional area of the specimen. Elongation was calculated by dividing film elongation at rupture by the initial gauge length of the specimen and multiplying by 100. Tensile Strength and Elongation determinations for each type of film were replicated four times with individually prepared films as the replicated experimental units and ten sampling units (specimens) tested from each film replicate.

2.7. Water Vapor Permeability

The mean thickness value was used as the specimen thickness in Water Vapor Permeability calculations. Prior to testing, all film specimens were conditioned at 25°C and 50% RH for two days. Four individually cast film specimens were tested from each type of film. Water Vapor Permeability (WVP) ($g \times m/m^2 \times s \times Pa$) was calculated as:

$$WVP = (WVTR*L)/\Delta p \tag{1}$$

where WVTR was the measured water vapor transmission rate (g/m^2×s) through a film specimen, L was the mean film thickness (m), and Δp was the partial water vapor pressure difference (Pa) across the two sides of the film specimen.

WVTR was determined gravimetrically using a modified ASTM Method E 96–80 (ASTM 1989b). Film specimens were mounted on poly(methyl methacrylate) cups filled with distilled water up to 1 cm from the film underside. Design of the cups was described by Gennadios et al (1994b). The cups were placed in an environmental chamber set at 25^{0}C and 50% RH. A fan was operated within the chamber creating an air velocity of 198 m/min over the surface of the cups to remove the permeating water vapor. Weights of the cups were recorded six times at 1 hour intervals. Steady state was reached after 1 hour. Slopes of the steady-state (linear) portion of weight loss versus time curves were used to estimate WVTR. Because of the low water vapor resistance of protein-based films, actual RH values at the film undersides during testing were lower than the theoretical value of 100%. Actual RH values at the film undersides and film WVP values were calculated after accounting for the resistance of the stagnant air layer between the film undersides and the water surface in the cups (McHugh et al, 1993; Gennadios et al, 1994b). The mean of the initial and the final stagnant air gap heights was used in the calculations.

3. RESULTS AND DISCUSSION

3.1. Mechanical Properties

Table 1 shows the results for thickness, Tensile Strength and Elongation for wheat gluten films coated with polylactic acid (cf. similar measurement reported in Chapters 5, 6 and 14). Mean thickness for all PLA-coated films was 0.156±0.007 mm. Though the average film thicknesses for wheat gluten protein remained in the range of 0.12 to 0.14 mm, the thickness for PLA-coated films increased with an increase in the PLA concentration in the coating solution. Tensile Strength increased for films with up to 1% PLA concentration in coating solutions then decreased with increasing concentration. Mean Tensile Strength values ranged from 3.09 MPa for the control to 4.18 MPa for 1% PLA concentration in the coating solution. There was a significant drop in Tensile Strength values in samples with 2% to 8% PLA concentration in the coating solution. Some of these differences maybe due to the different thickness of the interior wheat gluten layer. Absolute

Table 1. Mean values for thickness, Tensile Strength (TS), and Elongation (E) for wheat gluten film coated with polylactic acid (PLA)*

PLA Conc.(%)	Thickness (mm)	Tensile strength (MPa)	E (%)
0	0.15 ±0.01	3.09 ±0.26a	170 ±22.45b
0.5	0.16 ±0.01	3.60 ±0.45b	167 ±21.95b
1	0.14 ±0.01	4.18 ±0.50d	097 ±16.00a
2	0.15 ±0.01	4.03 ±0.10d	111 ±02.86a
4	0.15 ±0.01	3.94 ±0.66d	189 ±19.83b
8	0.17 ±0.01	3.83 ±0.16c	214 ±00.16c

*Any mean value in the same column followed by the same letter are not significantly different (P>0.05) according to the Duncan Multiple Range Test.

control of wheat gluten thickness was difficult to achieve, resulting in a compounding of error as shown for Tensile Strength.

Mean values for E, a measure of a film's extensibility, for the different types of film are also presented in the Table 1. Films coated with 8% PLA in coating solution showed a significant increase in Elongation over the control film. However, Elongation values decreased for films coated with up to 1% PLA concentration in the coating solution. Then Elongation showed a steady increase as PLA concentration in the coating solution was increased. Further the increase in the Elongation at higher PLA concentration in coating solution was due to effects of PLA. Relatively high values for the control and lower PLA concentrations in the coating solutions were attributed to moisture content. Water plasticizes hydrophilic films and improves film extensibility (Gennadios, 1994b; Gontard et al, 1993). In the two days allowed for conditioning, the films coated with lower PLA concentration may have absorbed greater amounts of water than those coated with higher PLA concentration.

3.2. Water Vapor Permeability (WVP)

WVP constants for the control and PLA-coated films, along with the calculated actual relative humidities (RH) conditions at the underside of the films during testing are reported in Table 2. WVP constant values of untreated wheat gluten films were 2–3 orders of magnitude less than those of typical polymeric packaging films (Gennadios et al, 1994a). The WVP constants (in $g \times m/m^2 \times s \times Pa$) for various polymeric films are: polyvinylidene chloride (0.7–2.4 x 10^{13}); high density polyethylene (2.4 x 10^{-13}); cast polypropylene (4.9 x 10^{13}); low density polyethylene (7.3–9.7 x 10^{-13}); ethylene vinyl acetate (2.4–4.9 x 10^{12}); polyester (1.2–1.5 x 10^{12}); and cellulose acetate (0.5–1.6 x10^{11}) (Briston, 1988). The WVP constant for wheat gluten film coated with 8% PLA was in the range of 0.94–1.24 x 10^{-11} $g \times m/m^2 \times s \times Pa$, which is comparable to cellulose acetate films.

Significant decreases in the WVP were observed with the increases in the PLA concentration in the coating solutions. Differences among the means were significant at the 95% level of confidence.

Figure 1 shows water vapor permeability of wheat gluten films coated with various amounts of PLA. An exponential line was fitted to the data to show the dramatic drop in WVP with increasing PLA concentration in coating solution. An increase in the calculated

Table 2. Changes in water vapor permeability (WVP) of wheat gluten films due to polylactic acid polymer (PLA) coatings

PLA Conc. (%)	Thickness (mm)	WVP (x 10^{-9} $g \times m/m^2 \times s \times Pa$)	RH inside cup (%)
0	0.14 ±0.08	2.56 ±0.06d	75.4 ±0.2a
0.5	0.15 ±0.50	2.33 ±0.25cd	77.1 ±0.3b
1	0.15 ±0.52	1.93 ±0.68c	79.9 ±3.8bc
2	0.12 ±0.48	0.93 ±0.15b	85.7 ±0.7d
4	0.16 ±0.16	0.39 ±0.03a	94.4 ±0.5e
8	0.20 ±0.57	0.12 ±0.01a	98.5 ±0.3f

WVP values are mean of six replicates plus/minus standard deviation. Any two WVP means followed by the same letter are not significantly (P> 0.05) different according to the Duncan's Multiple Range Test.
Actual RH values (means of six replicates plus/minus standard deviation) at the under side of the films calculated as described by Gennadios et al (1994) to account for resistance of stagnant air between film and water surface in testing cups. RH outside the cup was maintained at 50%.

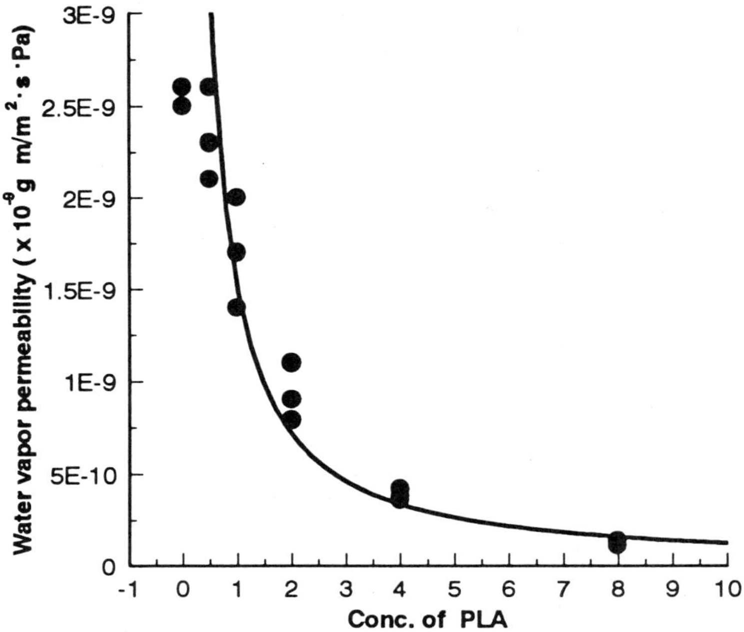

Figure 1. Water vapor permeability of wheat gluten coated with 0–8% concentrations of polylactic acid polymer (PLA) in coating solutions. An exponential regression line (r = 0.98) was fitted to the data.

actual RH's at the underside of the films was observed as the PLA concentration in the coating solution increased. Therefore, RH gradients (gradients increased with increasing concentration of PLA in solutions applied to films) applied across the films specimens during testing were not equal for all of the film types. Consequently, expected WVP values for PLA coated films would most likely have been even lower if equal RH gradients conditions had been applied across the films. The observed decrease in the WVP values with increasing amounts of PLA in coating solutions was attributed to the hydrophobicity of PLA. Additional study of PLA-coated wheat gluten films at different absolute humidities but at the same relative gradients is warranted, to aid understanding the influence of PLA on WVP of PLA-coated wheat gluten films.

3.3. Mulch Application

The total cost of a biopolymer mulch application is a major concern when attempting to develop commercial applications. Considering a current price of \$1.65 per kg for wheat gluten, a film thickness of 0.132±0.006 mm, 10 g of wheat gluten used for making a 23 x 33 cm piece of film (i.e. 1 g protein per 1.02 cm^3), volume of film required for 1 ha is (10,000 m^2 x thickness(m)) or 1.35 m^3, 1325 kg of wheat gluten is required to make a film to cover 1 ha of land. With our preliminary results, the cost for covering 1 ha of land with wheat gluten film would be \$2,286. Considering variable thickness of PLA coated films, PLA concentration, and PLA density (1.25 kg/m^3), the PLA requirement will be in the range 68 to 395 kg/ha. PLA coating costs would be in the range \$136 to 790 per hectare, depending on thicknesses of coating and anticipating that the commercial production price of PLA will be approximately \$2 per kg.

4. CONCLUSIONS

Tensile strengths of PLA-coated films increased for films of up to 2% PLA in the coating solutions, after which decreases in the Tensile strength were observed. PLA coating significantly reduced WVP values, indicating resistance to water vapor transmission through films. Initial estimates indicate that 1,325 kg of wheat gluten would be required to cover one hectare of land with WG-PLA film with a film thickness of 0.132±0.006 mm. The cost for wheat gluten would be $2,286 using a current price of $1.65 per kg. PLA coating costs would be in the range $136 to 790 per hectare, depending on thicknesses of coating and anticipating the commercial production price of PLA to be approximately $2 per kg.

REFERENCES

Astm (1989a) "Standard methods for tensile properties of thin plastic sheeting." D 882–88. Annual Book of ASTM Standards. American Society for Testing and Materials. Philadelphia, PA. 8.01 324–332

Astm (1989b) "Standard test methods for water vapor transmission of materials." Annual Book of ASTM Standards. American Society for Testing and Materials. Philadelphia, PA. Elongation 96–80 745–754

Briston JH (1988) "Plastic Films." 3 edn. John Wiley & Sons, Inc. New York. 434

Dennenberg RJ, Bothsst RJ and Thomas P (1978) "A new biodegradable plastic made from starch graft poly(methyl acrylate) copolymer." J. Appl. Polym. Sci. 22, 459

Doane WM (1988) "Proceedings of the first annual corn utilization conference." National Corn Growers Association. St. Louis, MO

Gennadios A, Brandenburg AH, Park JW, Weller CL and Testin RF (1994a) "Water vapor permeability of wheat gluten and soy protein isolate films." Industrial Crops and Products 2, 189

Gennadios A, Weller CL and Gooding CH (1994b) "Measurement errors in water vapor permeability of highly permeable hydrophilic edible films." J. Food Eng. 21, 395

Ghorpade VM, Gennadios A, Hanna MA and Weller CL (1994) "Soy protein/polyethylene oxide films." Cereal Chemistry, 72(6), 559

Gontard N, Guilberts S and Cuq J-L (1993) "Water and glycerol as plasticizers affect mechanical and water vapor barrier properties of an edible wheat gluten film." J. Food Sci. 58, 206

Griffin GJL (1977) US Patent 4016117

Guilbert S (1986) "Technology and application of edible protective films." In "Food Packaging and Preservation. Theory and Practice." M Mathlouthi, ed. Elsevier Applied Science Publishers Ltd. London, England. 371–394

Kester JJ and Fennema OR (1986) "Edible films and coatings, a review." Food Technol. 40(12), 47

Krochta JM (1992) "Control of mass transfer in foods with edible-coatings and films." In "Advances in Food Engineering." eds RP Singh and MA Wirakartakusumah. CRC Press Inc., Boca Raton, FL Ch-39. 517–538

McHugh TH, Avena-Bustillos R and Krocha JM (1993) "Hydrophillic edible films, modified procedure for water vapor permeability and explanation of thickness effects." J. Food Sci. 58, 899

Otey FH and Westhoff RP (1984) US Patent 4,454,268

Otey FH, Mark AM, Mehltretter CL and Russell CR (1974) "Starch based film for degradable agricultural mulch." Ind. Eng. Chem. Prod. Res. Dev. 13, 90

Otey FH, Westhoff RP.and Russell CR (1977) "Biodegradable films from starch and ethylene-acrylic acid co-polymer." Ind. Eng. Chem. Prod. Res. Dev. 16, 305

Otey FH, Westhoff RP and Doane WM (1980) "Starch based blown films." Ind. Eng. Chem. Prod. Res. Dev. 19, 592

Otey FH, Westhoff RP and Doane WM (1987) "Starch based blown films 2." Ind. Eng. Chem. Prod. Res. Dev 26, 1659

Park HJ, Bunn JM, Testin RF, Verango PJ and Edie DD (1993) "Characteristics of corn zein filled polyethylene films." Presented at the Annual Meeting of the Institute of Food Technologists, July 10–14. Chicago, IL. 822

16

ON-LINE MEASUREMENT OF BRAN IN FLOUR BY IMAGE ANALYSIS

Martin B. Whitworth,[1] Tony D. Evers,[1] and Christopher J. Brock[2]

[1]Campden & Chorleywood Food Research Association
Chipping Campden, Glos., GL55 6LD, United Kingdom
[2]Parascan Technologies Ltd.
Unit 8, Padgets Lane, South Moons Moat Industrial Estate,
Redditch, Worcs., B98 0RA, United Kingdom

1. INTRODUCTION

White flour contains microscopic specks of bran, the quantity of which must be controlled to maintain consistent flour quality. The two measurements most commonly made to characterise this aspect of flour quality are grade colour and ash value, based respectively on measurement of the total reflectance of a slurry made by mixing the flour with water, and of the mass of ash residue produced by incinerating the flour. These methods respond to bran by exploiting its darker colour and higher mineral concentration than other wheat tissues. However, they are also influenced by the colour and mineral content of these other tissues, which are variable among wheats, and prevent an absolute and universal measurement of bran content from being derived.

2. BRANSCAN 1000

We have developed a new method for measuring the bran content of flour and other mill stocks, based on image analysis. Specks of bran are individually identified within an image of a flour sample, allowing their quantity to be measured independently of the specific colours of the individual tissues. Previous applications of image analysis to such measurements have been based on fluorescence of the bran components (Harrigan, 1995; Symons and Dexter, 1996), and have required delicate microscopy equipment, which is inappropriate for use in dusty environments, and which could not be used on-line. The system described in this paper uses a different principle, based on the low reflectance of bran relative to other wheat tissues in visible light, and does not require a microscope. The system has been incorporated into a commercial instrument (Branscan 1000 - available from Parascan Technologies Ltd.); one model is suitable for laboratory use, and another (Figure

Cereals: Novel Uses and Processes, edited by Campbell *et al.*
Plenum Press, New York, 1997

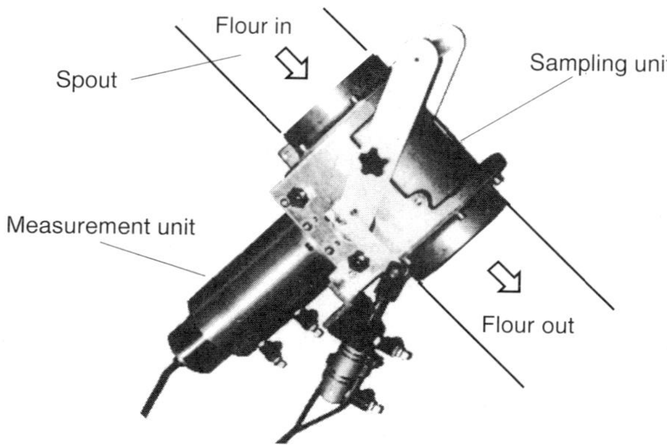

Figure 1. Branscan 1000 on-line measurement system, designed for connection to a flour spout. The majority of the spout cross section remains unobstructed.

1) incorporates automatic on-line sampling, allowing continuous monitoring of product quality within mills.

2.1. Principle of Operation

Both models of Branscan 1000 incorporate identical optical systems, housed in a sealed measuring head. This incorporates a flat window at one end, against which a flour sample is compressed. The analysis is non-destructive, and requires no special sample preparation. In the off-line model, a sample of < 200g is presented in a tray and compressed manually; in the on-line version, the sample is collected automatically from a flour stream and compressed by a pneumatic piston. After measurement, the sample is returned to the flow and the window is cleaned by compressed air. The flour is uniformly illuminated by a low power light source and is imaged by a video camera, both mounted within the measuring head. The image is digitised and is processed by a IBM compatible PC computer to identify bran specks (Figure 2), which are darker than their surroundings. The computer also controls the sampling system and is capable of controlling up to 8 instruments on-line, allowing many flour streams to be measured concurrently with no reduction in the sampling rate. Several hardware configurations are possible, including a sealed computer unit for mill floor use as an alternative to a desk-top computer, and options such as an audible alarm for bran content exceeding a chosen specification. The image-processing algorithm is designed to be insensitive to variations in the colour and brightness of bran specks and of endosperm and responds solely to the quantity of darkly contrasted tissues. Although the lighting is in fact highly uniform, the algorithm also has the capability to correct for substantial degrees of non-uniformity, increasing the robustness of analysis. The total area of bran specks is measured, and expressed as a percentage of the total image area. By calculating the mean of several presentations, sampling errors due to the random distribution of bran particles in flour can be reduced to any desired level. Ten samples are considered sufficient for most purposes and typically yield results accurate to about 0.1% bran content, but greater numbers of images can be averaged if higher accuracy or more extensive sampling is required. In the on-line system, samples are collected at a rate of 3 per minute, and results are presented as a graph of a moving average on a computer screen, and as a continuous printout; in the off-line model, as many as

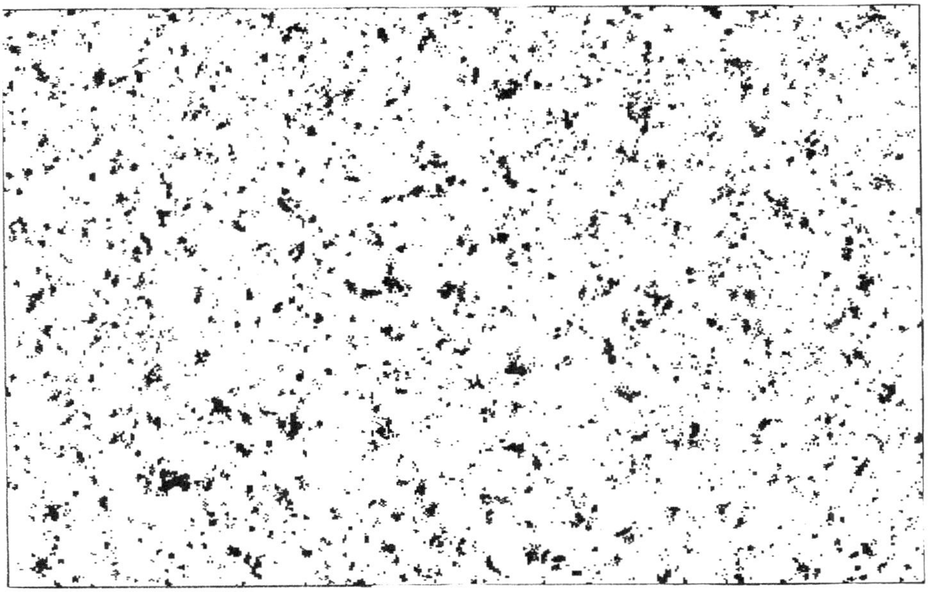

Figure 2. Processed image of bran specks in flour.

20 images per presentation can be taken automatically in under 2 minutes. The standard error is also estimated from the standard deviation of measurements of individual images.

In addition to measuring total bran content, Branscan also has the capability to count individual specks. Such types of measurement can provide an objective alternative to current manual speck counting procedures used in the assessment of durum wheat semolina quality, which suffer from considerable variation between operators (Symons et al, 1995). As with the system described by Symons et al, the measurement sensitivity can be adjusted by specifying the minimum size of speck to be counted. In an extension of this principle, the possibility also exists for using Branscan to measure the size distribution of bran particles, as has been demonstrated in a prototype instrument, which is capable of subdividing the total bran measurement into fine, medium and coarse categories.

3. PERFORMANCE CHARACTERISTICS

3.1. Calibration against Test Cards

Although optimised for mill products and configurable in several ways (for example, to alter the sensitivity to fine bran particles), the processing algorithm used by Branscan has been designed at a fundamental level to achieve accurate measurement of dark areas against a lighter background, independently of user settings. This aspect of performance is tested by measurement of black dots printed on a white background, and has proved robust, even when tested with different lighting and camera systems. Figure 3 shows Branscan 1000 measurements of several such test cards, plotted against the true percentage coverage of the dots, by area.

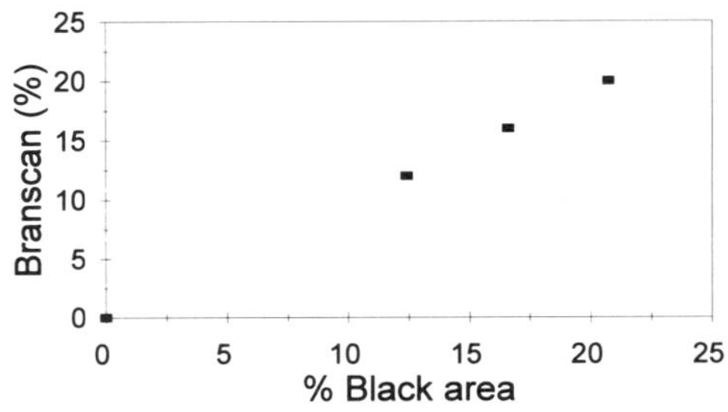

Figure 3. Branscan measurements of black printed dots, plotted against the percentage area occupied by the dots.

3.2. Specificity for Individual Wheat Tissues

Branscan 1000 operates in the visible waveband and is sensitive to the coloured components of bran (primarily the pericarp-testa), which are most relevant to the visual appearance of speckiness and of dark colour in flour. The sensitivity of the reading to these tissues is confirmed in Figure 4, which shows a greater response to the addition of pericarp-testa to patent flour than to the addition of aleurone, which has a lesser colouration. This higher sensitivity to pericarp-testa is also true of grade colour, which also operates in the visible waveband. The ash test, however, generally shows a higher sensitivity to aleurone, which has a higher mineral concentration. Some emerging milling technologies, such as those based on abrasive debranning processes, have the capability to produce white flours enriched with aleurone (e.g. Dexter and Wood, 1996; see also Chapters 29 and 32), and therefore with high ash values. Such flours perform better in baking tests than would be expected from their ash values on the basis of experience with conventionally milled flours. Tests which respond primarily to pericarp-testa may therefore prove more universal indicators of baking quality.

The ability of Branscan to measure bran content independently of endosperm colour has been confirmed by measurements of two chlorinated (1500 ppm) cake flours and of unchlorinated control samples of the same flours. The results in Table 1 reveal a reduction in grade colour due to the bleaching effect of the chlorine, but an unchanged (to within sampling error) Branscan reading, due to the unchanged bran content.

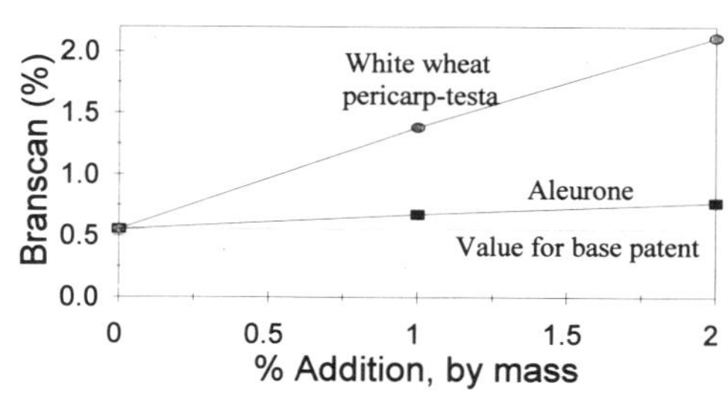

Figure 4. Response of Branscan to addition of pericarp-testa or aleurone to a patent flour.

Table 1. Measurement of chlorinated cake flours and unchlorinated control flours

	Grade Colour		Branscan (%)		Ash (%)	
	Untreated	Chlorinated	Untreated	Chlorinated	Untreated	Chlorinated
Flour 1	-3.0	-3.9	0.498	0.468	0.39	0.39
Flour 2	-2.7	-3.6	0.436	0.466	0.36	0.37

3.3. Sensitivity to Bran Addition

The sensitivity of Branscan 1000 measurements to different levels of bran content, and its ability to discriminate similar quantities, have been tested by measurement of sub-samples of a patent flour to which various levels of bran (75–125μm) had been added. Although a small initial bran content can be expected, even in a high quality patent flour, and the absolute bran content is therefore uncertain, it is possible to measure the increase in readings above the baseline value for the pure patent flour. Studies conducted previously (Evers, 1993) with an early prototype of Branscan demonstrated its superior ability to discriminate different levels of red bran addition to several base flours than other methods which showed a greater spread of values due to the variable endosperm properties of the base flours. Subsequent developments have further improved Branscan's sensitivity to small bran particles, such that particles as small as a single pixel (~25μm) can be detected, provided that they are sufficiently coloured. The sensitivity to poorly contrasted particles has also been improved, such that flours containing white wheat bran can also be measured, with only a slightly lower sensitivity. Results are shown in Figure 5 of the response to various levels of addition of either red or white wheat bran to a single patent flour. The response is close to linear over a wide range of bran levels, particularly at the lower end. Unlike grade colour, and in common with ash, the instrument can therefore easily be used to blend flours to a chosen specification.

3.4. Comparison with Ash and Grade Colour for Commercially Milled Flours

In addition to characterising Branscan's performance against flours of controlled composition, it is also important to evaluate its performance in commercial practice and to demonstrate how its measurements relate to the more familiar ones used to measure bran

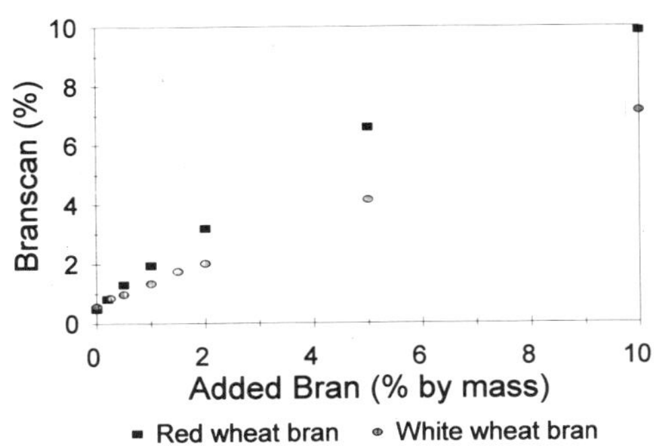

Figure 5. Response of Branscan 1000 to addition of red or white bran to a patent flour.

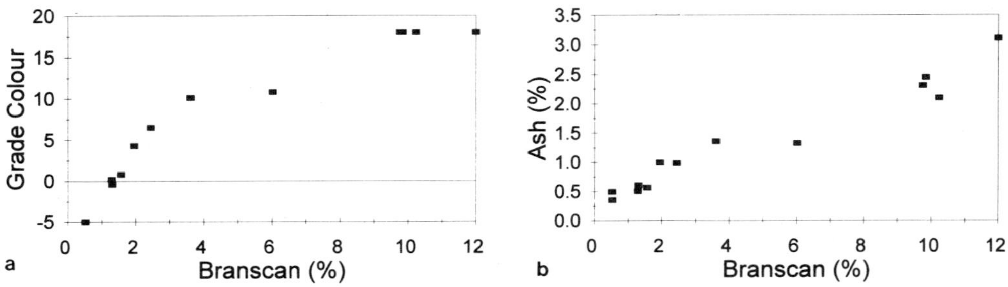

Figure 6. Comparison of Branscan 1000, grade colour, and ash measurements for single grist flour streams and blends from a single commercial mill.

contamination. It has been shown previously (Whitworth, 1994), for an early Branscan prototype, that when cumulative curves of bran content were plotted against extraction rate in a similar manner to ash curves for several pilot milled wheat varieties, Branscan gave a less variable baseline reading at low extraction and thus provided a more universal prediction of extraction rate.

The operational reliability of Branscan 1000 has been favourably tested on-line in a commercial mill, and off-line measurements have also been made of commercially milled flours and other mill stocks from many mills worldwide. Figure 6 shows comparisons of Branscan, grade colour, and ash measurements for a set of single grist flour streams from one mill. Figure 6a reveals the limitations in the measuring range of the colour grader, particularly with low-grade streams, several of which had off-scale grade-colour values of 18, whereas both ash and Branscan 1000 values of all samples (Figure 6b) increased with bran content and had a correlation coefficient of 0.975. When just the high and middle grade streams were compared, a high correlation coefficient of 0.995 was obtained between Branscan 1000 and grade colour, since both of these tests respond primarily to pericarp-testa; because ash responds more strongly to aleurone, the correlations with this are poorer. Grade colour and ash measurements are affected by variation in endosperm whiteness and mineral content respectively, which confound bran estimation. Therefore, although the above correlations are typical for single grist sample sets, correlations with each other and with Branscan, which responds more specifically to bran content, are poorer when flours from several grists are compared.

4. CONCLUSIONS

It has been shown that image analysis can be used to identify specks in flour samples and that this provides a useful measure of bran content. The measurement shows a linear response to addition of bran or to the proportions of flours in a blend. It is based on the relative areas of bran and endosperm in an image, and unlike ash or grade colour, is insensitive to the variable properties of these tissues. The measurement is based on contrast in visible light, and is therefore sensitive primarily to pericarp-testa, showing a good correlation with grade colour for flours milled from a single grist. The method has been developed as a commercial instrument (Branscan 1000). In addition to a laboratory instrument, a model suitable for on-line use has been developed, incorporating an automatic sampling system, which has been successfully tested in prolonged use in a commercial mill. The analysis software can be configured to measure either flour or semolina and, in addition to measuring bran content by total area, it is also possible to count individual specks, suitable for durum semolina testing. Size measurement of specks has also been demonstrated in a prototype.

REFERENCES

Dexter JE and Wood PJ (1996) "Recent applications of debranning of wheat before milling." Trends in Food Science & Technology, 7(Feb), 35–41

Evers AD (1993), "On-line quantification of bran particles in white flour." Food Science and Technology Today 7(1), 23–26

Harrigan K (1995) "Flour Power: Microscopic Image Analysis in the Food Industry." Cereal Foods World 40(1), 11–14

Symons SJ and Dexter JE (1996) "Aleurone and Pericarp Fluorescence as Estimators of Mill Stream Refinement for Various Canadian Wheat Classes." J. Cereal Science 23(1), 73–83

Symons SJ, Dexter JE, Matsuo RR and Marchylo BA (1995) "Rapid Instrumental Estimation of Bran Specks in Durum Semolina." AACC Annual Meeting, San Antonio, Cereal Foods World 40(9), 648 (abstract only)

Whitworth MB (1994) "Under the Spotlight." International Milling Flour & Feed 188(5) Supplement, 10–13

17

PRETREATMENT OF AGRICULTURAL CROP RESIDUES FOR CONVERSION TO HIGH-VALUE PRODUCTS

Anette Skammelsen Schmidt and Anne Belinda Bjerre

Environmental Science and Technology Department
Risø National Laboratory
PO Box 49, DK-4000 Roskilde
Denmark

1. INTRODUCTION

Lignocellulosic biomass residues from agricultural crops, e.g. straw and sugar-beet pulp, as well as new alternative industrial crops such as flax, kenaf, hemp, miscanthus and willow, are potential raw materials for production of several high-value products, including energy, ethanol, enzymes, xylitol and biofibres. These renewable raw materials look promising to replace environmentally unfriendly fossil hydrocarbons, and hence, creating "green" products.

The lignocellulose consists of three major constituents: cellulose, hemicellulose and lignin. The cellulose is a linear high molecular weight polymer of D-glucose units and, due to β-1,4 linkages, is a highly crystalline material resistant to enzymatic hydrolysis. Considering its physical structure, cellulose is most suitable for use as fibre, e.g. in paper making and fibreboard or as carbohydrate source for fermentation after hydrolysis to D-glucose (Büchert, 1990). The amorphous hemicellulose is composed of shorter chain polymers that provide linkage between lignin and cellulose (Fan et al, 1982). Lignin is probably the most complex component, with a three dimensional phenylpropane polymer structure held together by ether and carbon-carbon bonds (Fan et al, 1982). Cellulose and hemicellulose (70–80% of the dry weight) are intimately associated with lignin in the plant cell wall (Fengel and Wegener, 1989; Viikari et al, 1991).

The two polysaccharides are not directly available for bioconversion, as the lignin component functions as a physical barrier, which must be overcome by pre-treatment. Several pre-treatment processes have been developed in order to break down lignin and open the crystalline structure in cellulose, e.g. high-pressure steaming, alkaline or acid hydrolysis, gas treatment (chlorine dioxide, sulphur dioxide, ozone), hydrogen peroxide treatment, organo-solvent treatment, steam explosion, wet oxidation and biological treatment (Fan et al, 1982; Hörmeyer et al, 1988; McGinnis et al, 1983). The combination of the wet oxidation process (water, oxygen pressure, elevated temperature) and alkaline hydrolysis has proven to be an efficient method for wheat straw fractionation (Bjerre et al, 1996). By this treatment the hemicellulose was solubilized, the lignin partially degraded and the compact cellulose structure opened, thereby making the polysaccharides accessible for bioconversion or purifying the cellulose fibre fraction for high quality fibre-based products, as illustrated in Figure 1 (Chapters 20, 21, 25 and 27 describe similar integrated systems). The cellulose-rich fibres derived from wet oxidation of flax and straw were bleached in the process due to the presence of oxygen (Bjerre and Pallesen, 1994).

In this study two different raw materials, wheat straw and flax, were used to evaluate their potential for generating high-value products such as fermentable carbohydrates and high quality fibres, respectively. The effects of reaction temperature, alkaline addition and oxygen pressure on wheat straw and flax fractionation were evaluated. Preliminary fermentation of the aqueous hemicellulose-rich fraction to enzymes was carried out. The tensile strength of the flax fibres was also examined.

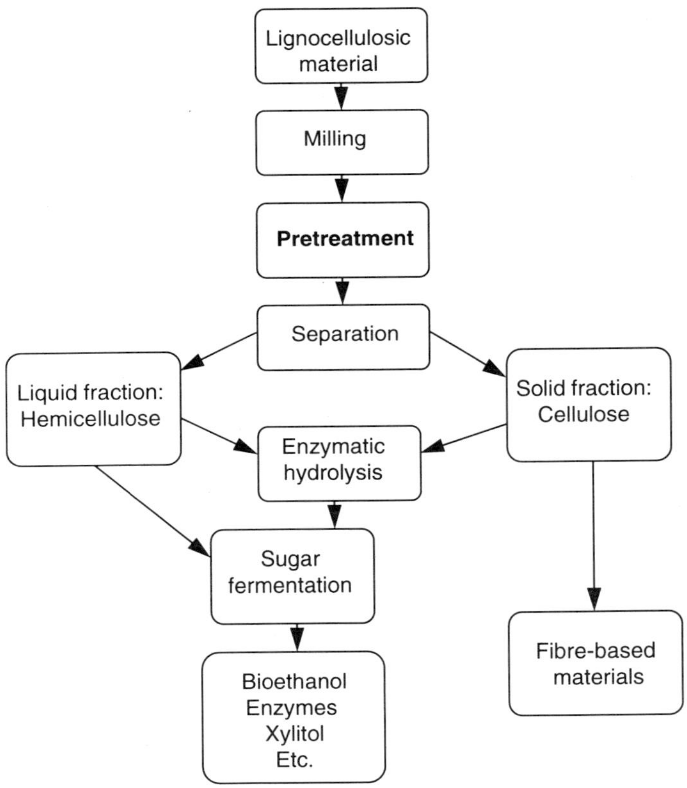

Figure 1. Flow diagram of possible utilization paths of agricultural crop residues in an integrated physical/chemical and biological treatment.

2. MATERIALS AND METHODS

2.1. Raw Materials

The wheat straw and flax were grown at Risø National Laboratory and Department of Plant Production, Denmark, respectively.

2.2. Pre-Treatment

The wet oxidation was carried out in a loop-reactor constructed at Risø National Laboratory (Bjerre and Sørensen, 1992; Bjerre et al, 1996). The wheat straw or flax was mixed with Na_2CO_3 and water, before increasing oxygen pressure and heating the suspension. After the reaction, the suspension was filtered to separate the solid cellulose-rich fraction (filter cake) from the liquid hemicellulose-rich fraction (filtrate). The pH of the filtrate was measured and the filter cake was dried and weighed. The composition of the two fractions was analysed together with the quality of the fibres.

2.3. Analysis

The gravimetric method of Goering and van Soest (1970) was used to determine the chemical composition (hemicellulose, cellulose and lignin) of the solid fractions. The convertibility of the cellulose to glucose was determined by a mixture of two enzymes: Celluclast and Novozym 188, kindly provided by Novo Nordisk A/S, Bagsværd, Denmark. The sample was suspended in 0.2 M acetate buffer (pH = 4.8) and hydrolysed by the enzymes for 24 hours at 50°C (Bjerre et al, 1996). The concentration of D-glucose in the filtrate was determined by HPLC.

The hemicellulose in the wet oxidation filtrate was hydrolysed by acid hydrolysis of 4% w/v H_2SO_4 at 121°C for 10 minutes. The samples were purified by combined precipitation and ion exchange treatment (Bjerre et al, 1996). The monosaccharides were quantified by HPLC cation exchange (Aminex HPX-87H column (Biorad)) with 0.004 M H_2SO_4 as eluent. Furfural and hydroxymethylfurfural were measured by HPLC (Nucleosil 5C-18 column) with a linear eluent gradient of methanol (10–90 %) at pH 3 (Bjerre et al, 1996).

The cellulose-rich flax fibres were processed into sheets of paper according to CPPA standard methods and the tensile strength of both dry and wet paper sheets determined by the zero-span tensile test method (Pallesen, 1996).

2.4. Fermentation

Aspergillus niger IBT 13099 was kindly supplied by DTU, Denmark. Stock cultures were maintained on PDA at 30°C. *A. niger* was grown at 30°C for 114 hours in submerged

Table 1. Chemical composition (% w/w of dry weight) of wheat straw and flax

Raw material	NCWM[a] (%w/w)	Hemicellulose (%w/w)	Cellulose (%w/w)	Total lignin (%w/w)	Ash (%w/w)
Wheat straw	18.8	32.8	38.0	8.9	1.4
Flax	14.1	12.9	68.3	4.8	trace

[a]NCWM = Non-cell wall material (pectin, proteins, etc.)

culture of wet oxidised wheat straw substrate supplemented with several salts (pH 5.5, 100 rpm) (Bjerre et al, 1996). β-Xylosidase activity was measured by incubating the intact mycelia with p-nitrophenyl-β-D-xylopyranoside (Bjerre et al, 1996; Stålbrand et al, 1992). The released p-nitrophenol was measured spectrophotometrically at 410 nm. The activity was defined as the release of one μmol of p-nitrophenol/minute.

3. RESULTS AND DISCUSSION

3.1. Wheat Straw - The Hemicellulose-Rich Fraction

Pretreatments have mainly been optimised with respect to the hydrolysis yield of the cellulose fraction, although up to 35–40% of the lignocellulose may be hemicellulose, mainly as xylan, especially in annual plant residues (Fan et al, 1982). Therefore, this section concentrates on the potential solubilization of the hemicellulose fraction by the wet oxidation process.

The temperature was previously found (Schmidt and Bjerre, 1996) to be the most important process parameter in wet oxidation affecting the fractionation of wheat straw. A temperature of 185°C gave the highest concentration of solubilized hemicellulose in the filtrate (Figure 2) together with the highest convertibility of the cellulose to glucose (Schmidt and Bjerre, 1996). In particular, the temperature affected the concentration of xylose and arabinose but not the concentration of glucose, in accordance with previous findings (Schmidt and Bjerre, 1996). At 185°C nearly 3 times more hemicellulose was in solution than at 150°C. Surprisingly, slightly more hemicellulose was removed from the solid fraction at 185°C (81% w/w) than at 200°C (76% w/w) but still significantly lower

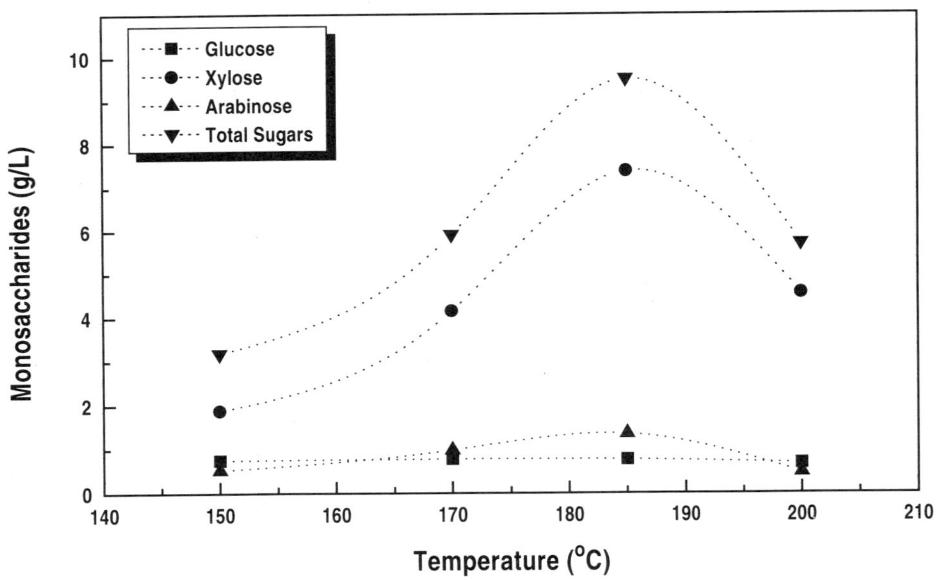

Figure 2. The hemicellulose concentration (measured as monosaccharides after acid hydrolysis) in the filtrate obtained by the wet oxidation process (60 g/L wheat straw, 6.5 g/L Na_2CO_3, 12 bar O_2 and 15 minutes) as a function of the reaction temperature.

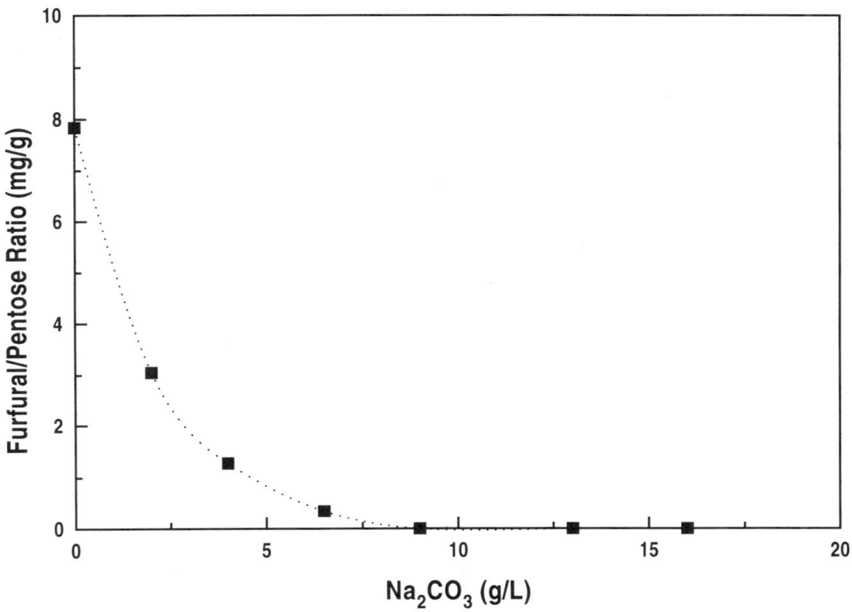

Figure 3. The ratio of furfural formation over the hemicellulose concentration (measured as pentoses) in the filtrate obtained by wet oxidation (60 g/L wheat straw, 12 bar O_2, 185°C and 15 minutes) as a function of the sodium carbonate concentration.

amounts were obtained in the filtrate at 200°C. This was probably due to a higher degradation of hemicellulose at 200°C than at 185°C. At 200°C slightly more hemicellulose remained in the solid fraction. At 185°C the recoveries of both hemicellulose (65%) and cellulose (99%) were significantly higher than at 200°C (51% and 64%, respectively). Hence, at the given conditions (60 g/L wheat straw, 6.5 g/L Na_2CO_3, 12 bar O_2 and 15 minutes) the optimal temperature was determined to be 185°C.

The sodium carbonate addition had a profound effect on the formation of furfural in the wet oxidation process (Figure 3) as only minimal effect was obtained on the amount of solubilized hemicellulose (Schmidt and Bjerre, 1996). At high levels of sodium carbonate no furfural was formed, whereas at low level of carbonate some furfural was formed. The concentration of furfural exponentially increased with the decrease in carbonate. In the absence of carbonate the highest concentration of furfural was obtained, in accordance with the formation of furfural in other pre-treatment processes (Büchert, 1990; Von Sivers et al, 1994). However, a low concentration of furfural was not expected to inhibit microbial growth significantly (Bjerre et al, 1996; Schmidt and Bjerre, 1995) or ethanol production, as has been shown for some thermophilic anaerobic bacteria (Ahring et al, 1996). Therefore, the carbonate addition could possible be as low as 4 g/L giving about 1 mg furfural per g soluble hemicellulose, a compromise between economics and inhibitor generation. Hydroxymethylfurfural was not observed in any wet oxidised filtrates.

Pre-treatments were carried out without oxygen and/or without carbonate addition in order to evaluate the effect of those two parameters in the wet oxidation process. When no carbonate was present a higher concentration of furfural was formed than when carbonate was added (more than a factor of 10 difference) (Figure 4) in accordance with Figure 3.

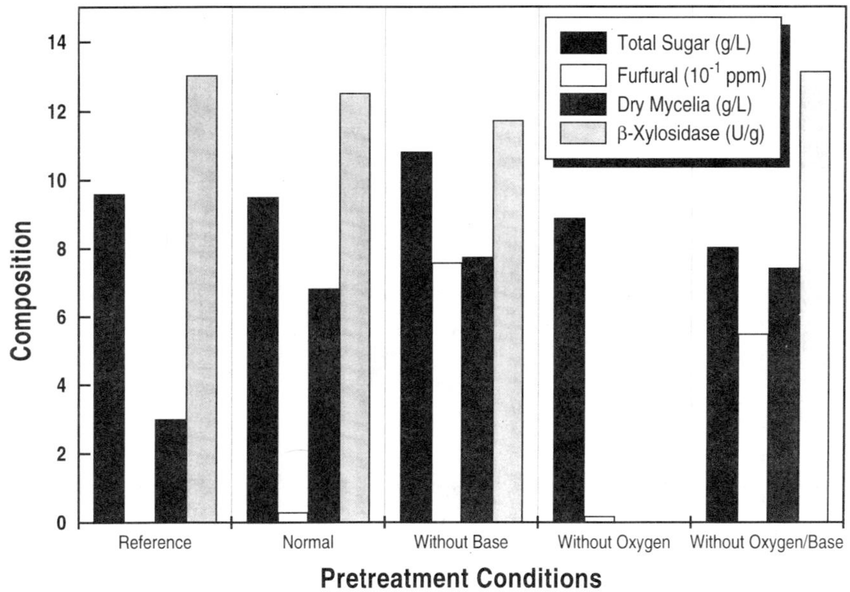

Figure 4. The effect of different hydrothermal pre-treatment conditions on the mycelia production (g/L) and b-xylosidase activity (U per dry mycelia) in *A. niger* fermentations. The total sugar (g/L) and furfural (10–1 ppm) was measured in the fermentation substrate. Normal: 60 g/L wheat straw, 6.5 g/L Na_2CO_3, 12 bar O_2, 185°C and 15 minutes. Without Oxygen: Normal without O_2. Without Base: Normal without Na_2CO_3. Without Oxygen/Base: Normal without O_2 and Na_2CO_3. Reference: monosaccharide substrate (0.8 g/L glucose; 7.4 g/L xylose; 1.4 g/L arabinose).

On the other hand, oxygen did not have a great effect on the formation of furfural (Figure 4). The presence of oxygen was important for the amount of produced carboxylic acids in the process (data not shown). This is probably related to the degradation of phenolic substances in the lignin network which are very susceptible to oxidation during the wet oxidation process (Devlin and Harris, 1984). These filtrates (without oxygen) also contained less furfural than the corresponding filtrates (with oxygen).

In order to examine the importance of different potential inhibitors, fermentation with *A. niger* was carried out. *A. niger* grew better in wet oxidised wheat straw (Normal) than in the reference substrate (Figure 4). The pre-treatment without added carbonate also provided a suitable substrate for fermentation of *A. niger*, giving similar dry mycelia concentration and β-xylosidase activity. Hence, the furfural concentration was not found to inhibit *A. niger* growth or β-xylosidase production. However, no growth was observed in the wheat straw pre-treated without oxygen but with Na_2CO_3. Therefore, other inhibitors must be present in this filtrate, possible iso-butyric acid (only present in this substrate (data not shown)) or some lignin degradation products, which had not been oxidised. The presence of these other potential inhibitors and their effect on the fermentability of the hemicellulose-rich fraction is presently being investigated. This study indicates that a variety of high-value products such as ethanol, furfural, enzymes, xylitol, carboxylic acids and oligosaccharides could be produced from the available carbohydrates.

3.2. Flax Fibres - The Cellulose-Rich Fraction

The solid cellulose-rich fraction can be either converted to a fermentation substrate after enzymatic hydrolysis to glucose or used in fibre composites (Figure 1). Due to the very high concentration of cellulose in flax (Table 1) this crop looks promising for utilisation of the cellulose fibre fraction. Lignocellulosic fibres have received a lot of interest due to their properties for replacing traditional fibres such as asbestos and glass in composite applications such as thermoplastics or cement. The properties of importance are the strengthening and stiffening part of the composite, generating composites with better specific stiffness (modulus/density) (Lilholt, 1994). The low density of annually-grown lignocellulosic fibres provides structural superiority (Sanadi et al, 1995). Furthermore, flax has a specific stiffness similar to asbestos but higher than glass fibres (Lilholt, 1994). Therefore, in this study attention was focused on the potential purification of the cellulose fraction by wet oxidation leading to high quality fibres.

The temperature was found to be the most important parameter affecting the chemical composition of the wet oxidised flax fibres, although oxygen and carbonate also had an effect (Figure 5) as found for wheat straw (Schmidt and Bjerre, 1996). Hemicellulose is partly responsible for the strength of the fibres (the stiffness), however, it and the Non-Cell Wall Material such as pectin are not desired in fibres for cement composites, as these compounds are responsible for moisture absorption (Sanadi et al, 1995). In general, wet oxidation efficiently solubilized the hemicellulose, thereby removing it from the fibre fraction, as for wheat straw (Bjerre et al, 1996; Schmidt and Bjerre, 1996). The application of higher concentrations of oxygen and carbonate in process no. 5 (Figure 5) resulted in a removal of 91% of the Non-Cell Wall Material together with nearly all lignin (96%)

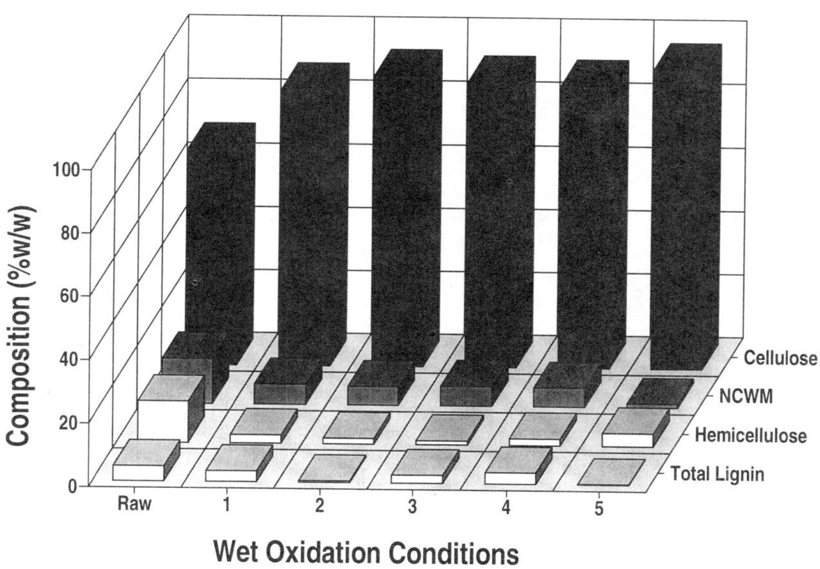

Figure 5. The chemical composition of the solid fraction (% w/w of dry weight) obtained by wet oxidation (10 g/L flax, 10 minutes) as a function of the process conditions. **Raw**: Raw material. **1**: 160°C, 1 g/L Na_2CO_3, 2.5 bar O_2. **2**: 180°C, 1 g/L Na_2CO_3, 5 bar O_2. **3**: 180°C, 2.5 g/L Na_2CO_3, 2.5 bar O_2. **4**: 160°C, 2.5 g/L Na_2CO_3, 5 bar O_2. **5**: 170°C, 5 g/L Na_2CO_3, 10 bar O_2 (Bjerre and Pallesen, 1994).

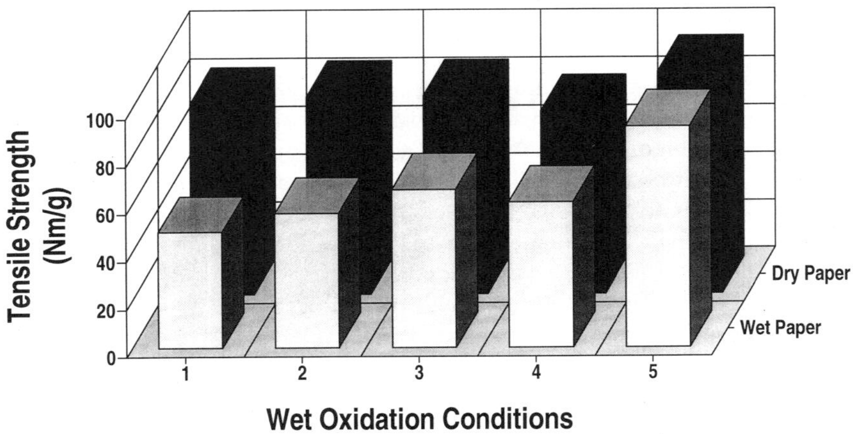

Figure 6. The tensile strength of flax fibres obtained by different wet oxidation pretreatments (conditions given in Figure 5) (Bjerre and Pallesen, 1994).

and hemicellulose. Hence, the solid fraction consisted of nearly pure cellulose (94.5% w/w).

The tensile strength of the fibre treated by conditions no.5 was very high, although the relative amount of hemicellulose in the fibres was higher than for the other 4 wet oxidation treatments (Figure 6). The reason why these conditions (no.5) gave a stronger fibre might be due to the very low content of Non-Cell Wall Material and lignin, and the hemicellulose adding to the strength. Due to the low hemicellulose content and the high tensile strength, wet oxidised flax fibres could be applied as, for example, composite material in cement or thermoplastics. Whether the achieved fibres actually will be suitable for reinforcing purposes in the composites is still to be tested.

4. CONCLUSIONS

Pre-treatment like wet oxidation provided an efficient fractionation of the two major carbohydrates in lignocellulose. In particular, wet oxidation was effective in view of achieving maximum solubilized hemicellulose from wheat straw without generating fermentation inhibitors, originating from the degradation of the carbohydrates during pretreatment. The sodium carbonate addition in the wet oxidation process prevented the formation of furfural, although at low carbonate concentration some furfural was formed. The furfural concentration in the hemicellulose-rich fraction obtained by different pretreatment conditions was not found to inhibit *A. niger* growth or β-xylosidase production. Preliminary studies of wet oxidation of flax indicated that this process was efficient at removing non-cell wall materials and lignin, generating high quality flax fibres with high tensile strength.

ACKNOWLEDGMENT

Funding was provided by the Danish Energy Ministry in the project "Development of chemical and biological processes for bioethanol production" (EFP 1383/94–0003).

REFERENCES

Ahring BK, Bjerre AB, Jensen K, Nielsen P and Schmidt AS (1996) "Pretreatment of wheat straw and conversion of xylose and xylan to ethanol by thermophilic anaerobic bacteria." Bioresource Technol. (submitted)

Bjerre AB and Sørensen E (1992) "Thermal decomposition of dilute aqueous formic acid solutions." IandEC Res. 31, 1574–1577

Bjerre AB and Pallesen BE (1994) "Production of flax fibers for industrial purposes; using combined wet oxidation and alkaline hydrolysis." 4th Oligo- and Polysaccharides Conference, Aussois, France, September 14–16

Bjerre AB, Olesen AB, Fernqvist T, Plöger A and Schmidt AS (1996) "Pretreatment of wheat straw using combined wet oxidation and alkaline hydrolysis resulting in convertible cellulose and hemicellulose." Biotechnol. Bioeng. 49, 568–577

Büchert J (1990) "Biotechnical oxidation of D-xylose and hemicellulose hydrolyzates by *Gluconobacter oxydans*." PhD thesis, Technical Research Centre of Finland

Devlin HR and Harris IJ (1984) "Mechanism of the oxidation of aqueous phenol with dissolved oxygen." Ind. Eng. Chem. Fundam. 23, 387–392

Fan LT, Lee Y-H and Gharpuray MM (1982) "The nature of lignocellulosics and their pretreatments for enzymatic hydrolysis." Adv. Biochem. Eng. 23, 157–187

Fengel D and Wegener G (1989) "Wood - Chemistry, Ultrastructure, Reactions." Walter de Gruyter, Berlin

Goering HK and van Soest PJ (1970) "Forage fiber analyses (apparatus, reagents, procedures, and some applications)." Agricultural Handbook 379, Agricultural Research Service - United States Department of Agriculture, USDA, Washington DC, 1–20

Hörmeyer HF, Schwald W, Bonn G and Bobleter O (1988) "Hydrothermolysis of birchwood as pretreatment for enzymatic saccharification." Holzforschung 42, 95–98

Lilholt H (1994) "Fiber-Reinforced Alloys." In "Mechanical Properties of Metallic Composites." (S Ochiai, ed.), Marcel Dekker, New York, 373–380

McGinnis GD, Wilson WW and Mullen CE (1983) "Biomass pretreatment with water and high pressure oxygen. The wet-oxidation process." Ind. Eng. Chem. Prod. Res. Dev. 22, 352–357

Pallesen BE (1996) "The quality of combine-harvested fibre flax for industrial purposes depends on the degree of retting." Ind. Crops Prod. 5, 65–78

Sanadi AR, Caulfield DF, Jacobson RE and Rowell RM (1995) "Renewable agricultural fibers as reinforcing fillers in plastics: mechanical properties of kenaf fiber-polypropylene composites." IandEC Res. 34, 1889–1896

Schmidt AS and Bjerre AB (1995) "Optimization of wet oxidation of wheat straw for the enzyme production by *Aspergillus niger*." GIAM X - Tenth International Conference on Global Impacts of Applied Microbiology and Biotechnology, Elsinore, Denmark, August 6–12

Schmidt AS and Bjerre AB (1996) "Process optimization of wet oxidation for production of fermentable carbohydrates from wheat straw." Bioresource Technol. (submitted)

Stålbrand H, Hahn-Hägerdal B, Réczey K and Tjerneld F (1992) "Mycelia-associated β-xylosidase in pellets of *Aspergillus* sps." Appl. Biochem. Biotechnol. 34/35, 261–272

Viikari L, Kantelinen A, Rättö M and Sundquist M (1991) "Enzymes in pulp and paper processing." In "Enzymes in Biomass Conversion." (GF Leatham and ME Himmel, eds.), ACS Symposium Series 460, ACS, Washington DC, 12–21

Von Sivers M, Zacchi G, Olsson L and Hahn-Hägerdal B (1994) "Cost analysis of ethanol production from willow using recombinant *Escherichia coli*." Biotechnol. Progr. 10, 555–560

INNOVATIVE USES OF CEREALS FOR FRUCTOSE PRODUCTION

Maria Grazia D'Egidio,[1] Cristina Cecchini,[1] Claudio Corradini,[2] Virgilio Donini,[2] Vito Pignatelli,[3] and Tommaso Cervigni[4]

[1]Istituto Sperimentale per la Cerealicoltura
via Cassia 176, 00191 Roma, Italy
[2]Istituto di Cromatografia del C.N.R.
Area della ricerca di Roma- 00016 Monterotondo(Roma), Italy
[3]ENEA INN BIOAG, C.R.
Casaccia, via Anguillarese 301, 0060 Roma, Italy
[4]C.R.A.
via Borgorose 15, 00189 Roma, Italy

1. INTRODUCTION

The EC policies, because of the well-known surplus situation in cereals production, are presently re-directed towards reducing the cultivation area and growing alternative plants for energy or industrial purposes.

Cereal crops, used by man particularly for grain consumption, produce and store for much of their growing cycle significant amounts of water soluble carbohydrates (WSC), composed of monosaccharides, sucrose and fructans. In such a frame, a new perspective can be proposed: the utilization of cereal crops for biomass production to be converted into products (fructose and fructose polymers) with high added value for new food and non-food primary products.

This innovative destination offers the advantage of keeping a not negligible portion of the labour force active in the sector, along with the benefit of preserving the quality of the environment, the possibility of ensuring farmer's income and of reducing the inputs of agrochemical and other energy-intensive production factors.

Moreover this new perspective is particularly attractive because currently most of the fructose for industrial production is obtained from corn starch by a complex process of enzymatic hydrolysis, followed by isomerisation and separation. Only few industries in Northern and Central European countries extract fructose from fructan-synthesizing plants, using as raw material roots from chicory (*Cichorium intybus*). The advantage in the use of cereals as the raw material for a direct fructose production is that they are well

Cereals: Novel Uses and Processes, edited by Campbell *et al.*
Plenum Press, New York, 1997

known and diffused crops, and the only change in the standard cultivation process is to anticipate the harvesting date.

Previous investigations (Kuhbauch and Thome, 1989; Pollock, 1991; D'Egidio and Cervigni, 1992) found that large quantities of fructose polymers are stored in stems of cereals (C_3 type) after anthesis, and that maximum accumulation occurs two-three weeks after flowering, at the physiological stage of milky phase.

The aim of this work is the evaluation of the productive potential for cereal crops in the perspective of using them as raw material for industrial fructose production.

2. MATERIALS AND METHODS

Cultivars of durum wheat, bread wheat and barley were grown on 10 m^2 plots according to a randomized block design and a sowing density of 450 seeds/m^2. Samples were collected on two segments of a 25 cm length on the same row. The stems together with the sheats and separately the ears were dried in a stove at 30°C for several days. The material was milled by a Cyclotec-Tecator (PBI) before performing the chemical and chromatographic analyses.

Chemical analyses were carried out following conditions reported by D'Egidio et al (1993). The chromatographic separations were performed with Dionex equipment (Sunnyvale, CA, USA) as reported in a previous work by Corradini et al (1995). Glucose, fructose, sucrose, raffinose, melezitose and trehalose standards were from Sigma (St. Louis, MO, USA). The trisaccharide 1-kestose was from Tokyo Kasei (Tokyo, Japan).

Chromatographic conditions for the kernel analyses were slightly varied in respect to those used for the stems, in order to facilitate the separation of mono-disaccharides, trisaccharides, and fructan oligosaccharides. Isocratic elution was protracted until the first 15 minutes, then a linear sodium hydroxide-sodium acetate gradient was performed from initial conditions (5 mM sodium acetate in 60 mM sodium hydroxide) to 200 mM sodium acetate in 120 mM sodium hydroxide in 45 minutes.

3. RESULTS AND DISCUSSION

Figure 1 reports the average content of WSC, sugars and fructans for different cereal crops cultivated in Central and Southern Italy and harvested at milky phase. The WSC production was highest for durum wheat; environmental conditions also influenced the WSC accumulation, with the average content for the three cereal crops generally lower in South Italy. This behaviour was due to a decrease in the level of fructans, as the sugar content did not change considerably with variations of the environment.

These results are probably due to an acceleration of synthesis processes depending on the higher temperatures of Southern regions, which determine a reduction in the storage of photosyntate for later utilization. These findings show that fructans represent the storage fraction for the plant metabolism. A strong relationship between WSC and fructan content is presented in Figure 2. The accumulation of fructans in the stems of cereal crops starts when the WSC content reaches a value of about 10 % and is not linked to species and environment.

The superiority of durum wheat in this accumulation process was verified in different years (D'Egidio et al, 1996) and is not yet explained; an hypothesis could be the higher photosynthetic capacity of this crop. In this regard, Austin et al (1982) found tetraploid and diploid wheats to have responses to photoperiod and temperature markedly

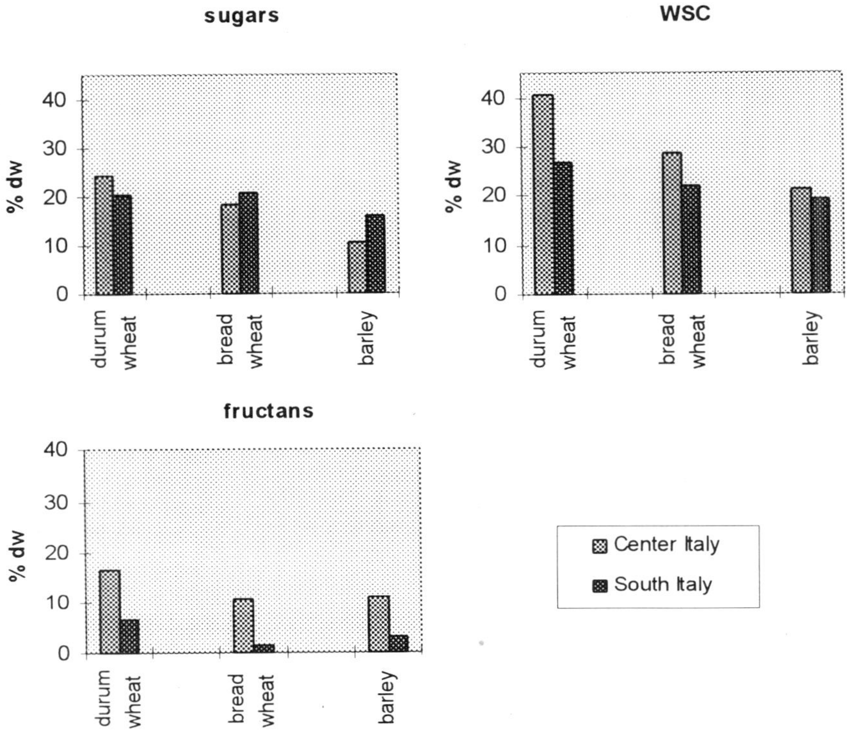

Figure 1. Average content of WSC, sugars and fructans in the stems at the milky phase of durum wheat, bread wheat and barley (Central and Southern Italy).

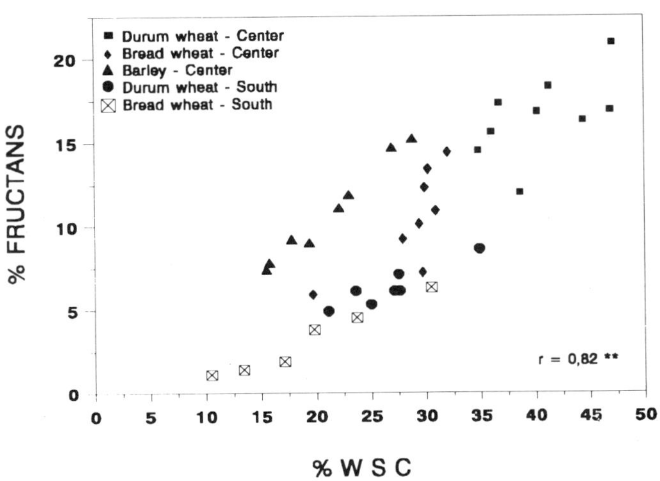

Figure 2. Correlation between WSC and fructan content in different genotypes and environments.

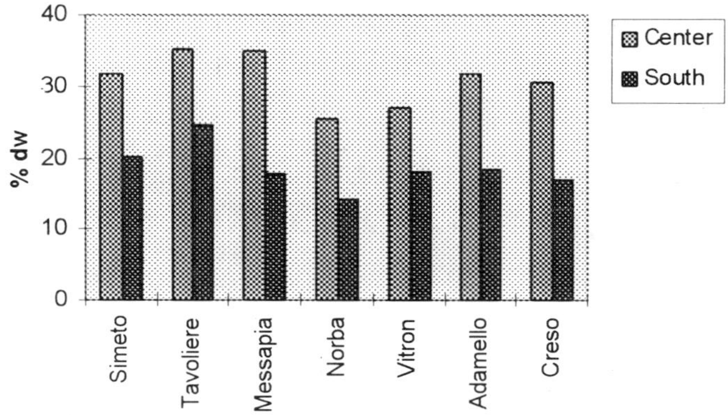

Figure 3. Total fructose content in stems of durum wheat cultivars at milky phase (Central and Southern Italy).

different from the hexaploid ones, and these may be factors which affect dry matter production.

The average fructose content expressed as percentage of the WSC is about 70%. In Figure 3 the fructose content in the stems of different durum wheat genotypes is shown; there is evidence that this amount is genotype-dependent and that the early maturing cultivars seem to be the most profitable.

In Figure 4, for the same cultivars, an estimated fructose yield per hectare is presented together with the yield of grains; immature grains are obtained as a co-product from harvesting at the milky phase.

On a durum wheat cultivar (Duilio) the accumulation of WSC and fructans was followed from anthesis to 14 days after milky phase, both in the stems and when possible in the grains. The results reported in Table 1 confirm for the stems the maximum accumulation of these compounds at milky phase; as regards the grains, the milky phase was characterised by high levels of WSC and fructose.

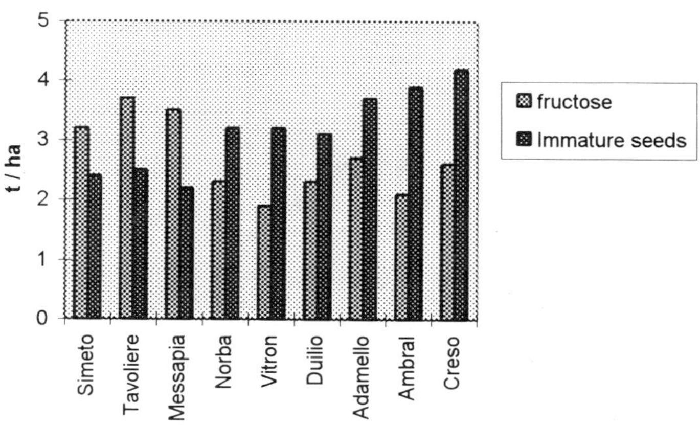

Figure 4. Estimated yield of total fructose and immature seeds in cultivars of durum wheat harvested at milky phase.

Table 1. WSC and fructan contents in durum wheat stems and immature seeds

	Stems		Immature seeds	
Sampling date	% WSC	% Fructans	% WSC	% Fructans
11/5 (flowering)	28.9	12.3	-	-
18/5	30.0	12.9	n.d.	n.d.
25/5 (milky phase)	34.5	17.1	20.4	15.5
1/6	20.4	1.7	3.0	1.9
8/6	9.5	1.2	3.7	2.0

These findings and previous results on a preliminary characterisation of this material (D'Egidio et al, 1995) point out very interesting properties of immature wheat grains; a more equilibrated amino acid composition (3.7% lysine against 2.3% of mature seeds) and a high level of fructose polymers could suggest the utilization of this material as functional food.

3.1. HPLC Analyses

The composition of water soluble carbohydrates in durum wheat stems of cultivar Duilio was studied also by anion-exchange HPLC coupled with pulsed amperometric detection (PAD), using an elution gradient well suited to separate glucose, fructose, sucrose and fructan oligosaccharides.

Glucose, fructose and sucrose were generally the dominant individual saccharides. The absence of purified standards for fructan oligomers limited the identification to the trisaccharide 1-kestose. Although this trisaccharide was present in all samples evaluated, it was not the dominant fructan, of the unidentified oligosaccharides. An unknown oligosaccharide with retention time approximately 16.85 minutes was found in all samples at concentrations higher than any other. It was eluted later than 1-kestose, but before any other fructan. Chromatograms of neutral sugars and fructans present are shown in Figure 5. Carbohydrate peaks corresponding to glucose, fructose, sucrose, 1-kestose and other

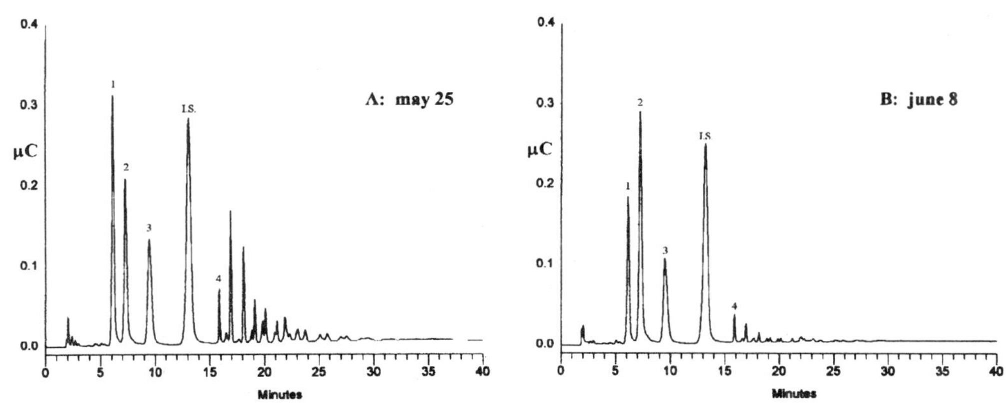

Figure 5. HPLC separation of glucose (1), fructose (2), sucrose (3), 1-kestose (4) and fructan oligosaccharides from 100 mg of stems of durum wheat Duilio (A: milky phase; B: 14 days after). I.S. = Internal Standard (melizitose).

unidentified fructan oligomers were observed in chromatograms A and B; in chromatogram B, corresponding to the extract at 14 days after the milky phase, as expected, the fructan oligosaccharide profile decreased dramatically. The trisaccharide melezitose was selected as the internal standard because it is not naturally present in our samples, is completely resolved from the other carbohydrates and is eluted near the peaks of interest.

Samples enriched in the fructan fraction by the method of Praznik et al (1992) were also separated by HPLC. From the chromatographic analysis (Figure 6), the qualitative profile seemed to be unchanged in all wheat stem samples, whereas a significant variation in the quantity of fructans present during the different physiological stages was observed.

Furthermore, fructan metabolism was examined in developing kernels. Figure 7 shows preliminary chromatographic data regarding accumulation of glucose, fructose, sucrose and fructan oligomers in kernels of Duilio from 7 days post-anthesis to 14 days after milky phase. Fructan content of the kernels decreased slightly from seven days post-anthesis (chromatogram 7A) to milky phase (chromatogram 7B); thereafter fructan content decreased dramatically 7 days after milky phase (chromatogram 7C). The fructan accumulation appears to be very active during the first two weeks after anthesis, while a rapid net degradation was observed a week after milky phase (Housley and Daughtry, 1987; Schnyder et al, 1988).

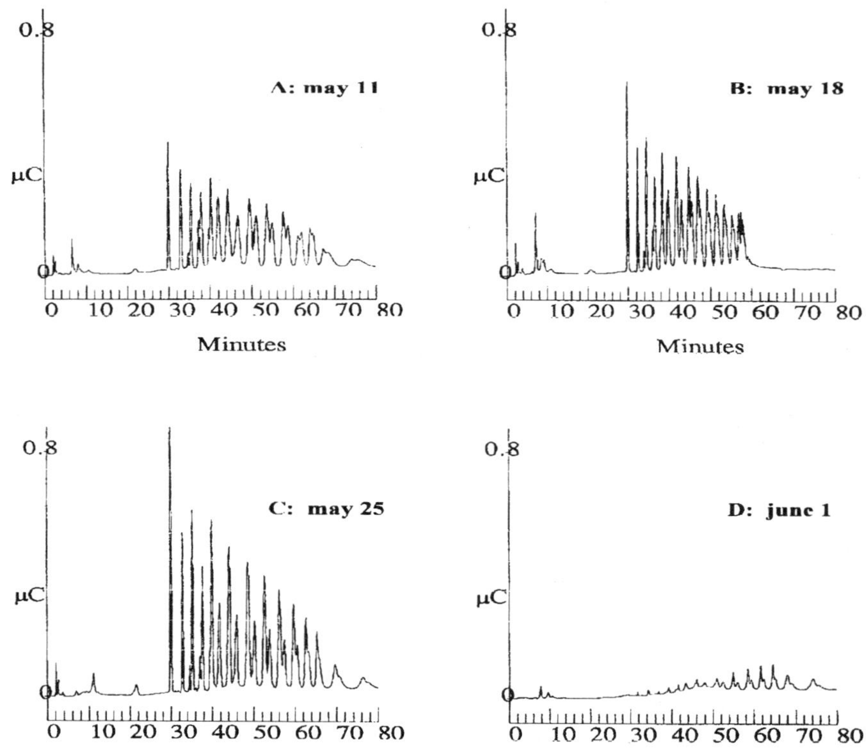

Figure 6. HPLC separation of samples of Duilio durum wheat stems, enriched in the fructan fraction by acetone precipitation (Praznik et al, 1982). A: anthesis; B: 7 days after; C: milky phase; D: 14 days after.

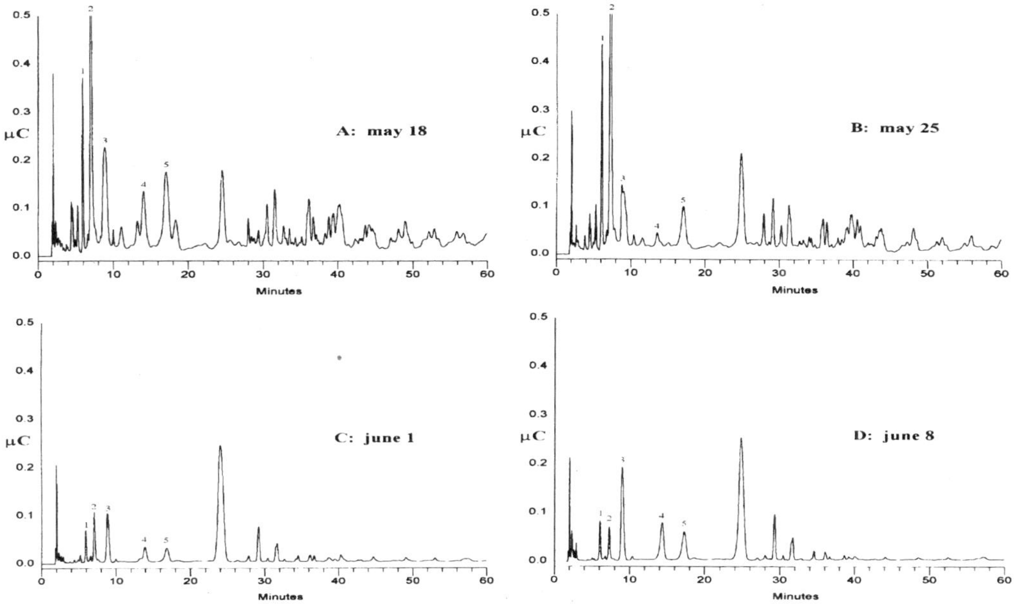

Figure 7. HPLC separation of glucose (1), fructose (2), sucrose (3), 1-kestose (5) and fructan oligosaccharides from 100 mg of immature grains of durum wheat Duilio. A: 7 days post anthesis; B: milky phase; C: 7 days after; D: 14 days after. Peak 4 has the same retention time as trisaccharide raffinose.

In chromatograms peak numbers 1, 2, 3 and 5 were identified as glucose, fructose, sucrose and 1-kestose respectively. Peak number 4 has the same retention time as the trisaccharide raffinose. This trisaccharide was not present in the WSC extract from stems.

4. CONCLUSIONS

The results obtained so far seem particularly encouraging and allow the suggestion of the harvesting of cereals at milky phase for the innovative use of these crops in fructose industrial production; early maturing genotypes, low nitrogen input and North-Central lands can be indicated as suitable for this purpose (D'Egidio et al, 1996). Another result of interest is that regarding the immature seeds, the co-product of early harvesting.

Figure 8 presents a plan for the complete utilization of cereal crops harvested at milky phase (see Chapters 17, 20, 21, 25 and 27 for similar whole crop utilization ideas). It is possible to harvest all the biomass, then to resort to forced hay-making (by mild heating and ventilation, or by solar panels as suggested by Borghi et al, 1984). After drying it is possible to proceed to a normal threshing, separating seeds from the straw. Alternatively it could be convenient to harvest separately ears and the rest of the plant by a planing machine equipped with bars at different levels from the soil. Seeds should be gently dried before use; straw instead can be directly utilized for the industrial transformation.

After extraction, the water soluble carbohydrates of the straw can be utilized for production of fructose and fructose-syrups, for production of fructans at low polymerisation degree, or for industrial fermentation processes (ethanol, citric acid etc.). Moreover imma-

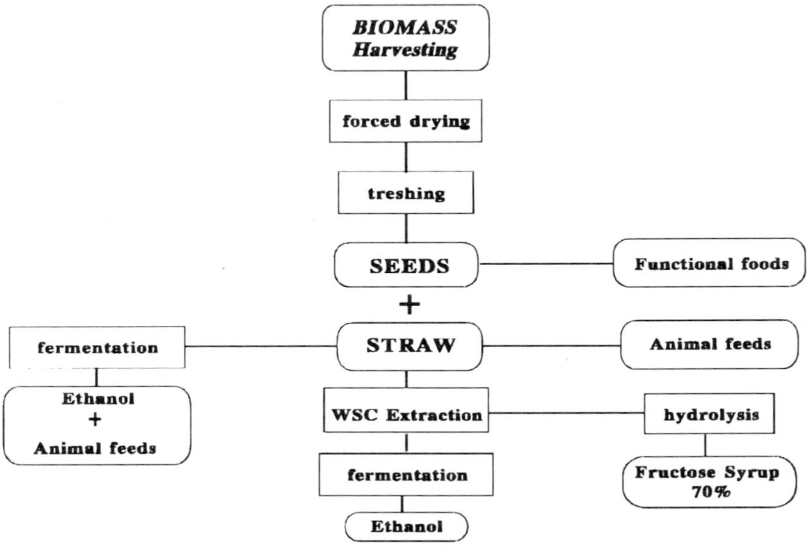

Figure 8. Utilization of cereal crops harvested at milky phase.

ture seeds can be utilized by simple industrial transformations as functional foods (Chapter 28 discusses some of the issues surrounding functional foods).

Recent *in vivo* and *in vitro* studies have been extensively pursued (Nilsson, 1988; Hidaka et al, 1991; Gibson and Wang, 1994; Gibson et al, 1996), finding remarkable biofunction and usefulness of fructo-oligosaccharides in human diet and in animal nutrition. Cereal crops (seeds and straw) appear to be an easy and suitable source of these compounds. In conclusion it is possible to affirm that cereals were the most ancient crops, but will also be the crops of the future.

REFERENCES

Austin RB, Morgan CL, Ford MA and Bhagwat SG (1982) "Flag leaf photosynthesis of *Triticum aestivum* and related diploid and tetraploid species." Ann. Bot. 49, 177–189

Borghi B, Canzi L, Rossi L and Facchini U (1984) "Raccolta anticipata di frumento tenero e di orzo ed essiccazione con enrgia solare." L'Informatore Agrario, 40(20), 51–53 (in Italian)

Corradini C, D'Egidio MG and Donini V (1995) "Separazione e caratterizzazione di carboidrati solubili e di fruttani presenti nelle spighe e negli steli di frumento duro." Atti del 2° Congresso Nazionale di Chimica degli Alimenti, 24–27 May 1995, Giardini Naxos, Italy (in Italian)

D'Egidio MG and Cervigni SE (1992) "Fruttani-carboidrati di riserva dei cereali a C_3: Variazioni negli steli di frumento duro durante la maturazione della cariosside; effetto della asportazione della spiga." Atti X Convegno Società Italiana di Chimica Agraria, 15–18 September, Roma, Italy, 47–51. (in Italian)

D'Egidio MG, Cervigni SE and Cervigni T (1993) "Water soluble carbohydrates in cereal stems at milky phase as raw material for industrial purposes." Proceedings of ICC Symposium on "Non food uses of cereals", 28–30 October 1993, Budapest, Hungary, 73–77

D'Egidio MG, Nardi S, Cecchini C and Calcagno C (1995) "Variabilità delle caratteristiche chimiche e tecnologiche della granella dei cereali durante la maturazione." Atti XII Convegno Società Italiana di Chimica Agraria, 19–21 September, Piacenza, Italy, 215–220 (in Italian)

D'Egidio MG, Cecchini C, Cervigni T, Donini B and Pignatelli V (1996) "Production of fructose from cereal stems and polyannual cultures of Jerusalem artichoke." Third International Symposium on "Industrial Crops and Products", 22–24 April 1996, Reims, France

Gibson GR and Wang X (1994) "Bifidogenic properties of different types of fructo-oligosaccharides." Food Microbiology, 11(6), 491–498

Gibson GR, Beatty ER, Wang X and Cuimmings JH (1995) "Selective stimulation of bifidobacteria in the human colon by oligofructose and inulin." Gastroenterology 108(4), 975–982

Hidaka H, Hirayama M and Yamada K (1991) "Fructooligosaccharides. Enzymatic preparation and biofunctions." J. Carboydrate Chemistry, 10(4), 509–522

Housley TH and Daughtry CST (1987) "Fructan content and fructosyltransferase activity during wheat seed growth." Plant Physiol. 83, 4–7

Kuhbauch W and Thome U (1989) "Non structural carbohydrates of wheat stems as influenced by sink-source manipolations." J.Plant Physiol, 134, 243–250

Nilsson U (1988) "Cereals fructans-preparation, characterization fermentation and bioavailability." First International Symposium on Fructan, 26–29 July 1988, Bonn, Germany, 31

Pollock CJ (1991) "Fructan metabolism in grasses and cereals." Ann. Rev. Plant Physiol. Plant Mol. Biol. 42, 77–101

Praznik K, Spies T and Hofinger A (1992) "Fructo-oligosaccharides from the stems of *Triticum aestivum*." Carbohydr. Res. 235, 231–238

Schnyder H, Ehses U, Bestajovksy J, Mehrhoff R and Kühbauch W (1988) "Fructan in wheat kernels during growth and compartmentation in the endosperm and pericarp." J. Plant Physiol. 132, 333–338

19

STRAW AS A FUEL

Damian Culshaw

ETSU
Harwell, Didcot, Oxfordshire OX11 ORA, United Kingdom

1. STRAW AVAILABILITY

A recent study has been carried out to determine the current production and uses for straw. This was conducted by the largest straw contracting and merchanting organisation in the country and funded by the Department of Trade and Industry through the New and Renewable Energy Programme managed by the Energy Technology Support Unit, ETSU (Anon, 1995a)

The total straw produced in England, Wales and Scotland is 12.5 Mtonnes/year. The split between the different crops is shown in Table 1 and Figure 1.

Table 2 shows the regional variation. It is clear that the majority of the surplus is in the east of the country in the grain growing areas, and that the deficit is in the western livestock areas. There is a significant trade in straw from west to east. The 'straw produced' figure has been calculated by multiplying the crop area with the straw yield based on many years of experience.

2. STRAW USE IN BRITAIN

Figure 2 shows the proportions of straw used in each application, while Table 3 reports the quantities. For the livestock applications, this was calculated by multiplying the amount of straw used by each animal by the number of animals. Data on the number of animals has been obtained from the Ministry of Agriculture's census statistics which is presented on a county basis. The amount of straw used in industrial applications is an estimate based on the report author's knowledge of the industry.

3. FUEL PROPERTIES

The properties of straw as a fuel are compared with those of coal in Figures 3–6. The energy content of straw is lower than that of coal and depends on the moisture content. In comparison with most other biomass fuels, including wood, straw has a low mois-

Cereals: Novel Uses and Processes, edited by Campbell *et al.*
Plenum Press, New York, 1997

Table 1. Total straw produced in the UK, by crop

Crop	Straw (tonnes/year)
Wheat	6,859,842
Winter Barley	2,508,382
Spring Barley	1,848,376
Oats	364,889
Other cereals	61,081
Oil Seed Rape	856,798
Total	12,499,368

ture content and hence a high energy value. The typical moisture content of straw, as delivered, is 15%, at which the energy value is around 14 GJ/tonne. Based on recent data (Anon, 1994), the mean energy value of straw, based on 137 different samples of cereal straw from all over the country, was found to be 18.3 GJ/tonne (oven dry). The typical energy range for coal is from 25 to 28 GJ/tonne, although lignite coals have a lower value and anthracites a higher one (Anon, 1994).

Straw is generally low in sulphur (Figure 4) and since it contains high levels of calcium, it emits very low levels of acid gases when burned. There is even evidence that it can reduce the acid gas emissions from a unit being fired on high sulphur coal co-fired with straw (Anon, 1991).

When considering CO_2 emissions, straw can be regarded as largely carbon neutral. The carbon released into the atmosphere by burning straw is 'within the carbon cycle', since the plants which grow to produce it take the carbon from the atmosphere in the first place. Figure 6 is based on a recent study of the carbon and energy balance for straw (Anon, 1995b). The analysis quantifies the amount of fossil fuel energy used (and hence CO_2 released) in the process of gathering and converting straw to electricity. This report shows that for every MJ of electricity generated from coal, 299 grams of CO_2 are released into the atmosphere, while the equivalent figure for straw is between 7.7 and 15.8 grams.

As well as the low levels of gaseous pollutants, straw is also a sustainable resource, unlike coal which is being depleted.

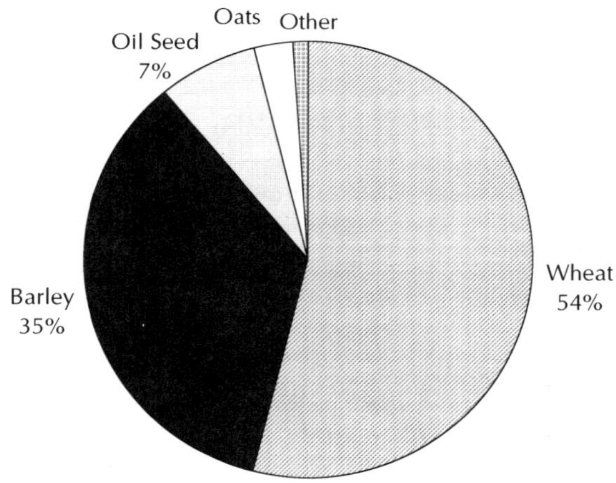

Figure 1. Total straw produced in the UK, by crop.

Table 2. Regional variation in straw production in the UK

Region	Straw produced (tonnes/year)	Straw surplus (tonnes/year)
Scottish Highlands & Islands	656,009	254,325
East Scotland	906,645	557,625
West Scotland	225,860	-404,261
North England	575,418	-93,248
Yorks. and Humbs.	1,490,727	614,397
East Midlands	1,938,048	1,320,372
East Anglia	1,839,767	1,468,518
South East England	2,351,587	1,602,821
South West England	1,247,313	-384,358
West Midlands England	949,706	213,177
North West England	119,514	-354,212
Wales	198,775	-858,021
Total	12,499,369	3,937,135

4. STRAW AS A FUEL NOW

Technology for using straw for domestic and other rural applications continues to develop. The whole-bale batch fired boilers manufactured in Britain have been improved significantly. They can now accept all bale sizes and can operate efficiently and cleanly at up to 0.4 MW, larger if a number of boilers are coupled together. These boilers can be cost effective for hot water heating under the right circumstances.

There is now interest in electricity generation from straw. This has been stimulated by the Government's premium pricing mechanism known as the Non Fossil Fuel Obligation (NFFO); it is designed to encourage the use of renewable energy for electricity generation. The government's declared policy in the development of renewables is '*to stimulate the development of new and renewable energy technologies where they have prospects of being economically attractive and environmentally acceptable*'. They aim to '*work towards a figure of 1500 MW of new renewable electricity generating capacity by the year 2000*'.

Table 3. Straw usage in the UK

Application	Amount used (tonnes/year)
Beef	1,465,281
Other Cattle	2,618,509
Pigs	674,712
Sheep	605,968
Other Animals	3,972
Crop Protection	74,000
Feed Compound	110,000
Mushroom	400,000
Other Agro-industrial	210,500
Farm Fuel	170,000
Dairy	2,399,291
Total	8,732,233

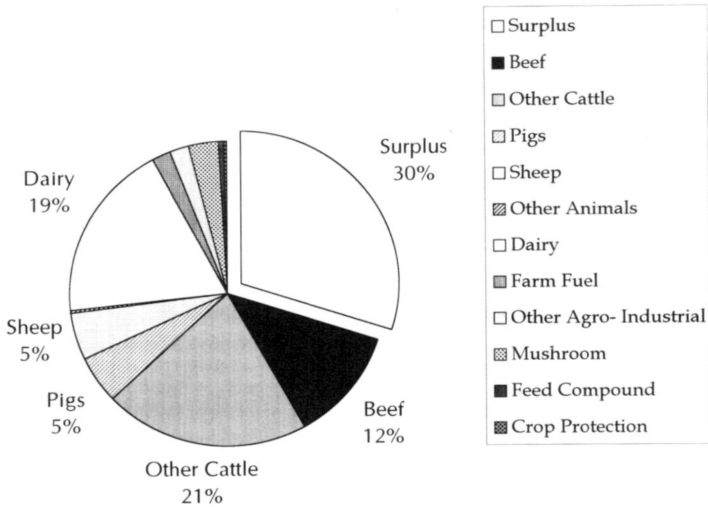

Figure 2. Straw usage in the UK, showing a surplus of supply over demand.

Under the third round of the NFFO, a project at Ely in Cambridgeshire was granted a contract to generate electricity from straw; building work on this plant is expected to start late in 1996. The plant has a contract to generate 31 MW of electricity and expects to use 180,000 tonnes of straw per year. There are also plans to use the waste heat from the plant in greenhouses used to grow tomatoes.

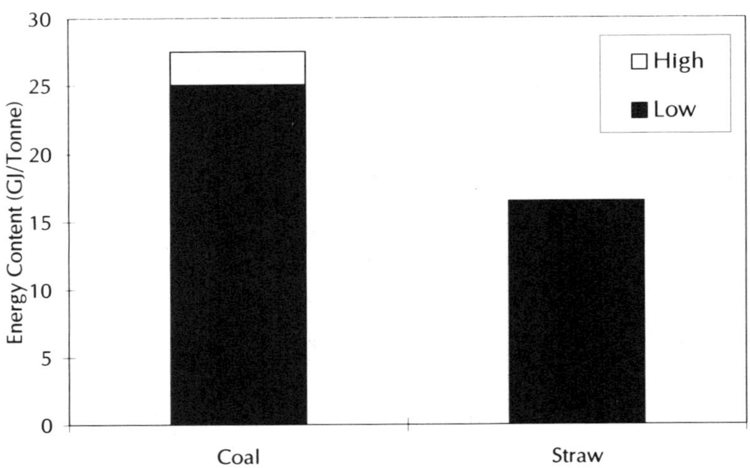

Figure 3. Energy content of straw compared with coal (based on data from Anon, 1995b).

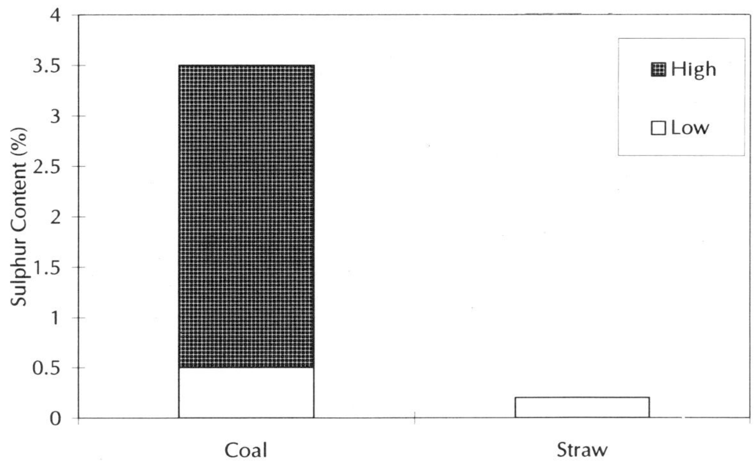

Figure 4. Sulphur content of straw compared with coal (based on data from Anon, 1995b).

5. STRAW FOR FUEL IN THE FUTURE

The NFFO is the government's main instrument for pursuing the development of electricity generating capacity from renewable energy sources. Straw could, in principle, be included under future rounds of NFFO and is attractive because:

- the resource could make a significant contribution to the government's target of 1500 MW of new renewables-generated electricity by the year 2000;
- there are no significant environmental drawbacks to the use of straw as a fuel, and no change in land use will be needed to produce fuel;

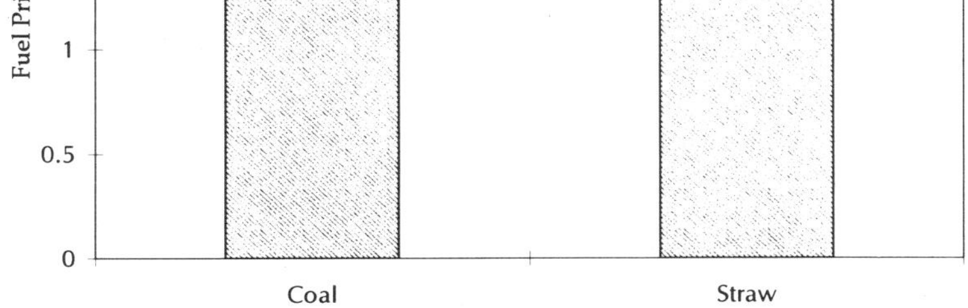

Figure 5. Fuel price of straw compared with coal (based on data from Anon, 1995b).

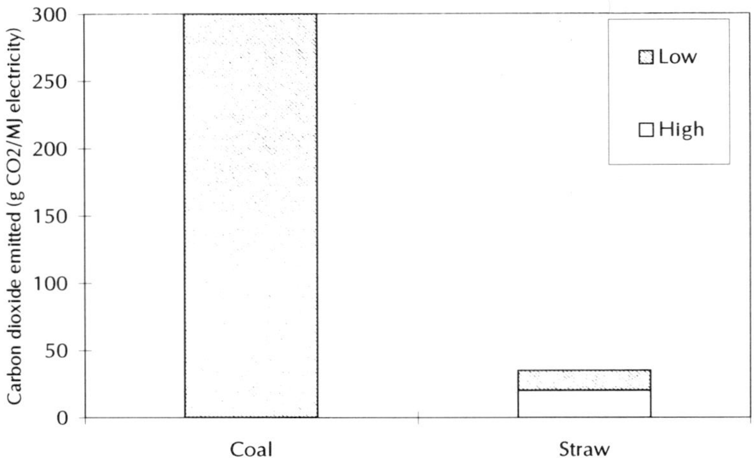

Figure 6. Carbon dioxide emission from straw compared with coal (based on data from Anon, 1995b).

- straw can be used now with little technical risk, since it is already used on a large scale for other purposes, and the conversion technology is well developed within Europe;
- the cost of power generation from straw is currently lower than that from woody biomass fuels; there is the potential to reduce costs further through the development of advanced conversion techniques (such as pyrolysis and gasification), by co-firing with fossil fuels, and by streamlining the fuel collection techniques.

Whether further straw for electricity generation will be supported in the future is at present unclear. Currently around 170,000 tonnes/year is used for heating in agriculture-related applications; this could grow in the future.

REFERENCES

Anonymous (1991) "Straw Ash Characteristics - ETSU B1242, Babcock Energy Ltd.
Anonymous (1994) "The Analysis of Straw - ETSU B/M3/00388/39/REP, ADAS
Anonymous (1995a) "Non- Energy Markets for Straw." ETSU Report B/M4/00487/16/REP, Northern Straw
Anonymous (1995b) Energy and Carbon Analysis of Using Straw as a Fuel - ETSU B/M4/00487/01/REP
These references are available from the Renewable Energy Enquiries Bureau, ETSU, Harwell, Oxfordshire, OX11 ORA, UK; Tel. +44 (0)1235 432450; Fax. +44 (0)1235 433066

20

FOOD AND NON-FOOD USES OF IMMATURE CEREALS

Rolf Carlsson

Institute of Natural Sciences
Kalmar University
Box 905, S-391 29 Kalmar, Sweden

1. INTRODUCTION

In the past the aim of agriculture has been primarily to produce more and better food. However, during the last decade, production of industrial (non-food) raw materials from crops has been strongly advocated (Carlsson, 1995). The two demands for food and non-food materials require optimal utilisation of every potential source of plants and lands (Carlsson, 1994). Both dry crop and green crop fractionation of cereals have been developed for multipurpose uses for the food and non-food industries.

Current "traditional" agriculture limits what it is presently possible to obtain from agricultural lands (Jones, 1977). Only a part of the available photosynthetic production during a growth season is utilised for production of agri-commodities; only a minor part of the material that can be harvested is used. The alternative is an ecologically superior agriculture based on the primary production by photosynthesis in green plants during the whole growth season (Carlsson, 1985a). Such an agricultural system would be based on *green* crops, along with a renewed and expanded use of nitrogen-fixing plants or plants with nitrogen-fixating root zones.

By primary production of photosynthesis in green plants more than 20 tons of dry matter and 3 tons of protein per ha in temperate climates, and 80 tons of dry matter and 6 tons of protein per ha in tropical areas can be obtained per year (Carlsson, 1985a).

2. GREEN CROP FRACTIONATION

Green crops are now used primarily as forage and a source of leafy vegetables. Very seldom is the green crops' potential for production of food and non-food raw materials utilised directly. Green crops can be used for simultaneously manufacturing several food and non-food industrial products, by a process called green crop fractionation or wet-fractionation of green crops (Carlsson, 1994; see Figure 1). At Rothamsted Experimental Sta-

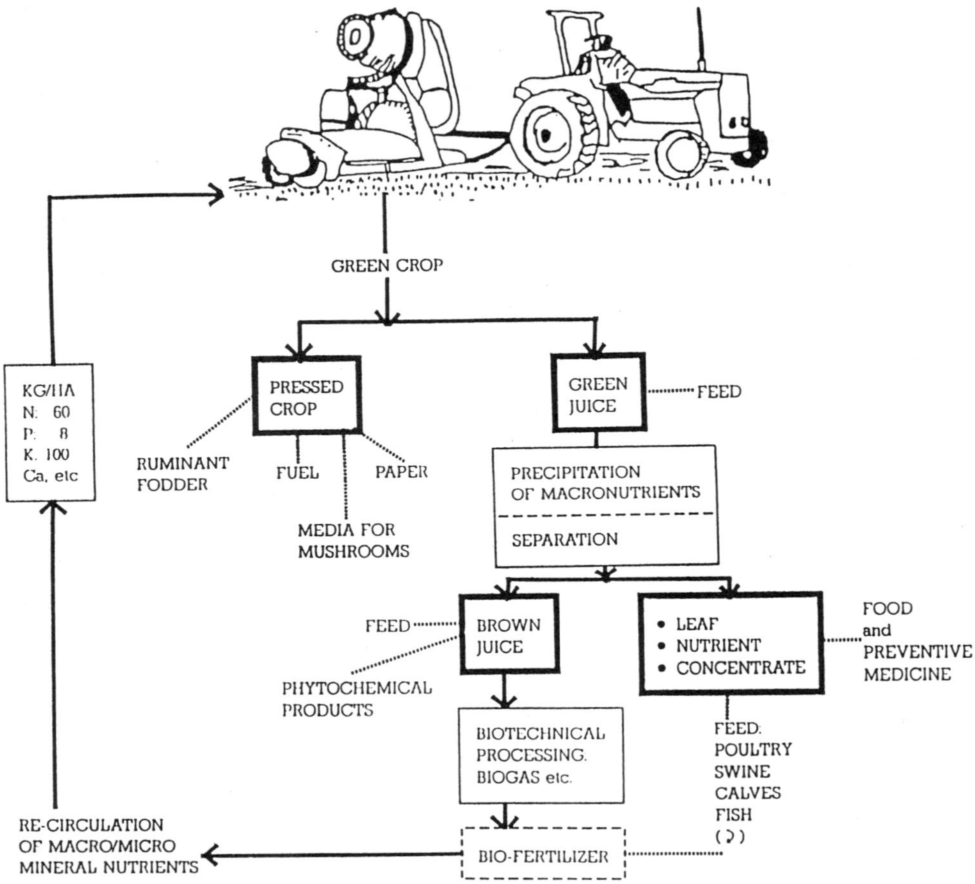

Figure 1. Wet-fractionation of green crops for multipurpose use.

tion, Professor NW Pirie and his colleagues (Pirie, 1978, 1987) have been developing green crop fractionation, as have many other scientist from different places in Europe, North and South America, USSR, Asia, Australia and New Zealand.

Green crop fractionation is now being studied and developed in about 80 countries, the majority of which are in the tropics (Carlsson, 1982, 1985b, 1993a). Presently, Argentina and other Latin American countries, China (PRC), and Russia, plus former USSR countries, are increasing their green crop fractionation activities. Several hundreds of temperate and tropical plant species have been investigated for green crop fractionation (Telek and Graham, 1983; Carlsson, 1989a).

Reviews on green crop fractionation and production of leaf nutrient concentrate (LNC) (synonyms: "leaf protein concentrate", "leaf protein" and "leaf concentrate" in British literature) have been given by Pirie (1971, 1987), Wilkins (1977), Costes (1981), Telek and Graham (1983), Singh (1984, 1996), Tasaki (1985), Fantozzi (1989), Carlsson (1992, 1996) and Ostrowski-Meissner (1993a). By green crop fractionation, a fibre-enriched pressed crop (PC) and an expressed green juice (GJ) are produced in the first fractionation step (Figure 1). The green juice can be fractionated by heat, acid, anaerobic fermentation, and centrifugation into a LNC and a brown de-proteinized juice (BJ), in a second step.

The LNC consists of a mixture of chloroplastic and other organelle membranes plus denaturated soluble chloroplastic and cytoplasmic proteins. The BJ, based on vacuolar substances, is enriched in water-soluble constituents, such as sugars, organic acids, low-molecular weight nitrogenous substances, glycoside phytochemicals, and mineral ions (potassium, sodium, calcium, chloride, nitrate, sulphate, etc.).

By more advanced protein fractionation techniques, the proteins of the GJ can be separated into a chloroplastic membrane, lipid-enriched green protein concentrate (GLPC), and two concentrates of soluble, whitish proteins: Rubisco protein isolate and a protein isolate of all other pooled soluble proteins, called FI and FII proteins, respectively.

Each of the above mentioned fractionation raw products can be used for an environmentally friendly production of food and non-food industrial raw materials, using a sustainable agriculture based on the biological diversity of about 300,000 "higher" developed plant species.

3. OPTIONAL GREEN CROP FRACTIONATION PRODUCTS

Pressed crop (PC) from wet fractionated green plants can be transformed into possible industrial products (Carlsson, 1994, Figure 1), such as fibres for printing paper, construction and mobile boards, solid fuel, and biologically better tobacco products, or used as growth media for mushroom cultivation (Chanda and Das, 1993). The PC can be hydrolysed to fermentation media to produce liquid fuels. Mostly, the PC is used as fodder for ruminant animals, such as milk cows, beef cattle, horses, sheep, and goats. The PC, compared to hay, can be produced with a standardised composition, which is most valuable for its use as a quality-oriented, premium-priced class of agri-commodity (Ostrowski-Meissner, 1993b).

The leaf nutrient concentrate, LNC, is mainly used for non-ruminant feed to enhance the colour of chicken skin or egg yolk. It also produces tender meat in chickens, ducks and pigs. Pigs fed LNC give pork with increased contents of healthy oleic and linoleic fatty acids in the fat.

The original idea to use LNC to improve the human diet is persisting and spreading. The major consumption and development of LNC as food supplements in diets are presently taking place in tropical countries. However, there is a growing awareness of LNC's advantages for health food products in industrialized countries. In these countries, high-quality vegetable food proteins presently tend to substitute animal proteins. Several possible sources were earlier reviewed by Norton (1978). One reason for the new trend is the negative impact of the earlier surplus fertilization of arable lands, where plants are cultivated for animal forage. This use seems to have destroyed agricultural lands and polluted the environment in "developed" countries.

4. INDUSTRIAL PROCESSING

The largest industrial green biorefineries for green crop fractionation are situated in France, where PC plus condensed BJ are produced as ruminant pellets, along with LNC pellets for non-ruminants; de Mathan (1989) reports 120 tonnes lucerne processed per hour. A French patent exists for using BJ in ethanol fermentation industries instead of water, giving an additional yield of 400 L of ethanol per ha lucerne. Two commercial biorefineries are working in Denmark. The one in Nykoebing, Falster (50 tonnes/hour), produces the same products for the EU market as the ones in France. In collaboration with Biosystemer A/S (Aalsgaarde, Denmark), paper from PC of lucerne have been manufactured (Holm-Christensen, 1990; cf. Carlsson, 1993c). A second green biorefinery, the Institute of Biomass Utilization and Biorefinery (Olgod) specialises in fermentation of green juice, brown juice, silage effluents, and agro-industrial liquid by-products into lactic acid bio-degradable polymers, lysine, and other large-scale fermentation industrial products, e.g., ethanol (Kiel, 1993; see also Chapter 27). In New Zealand a green biorefinery is specialised in high-quality feed production (Singh, 1996). There is strong co-operation between Australia and Japan to develop industrial green crop fractionation technology for quality agro-commodities (Ohshima, 1993).

In the former USSR, large-scale green biorefineries with process capacities of 50 to 100 tonnes green biomass per hour exist. However, to what extent they are running today is not specified. Since about 5 years ago, the changes in the former USSR have caused a huge demand for locally produced animal fodder and feed in the newly formed countries. Wet-fractionation of green crops is apparently looked upon as a seemingly easy solution to fulfil such demands. Earlier working green biorefineries in USA, Italy, Spain, Japan, Hungary (the first commercial processing plant), Latvia, former Czecho-Slovakia, and Germany have at least temporarily or partly closed down, due to the economic market situation (Singh, 1996, Carlsson, unpublished 1996). Chapter 21 describes some of the activity in integrated processing systems in Latvia.

Green crop fractionation on a farm-scale is continuously persisting and being introduced, e.g. in the US, Tatarstan, and other former USSR states/regions. In the latter cases, often *Amaranthus* species are most appreciated (Information from proceedings of conferences of the European Amaranth Association; 1992: Olomouc and Nitra, Czechoslovakia; 1993: Tashkent, Uzbekistan, and Olomouc, the Czech Republic; Printed proceedings in limited circulation. See also Chapters 11 and 12).

The earlier green tobacco processing plant in Sweden for "biologically better" tobacco products and phytochemicals (Carlsson, 1988, 1993b), has been transferred to Kentucky in the USA for especially Rubisco protein processing (Sheen, 1994) . A review on white leaf protein products investigated and developed for medical and other purposes has been made by Carlsson (1985c). Green tobacco as a source for anti-oxidants/anti-cancer chemicals, i.e. carotenoids, have been studied in Sweden (Carlsson, 1988, 1993b) and in Japan (Layug et al, 1993a).

5. GREEN CROP WET-FRACTIONATION OF GRASSES AND IMMATURE CEREALS

Grass cheese production from British grasslands was suggested by Slade et al (1939). Extracted protein from grass should be used for human consumption, while the residue, the pressed crop, should be feed for ruminants. The same concept to produce human food based on LNC from grasses and other green crops was taken up at the beginning of the Second World War, as questions in the House of Commons in the UK in 1941 (Pirie, 1978). The advantages of green crop fractionation of grass for forage protein conservation and the uses of the fractionation products in both non-ruminant and ruminant feeding were elucidated by Wieringa, Jones, Maguire et al, and Wilkins (Griffiths and Maguire, 1982). The above mentioned utilisation of wet-fractionated grass is based on high yields of edible protein per ha and the very low price per kg protein (Jones, 1977; Ostrowski-Meissner, 1983).

6. GRAMINAE SPECIES FOR WET-FRACTIONATION: EFFECTS ON LNC YIELD AND QUALITY

Both temperate grass species, including green cereals (Carlsson, 1983, 1989, 1994), and tropical grasses (Telek and Martin, 1983) have been investigated for LNC production. Most temperate grasses are species with a C-3 photosynthesis, while grasses adapted to hotter climates have a C-4 photosynthesis. The type of photosynthesis affects the yield and quality of the LNC obtained (Carlsson, 1994). The plant cell structures differ, e.g. chloroplasts with high-quality Rubisco protein (FI protein) are abundant in all mesophyll cells in C-3 species, while the same type of chloroplasts are limited to the bundle sheet cells around the vascular tissue in C-4 species. The C-4 species contain many fibre cells and the cell walls are thicker. Thus, C-3 species are easy to disintegrate relative to C-4 species. Less energy is needed for processing C-3 species. Lush young green plant shoots of any crop are richer in nutrients and easier to disintegrate than more mature shoots.

For C-4 species the extraction rate of high-quality protein is severely reduced, due to frictional heating of native proteins during processing, and the formation of thick fibre layers with ultrafiltration effects. On the other hand, the yield of extractable protein per ha is much higher due to higher photosynthetic yields of C-4 species in tropical countries.

7. EFFECTS OF CULTIVATION AND HARVESTING ON WET-FRACTIONATION OF IMMATURE CEREALS

The yields of extracted leaf protein by large-scale processing from immature cereals and forage grasses per ha and season have been determined by Byers and Sturrock (1965), Arkcoll and Festenstein (1971), Cheeseman, Heath and King, Jones, Wilkins and others (Wilkins, 1977). The extractable yields varied from 1 to 2 MT per ha. Effects of species and cultivars (of, for example, wheat, barley, rye, maize, rye-grass and cocksfoot), regrowth cuts of 3 to 4 times per season, harvesting ages, and especially nitrogen fertilizers were investigated. The composition and uses of PC, GJ, LNC, and BJ are described. The PC contained on average over a season 15% crude protein, the GJ 33%, the LNC 55%, and the BJ 15%. Products from young plants, especially nitrogen fertilized ones (200 to 1,000 kg N/ha and season) had the highest protein contents. All products from the wet-fractionated crops were competitive to unprocessed ones in ruminant as well as non-ruminant feeding (Wilkins, 1977: Session 3). Process flow systems, energy balances, and cost analyses of different processing systems for green crop fractionation are described in "Leaf Protein Concentrates" by Telek and Graham (1983).

8. NUTRITIVE VALUE OF LNC FROM GRASSES AND OTHER CROPS

Yields and quality of LNC vary due to species, physiological development, cultivation conditions, harvesting and processing techniques (Carlsson, 1994). The occurrence of antinutritive factors in plants (Liener, 1980) influence the quality of LNC (Carlsson, 1994).

Species such as selected *Graminae* species (wheat, rye, barley, etc.), *Chenopodiaceae*, *Crucifearae* and *Solanacae* species show high nutritive values of LNC by *in vivo* assay, using rats. This is often related to a relatively high content of Rubisco protein and an absence of antinutritive secondary substances. On the other hand, most secondary substances can be used for pharmaceutical purposes (Carlsson, 1996).

9. DRY CROP FRACTIONATION OF IMMATURE CEREALS

Dry crop fractionation in agricultural refineries is based on an integration of agriculture and industry. All biomass from an extended season can be dealt with. Whole crops of immature plants are harvested and dried with the same techniques as in green crop dryers. Sometimes the harvested, immature crop of the cereal is preserved by different chemicals before the eventual drying, as used by Kockums Constraction Ltd. (Sweden), later taken up by Scandinavian Farming Ltd. (Sweden).

The dry fractionation of the whole crop optimizes the utilization of all botanical components of the biomass. Kernels, straw chips of internodes, and straw meal (leaves, ears, chaff, and nodes) are separated (Figure 2; Rexen, 1986). The fractions are suitable as raw materials for the starch industry, feed industry, cellulose industry, particle board industry and chemical industry, and also for use as a fuel. An Agricultural Development and

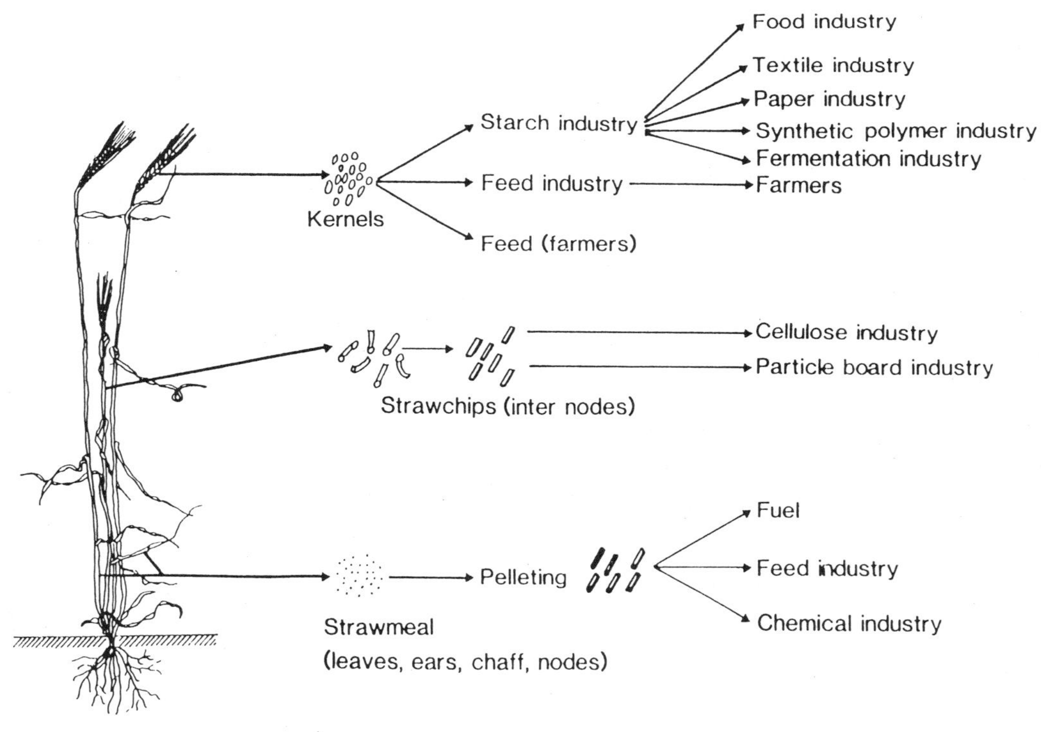

Figure 2. Dry crop fractionation of near-maturity cereals for multipurpose use.

Innovation Centre "Bioraf" has been working with dry crop fractionation at Bornholm in Denmark since the early 1990's as a Danish-South Swedish co-operation; Chapter 24 presents an evaluation of the profitability of the Bioraf-type biorefinery concept.

The whole crop harvesting system is claimed to be cheaper compared to traditional combine harvesting, both according to theoretical calculations and calculations based on practical experiences by Danish and Swedish companies.

10. OTHER USES OF IMMATURE CEREALS

Germinated grains with young green leaves are more nutritious than grains as such. The protein-, vitamin- and mineral-rich germinated grains are used as horse feed and for human health food products. Young cereal leaves are used both for production of healthy grass juices and as dried grass meal re-dissolved as health food drinks.

11. CONCLUSION

In global agriculture, cereals are the dominant crops in both temperate and tropical climates. Immature cereals can be used for food and non-food products. Vegetative plants are being processed by wet-crop fractionation, while near-maturity plants are dry-fractionated. In both cases multiple products are manufactured for industrial uses. Both fractionation techniques increase immensely the global utilization of cereals, even in areas where a mature crop for grains cannot be grown.

REFERENCES

Arkcoll DB and Festenstein GN (1971) "A preliminary study of the agronomic factors affecting the yield of extractable leaf protein." J. Sci. Food Agric. 49–56

Byers M and Sturrock JW (1965) "The yield of leaf protein extracted by large-scale processing of various crops." J. Sci. Food Agric. 16, 341–355

Carlsson R (1982) "Trends for future applications of green crops." Proc. EEC Conf. Forage Protein Conservation and Utilization, Dublin, Ireland, 57–81

Carlsson R (1983), in "Leaf Protein Concentrates." (Eds L Telek and HD Graham), AVI Publ Co., Westport Conn., USA, 52–80

Carlsson R (1985a) "An ecologically better adapted agriculture. Wet-fractionation of biomass as green crops, macro-alga, and tuber crops." Proc. 2nd Int. Conf. Leaf Protein Res., Nagoya, Japan, 19–23

Carlsson R (1985b) "Wet-fractionation of green crops in Europe. A short review on status and development." Proc. 2nd Int Conf. Leaf Protein Res., Nagoya, Japan, 93–100

Carlsson R (1985c) "White leaf protein products for human consumption. - A global review on plants and processing methods." Tobacco Protein Utilization Perspectives, Proc. Round Table Conf. at 1st Int. Congr. Food Health, (Ed. P Fantozzi), CNR/Italian National Research Council, Special Project IPBR, Subproject L, European Community Commission Agrimed. Project, 125–145

Carlsson R (1988) "New tobacco products and phytochemicals from selected, field-cultivated Nicotiana species." Proc. Tobacco Protein Utilization Perspectives, Agrimed Res. Programme: Agriculture, Symp. Perugia, Italy, Report: EUR 11923 EN, 54–61

Carlsson R (1989) "A tentative list of plants for commercial production of leaf protein concentrates." Proc. 3rd Int. Leaf Protein Conf., Pisa-Perugia-Viterbo, Italy, 350–353

Carlsson R (1992) "Green Crops for Multipurpose Use, and Pseudo-cereals from the Inca and Aztek Empires." (Book in Chinese; ed Song Yuhua), Chendu Xingguang Development Corporation for Appropiate TechnologyDCAAT, Chengdu, China (PRC), Oct. 1992, 142

Carlsson R (1993a) "Green crop fractionation in Europe, Leaf Protein Processing and Fractionation." Proc. 4th Int. Conf. Leaf Protein Res. New Zealand-Australia, 29–34

Carlsson R (1993b) "Wet fractionation of green tobacco for industrial raw materials in Sweden." Proc. 4th Int. Conf. Leaf Protein Res., New Zealand-Australia, 75–80

Carlsson R (1993c) "Pressed crop for possible production of paper crop." Proc. 4th Int. Conf. Leaf Protein Res., New Zealand-Australia, 69–74

Carlsson R (1994) "Sustainable production-Green crop fractionation: Effects of spceies, growth conditions, and physiological development on fractionation products." In "Handbook of Plant and Crop Physiology" (Ed M Pessarakli), Marcel Dekker, Inc., New York, USA, 941–963

Carlsson R (1995) "New industrial crops and products from agriculture in Europe." 20th Congr. Nordic Agri. Researchers Assoc., Poster Abstract No 16

Carlsson R (1996) "A global renewed review on green crop fractionation." 5th Int. Conf. Leaf Protein Res., Rostov-on-Don, Russia, 12

Chanda S and Das S (1993) "Cultivation of edible fungi Pleurotus sajor-caju on ligno-cellulosic by-product of LP technology." Proc. 4th Int. Conf. Leaf Protein Res. New Zealand-Australia, 141–146

Costes C ed. (1981) "Proteines foliares et Alimentation." Gauthiers-Villars, Bordas, Paris, France

Fantozzi P ed. (1989) Proc.3rd Int. Conf. Leaf Protein Res., Pisa-Perugia-Viterbo, Italy

Griffiths TH and Maguire MF eds. (1982) Proc EEC Conf. Forage Protein Conserva-tion and Utilization, Dublin, Ireland

Holm-Christensen B (1989) "The dehydration plant as producer for the cellulose industry." Proc. Dri-Crops 89, 4th Int. Green Crop Drying Congr., Cambridge, Agra Europe Ltd. London, UK, 91–94

Jones AS (1977) "The principles of green crop fractionation." In "Green Crop Fractionation." Occasional Symp.9, Brit. Grassl. Soc., Grassl. Res. Inst., Hurley, Maidenhead, UK 1–8

Kiel P (1993) "Bioconversion of agricultural residues." Beitroge zur Ikologischen Technologie, Band 1, Proc. Ecologic Bioprocessing.- Challanges in Practice, Verlag Gesellschaft fur /kologische Technologie und Systemanalyse e.V., Berlin, Germany, 147–152

Layug D, Ohshima M, Takabatake H, Ueda M, Okajima T and Yokota H. (1993) "The effect of anti-oxidants on extraction efficiency and the total carotenoid contents in LPC's from tobacco leaves." Proc. 4th Int. Conf. Leaf Protein Res., New Zealand-Australia, AFIC National Facilities, Sydney, Australia, 121–127

Liener IE (1980) "Toxic Constituents of Plant Foodstuffs." (2nd ed.), Academic Press, New York, USA

de Mathan O (1989) "Large scale commercial operations in Europe." Proc. 3rd Int. Leaf Protein Res., Pisa-Perugia-Viterbo, Italy, 36–49

Norton G ed. (1978) "Plant Proteins." Butterworths, London, UK

Ohshima M (1993) "The potential of crop processing and fractionation technology used for supplying quality stock feeds as an alternative to conventional agri-commodities imported to Japan." In "Crop Processing for Quality in Agri-Commodity Trade." Proc. Regional Asia Pacific Conference, AFIC National Facilities, Sydney, Australia, 15–25

Ostowski-Meissner HT (1983) "Protein extraction from grasslands." in "Leaf Protein Concentrates." (Eds L Telek and HD Graham), AVI Publ. Co., Westport, Conn., USA, 9–51

Ostrowski-Meissner HT ed (1993) Proc. 4th Int. Conf. Leaf Protein Res., New Zealand-Australia, AFIC National Facilities, Sydney, Australia

Ostrowski-Meissner HT (1993b) "Crop processing and fractionation technology as a tool in implementing quality to agri-commodity trade with Japan." In "Crop Processing for Quality in Agri-Commodity Trade." Proc. Regional Asia Pacific Conference, AFIC National Facilities, Sydney, Australia, 27–40

Pirie NW (1971) "Leaf Protein: its agronomy, preparation, quality and use." Blackwell Scientific Publications, Oxford/Cambridge, UK

Pirie NW (1978) "Leaf Protein and Other Aspects of of fodder fractionation." Cambridge Univ. Press, UK

Pirie NW (1987) "Leaf Protein and Its By-Products in Human Nutrition and Animal Nutrition." Cambridge Univ. Press, Cambridge, UK

Rexen F (1986) "New industrial application possibilities for straw" (Danish), Fytokemi i Norden, Stockholm, Sweden, 1986–03–06, Documentation of Svebio Phytochemistry Group, 12

Sheen SJ (1994) University of Kentucky, Lexington, Kentucky, USA, personal communication

Singh N ed (1984) "Progress in Leaf Protein Research." Proc. 1st. Int. Conf. Leaf Protein Res., Today and Tomorrows Printers and Publishers, New Delhi, India

Singh N ed (1996) "Green Vegetation Fractionation Technology." (To be published soon; Wageningen, the Netherlands)

Slade RE, Birkinshaw JH, and ICI (1939) "Improvements in or related to the utilization of grass and other green crops." Br. Pat. 511, 525

Tasaki I ed (1985) "Recent Advances in Leaf Protein Research." Proc. 2nd Int. Conf. Leaf Protein Res., Faculty of Agric., Nagoya Univ., Togo-cho, Aichi-ken, Japan

Telek L and Graham HD Eds (1983) "Leaf Protein Concentrates." AVI Publ. Co., Westport, Conn., USA

Telek, L and Martin, FW (1983) "Tropical Plants for leaf protein con-centrates." In "Leaf Protein Concentrates." Eds L Telek and HD Graham, AVI Publ. Co., Westport, Conn., USA, 81–116

Wilkins RJ ed. (1977) "Green Crop Fractionation." Occasional Symp. 9, Brit. Grassl. Soc., Grassl. Res. Inst., Hurley, Maidenhead, UK pe Ltd. London, UK, 91–94

21

A CLOSED BIOTECHNOLOGICAL SYSTEM FOR THE MANUFACTURE OF NONFOOD PRODUCTS FROM CEREALS

M. Bekers,[1] J. Laukevics,[1] N. Vedernikovs,[2] M. Ruklisha,[1] and L. Savenkova[1]

[1]Institute of Microbiology and Biotechnology
University of Latvia
Kronvalda boulevard 4, Riga, LV 1586, Latvia
[2]Institute of Wood Chemistry
Dzerbenes str. 27, Riga, LV 1006, Latvia

1. PRINCIPLES OF CLOSED BIOTECHNOLOGICAL SYSTEMS

The production of nonfood products from agricultural raw materials is an important challenge. Efficiency of grain processing for sustainable development depends to a large extent on the harmony of the proposed system. Such a system must include consideration of all steps in the cycle from preparation of the field through cultivation of the grain crop; harvesting; total biomass utilization during processing stages; treatment of wastes; and back to preparation of the field. A closed biotechnological system can be recommended for processing of agricultural raw materials. Ethanol production from potatoes in a closed system is presented in Figure 1. Lysine production is represented schematically in Figure 2, while details are given in Table 1. A semiclosed biotechnological system of leaf protein production from green biomass, integrated with straw and farm waste utilization, was realized in the Bauska region farm "Uzvara" in Latvia. This system includes lactic acid fermentation of juice and presscakes and methane fermentation of brown juice, together with pig manure and straw.

2. ECONOMICAL ASPECTS OF ETHANOL PRODUCTION FROM AGRICULTURAL RAW MATERIALS BY FERMENTATION

The final product yield from a land unit is one of the most important economical parameters. Figure 3 compares the ethanol yield from sugar beet, sorghum, potato, wheat and wood in Middle Europe. Up to 6000 L of ethanol can be produced from 1 ha of sugar beets. However, the cost of ethanol produced from sugar beet is 1.5 $US per litre of absolute alcohol, while that from wheat is only 0.65 $US (Table 2). This shows the impact of

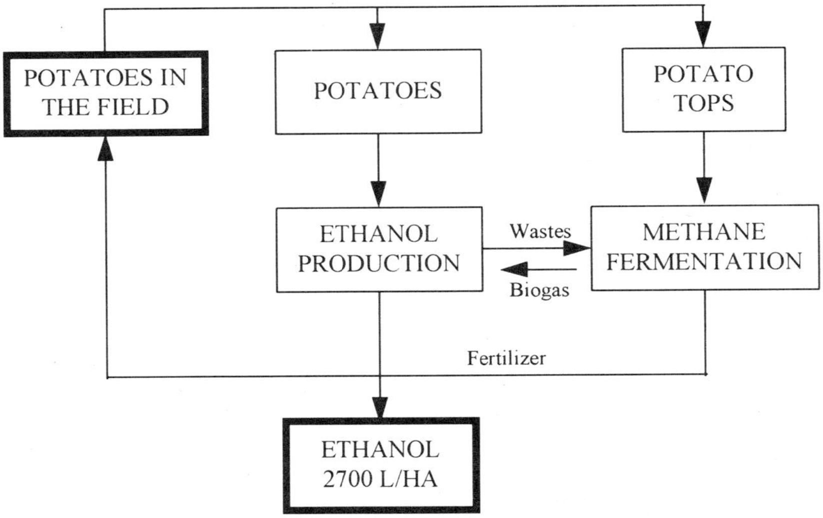

Figure 1. A closed system for ethanol production from potatoes.

technological expenses for raw material processing. A significant factor in the efficiency of any system is the ratio of energy output to input associated with the production of energy rich agricultural raw materials (Table 3). The output/input ratio increases from 1.14 to 2.71 in the case of wheat if straw is also used for energy-chemical production. The same effect can be achieved using Vedernikov's (private communication) method for straw processing to obtain furfural, ethanol, acetic acid and fuel granules. Vedernikov's

Table 1. L-Lysine production by auxotrophic *Corynebacterium* strains

Medium:	glucose or sucrose (molasses, starch, hydrolysate) NH_3, P_2O_5, K
Growth factors :	threonine (300-600 mg/l), methionone (~100-200 mg/l), biotin (0.3 mg/l); thiamin (0.2 mg/l); (source - corn steep liquor, hydrolysates of proteins etc.)
Fed - batch fermentation	48–72 hours
System productivity of L - Lysine	1.2 -1.6 g/L/hour
Lysine yield	0.4–0.48 g/g sugar
	L - Lysine concentrate, % DM
L - Lys.HCL	20–40
Bacterial biomass	2.0–2.5
Crude protein (Nx6.25)	45–50
Betaine	6–12
Reducing sugar	4–12
Thiamin (B_1) mg/kg	2–10
Riboflavin (B_2) mg/kg	80–160
Pantothenic acid (B_3) mg/kg	30–60
Folic acid (B_4) mg/kg	10–20
Pyridoxine (B_6) mg/kg	200–340
Nicotinic acid (PP) mg/kg	8–10
Ash	19–25

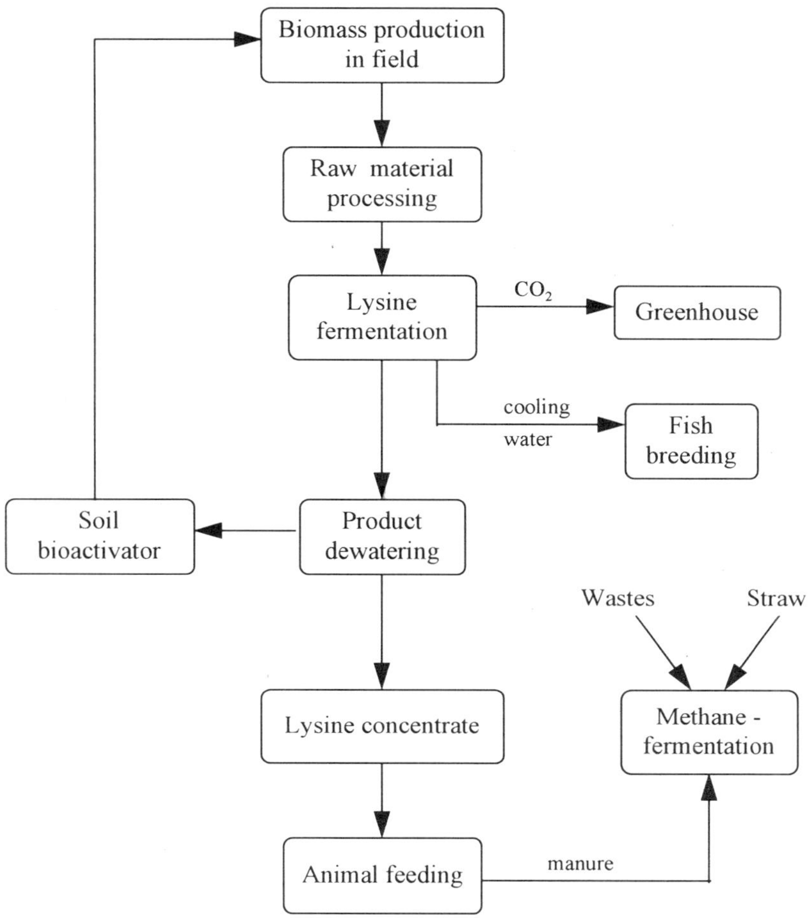

Figure 2. Closed systems in Lysine production.

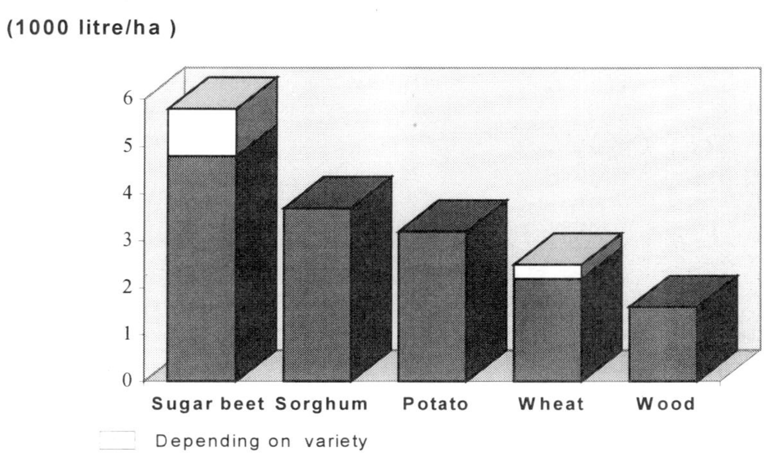

Figure 3. The potential yield of ethanol from different raw materials in terms of volume per area of planted crop.

Table 2. Cost of ethanol production from various raw materials and technologies

Raw material and technology used	Country	Cost ($US per litre absolute alcohol)	Source of information
Maize, traditional technology	USA	0.60	Zhang et.al. 1995
Grain, traditional technology	Latvia		Unpublished materials
Raw ethanol (88-90%)		0.60	
Rectificate (96%)		0.70	
Agricultural raw material in Europe	Europe		
Wheat		0.65	EC data
Maize		1.1	Riuiz 1994
Sugar beet		1.50	
Potatoes		3.45	
Corn + straw, hydrolysis and use of genetically engineered *Zymomonas mobilis*	USA	0.32	Zhang et.al. 1995
Wheat + straw, hydrolysis, with recovery of furfural, acetic acid, ethanol calculations		0.31	Vedernikov1995

proposed process for the production of products from wheat straw is illustrated in Figure 4. Silaging of straw with brown juice (the liquid residue after thermal protein separation from green juice), is an alternative possibility for the utilization of straw.

The batch process for ethanol production is generally practised by using amyloytic enzyme preparations for liquefaction and saccharification of starch, and the yeast *Saccharomyces cerevisiae* for ethanol fermentation. Typically, fermentation is carried out at between 29 and 35°C in a medium containing 16–24% sugars. This yields a product at around 8–12%, with the theoretical conversion of sugar to alcohol being 0.538 kg/kg. Typical productivities are in the range 1.3 to 2 g/L/hour. Fermentation processes using *Zymomonas mobilis* bacteria have also been developed in many laboratories, including our

Table 3. Energy output and input for the production of energy chemicals from renewable raw materials in Germany. Data are presented as energy equivalent (GJ per hectare of crop)

Raw material	Rape	Sugar beet	Potatoes	Wheat	Maize (CCM)
Input:					
farming	17.7	33.7	40.6	26.2	28.0
conversion	4.2	64.1	43.5	29.9	25.7
Total input	21.9	97.8	84.1	56.1	53.7
Output:					
rape oil methylester (RME)	47.8				
ethanol		101.8	64.1	44.5	46.7
pulp of thick stillage		32.3	12.2	10.1	12.0
biogas		14.7	11.2	9.6	10.2
leaves as fertilizer		11.0			
oil cake	28.2				
straw	(43.0)			(88.1)	
Total output	76.0 (119)	159.8	87.5	64.2 (152.3)	68.9
Output / input	3.47 (5.43)	1.63	1.04	1.14 (2.71)	1.28

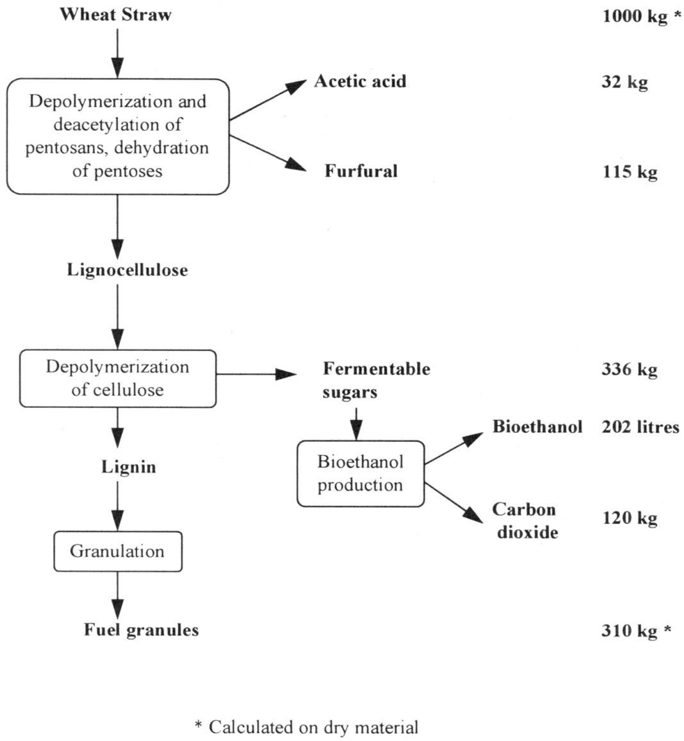

Figure 4. A proposed scheme for the manufacture of biothanol and other products from wheat straw.

institute. Table 4 gives details of ethanol production by *Z mobilis* for different modes of operation. An alternative process uses *Z mobilis* to produce ethanol, sweetener and levan (Bekers et al, 1990). Based on a 40 g/L sucrose substrate and fermentation at 25°C and pH 4.8, between 40 and 60 g/L of levan and 50 to 60 g/L ethanol can be produced. The process involves fermentation followed by centrifugation to remove biomass, then precipitation of levan by ethanol, distillation of ethanol and evaporation of the sweetener. Overall

Table 4. Ethanol production characteristics for *Zymomoncus mobilis* 113 fermentation under different modes of operation. D = dilution rate; P = product (ethanol) concentration; $Y_{p/s}$ = yield of product from substrate; Q_p = ethanol productivity; X = biomass concentration (dry weight basis)

Parameter	Batch culture 40 h	Continuous culture	Immobilized cell culture
Glucose concentration, g/L	220	100	150
D, h^{-1}	-	0.15	1.6
P (ethanol), g/L	109.5	46.3	60.0
Y p/s, g/g	0.50	0.46	0.44
Q_p, g/L.h	2.74	6.94	96
x, g/L	1.5	1.0	-

Table 5. Production of Poly-β-hydroxybutyrate (PHB) by *Azotobacter chroococcum* 23. Medium: glucose or sucrose (molasses), NH_3, P_2O_5, K Fed - batch fermentation 36 hours

Dry cell weight	110 g/L
PHB content	75 %
System productivity of PHB	Q = 2.29 g/L/ hour
PHB yield	$Y_{p/s}$ = 0.3 g/g glucose

yield of levan from sucrose is 14–17%, while sweetener is 40–65%. Ethanol and CO_2 are both yielded at between 7 and 21%.

3. PRODUCTION OF BIODEGRADABLE POLYMERS BY FERMENTATION

Synthesis of polyhydroxyalkanoates (PHA) has been realized using *Azotobacter* strains at laboratory scale, in the Institute of Microbiology and Biotechnology at the University of Latvia. Table 5 gives details of Poly-b-hydroxybutrate (PHB) production, while Table 6 details the main characteristics of the PHB produced. Up to 100 g/L of biomass was produced, with a PHA content of 75%. It planned to investigate PHA composites with wheat polymers in future studies.

4. THE ROLE OF METHANE FERMENTATION OF WASTES

The energy flow diagram for ethanol production from wheat (Figure 5) demonstrates that the energy available from biogas obtained from waste methane fermentation is as much as 21% of that associated with the ethanol itself. Thermophilic methane fermentation of agricultural wastes has been carried out in laboratory, pilot and industrial scale bioreactors. It was established that inactivation of pathogenic bacteria, gelmints and weed seeds occurs during the thermophilic process, and that pesticides are also biodegraded.

5. INTEGRATED CLOSED BIOTECHNOLOGICAL SYSTEM FOR PROCESSING OF AGRICULTURAL RAW MATERIAL

The general principles of a closed biotechnological system for nonfood production from agricultural raw materials are as follows:

Table 6. Main characteristics of PHB produced by *Azotobacter chroococcum* 23

Crystalline melting point, T_m, ^0C	181–185
Crystallinity, %	60–65
Glass transition temperature, T_g, ^0C	4.5
Molecular weight, M_w	200,000–400 000
Tensile strength, MPa	30–38
Extension to break, %	0–5

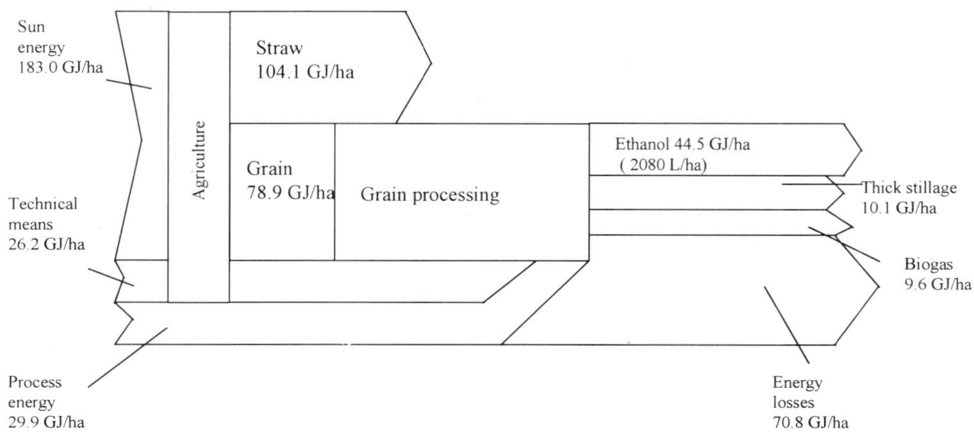

Figure 5. Energy flow diagram for ethanol production from wheat.

- usage of no-waste technologies with minimum energy consumption;
- optimized use of soil for biomass production with minimum mineral fertilizers and chemical inputs;
- maximum use of biological as opposed to chemical processes;
- utilization of wastes, preferably by methane fermentation to obtain biogas as a local energy source and to replace minerals in the soil from the liquid fraction;

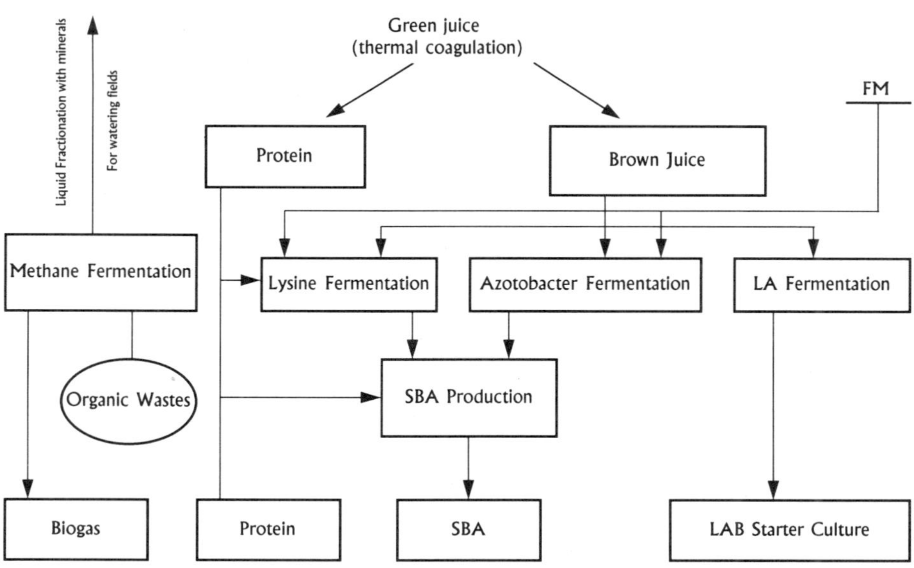

FM - Fermentation Medium from grain processing
LA - Lactic acid
SBA - Soil bioactivator
LAB - Lactic acid bacteria

Figure 6. Closed biotechnological system for agricultural biomass utilization.

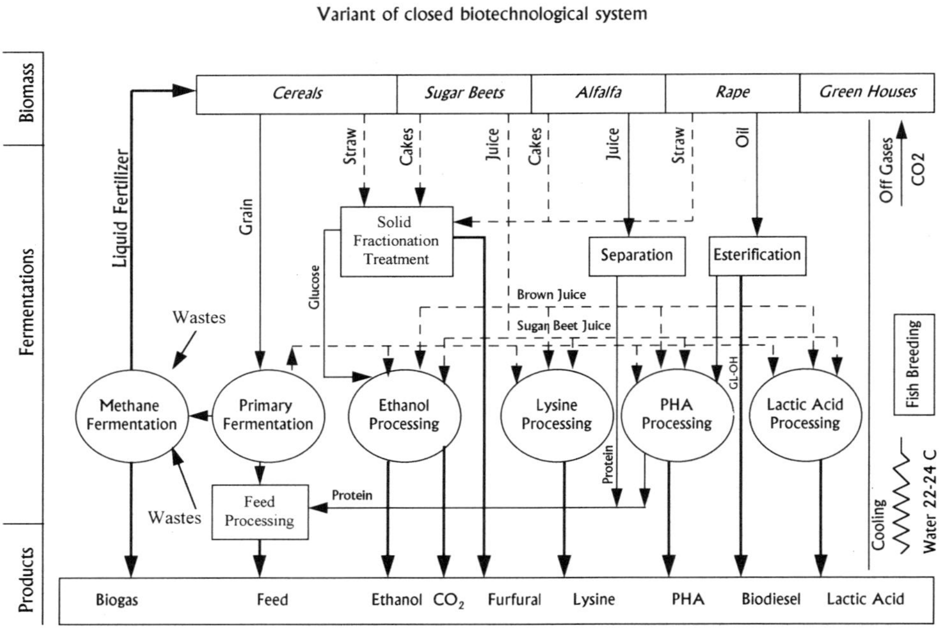

Figure 7. Variant of closed biotechnological system.

- the use of environmentally friendly technologies, including utilization of CO_2 from fermentation, in greenhouses and use of cooling water for fish breeding.

Such a closed system is represented schematically in Figure 6 and features a fermentation medium produced from grain as discussed in Chapter 25 as well as green juice (see Chapters 20 and 27). A more comprehensive process based on a variety of raw materials is proposed in Figure 7. In this process, cereal crops, sugar beet, alfalfa and rape are used to produce a range of products as well as to supply CO_2 to greenhouses and warm water to fish farms.

It is clear that closed biotechnological systems for processing of agricultural raw materials offer potential and may ultimately provide a guarantee for sustainable development.

REFERENCES

Beker M, Shvinka J, Pankova L, Laivenieks M and Mezhbarde I (1990) "A simultaneous sucrose bioconversion into ethanol and levan in *Zymomonas mobilis*." Appl. Biochem. Biotechnol., 24/25, 265–274

Ruiz Altisent M (Ed) (1994) "Application of biologically derived products as fuels or additives in combustion engines." EC Directorate-General, XII: Science, Research and Development, Madrid

Verernikov (1995) Private communication

Zhang M, Eddy C, Deanda V, Finkelstein M and Piccatagio S (1995) "Metabolic engineering of a pentose metabolism pathway in ethanologenic *Zymomonas mobilis*." Science 286, 240–243

22

REDUCTION OF THE ENVIRONMENTAL IMPACT OF WHEAT STARCH AND VITAL WHEAT GLUTEN PRODUCTION

George Svonja

Barr Rosin
Maidenhead, Berkshire, SL6 1BR, United Kingdom

1. INTRODUCTION

Traditionally wheat flour has been separated into its constituent components, principally starch and gluten, by screening and centrifugation. The main process used for many years was the Martin Process which used enormous quantities of water to achieve good yields and product quality.

This paper describes a new process which is based on a three phase decanter with hydrocyclone refining, allowing exceptional flexibility in raw material process capability and in choice and balance of products. This process uses very little water (2 to 2.5 m^3 / tonne of flour) in the wet process. Effluent is consequently reduced and effluent production can be as low as 1 m^3 / tonne of flour. The effluent can be treated in an anaerobic treatment plant to produce methane from the soluble solids in the wheat; alternatively the effluent stream can be evaporated to produce a concentrate which can be used for animal feed. In this case virtually 100% of the solids processed are recovered as saleable products.

The basic process is designed to handle wheat flour and is capable of processing a range of flours from baking quality down to some feed. The processing suitability of the flour is determined by the content and quality of gluten and so the raw material for the process can be second and third cut flours from the mill.

As far as the products are concerned, operating adjustments of the Three Phase Decanter ensure that control of the products and yields is available right at the beginning of the process. By adjusting the phases while the Three Phase Decanter is in operation, and monitoring the streams from the Three Phase Decanter, the quality and proportions of 'A' starch and 'B' starch can be fully controlled and adjusted to give the product mix required.

The particular advantage of the Three Phase Decanter is its ability to remove the pentosans and soluble proteins from the process right at the beginning of the washing operation, thus allowing the rest of the process to take place at a lower viscosity and with more precise separation than was possible with traditional processing. The lower viscosity

Cereals: Novel Uses and Processes, edited by Campbell *et al.*
Plenum Press, New York, 1997

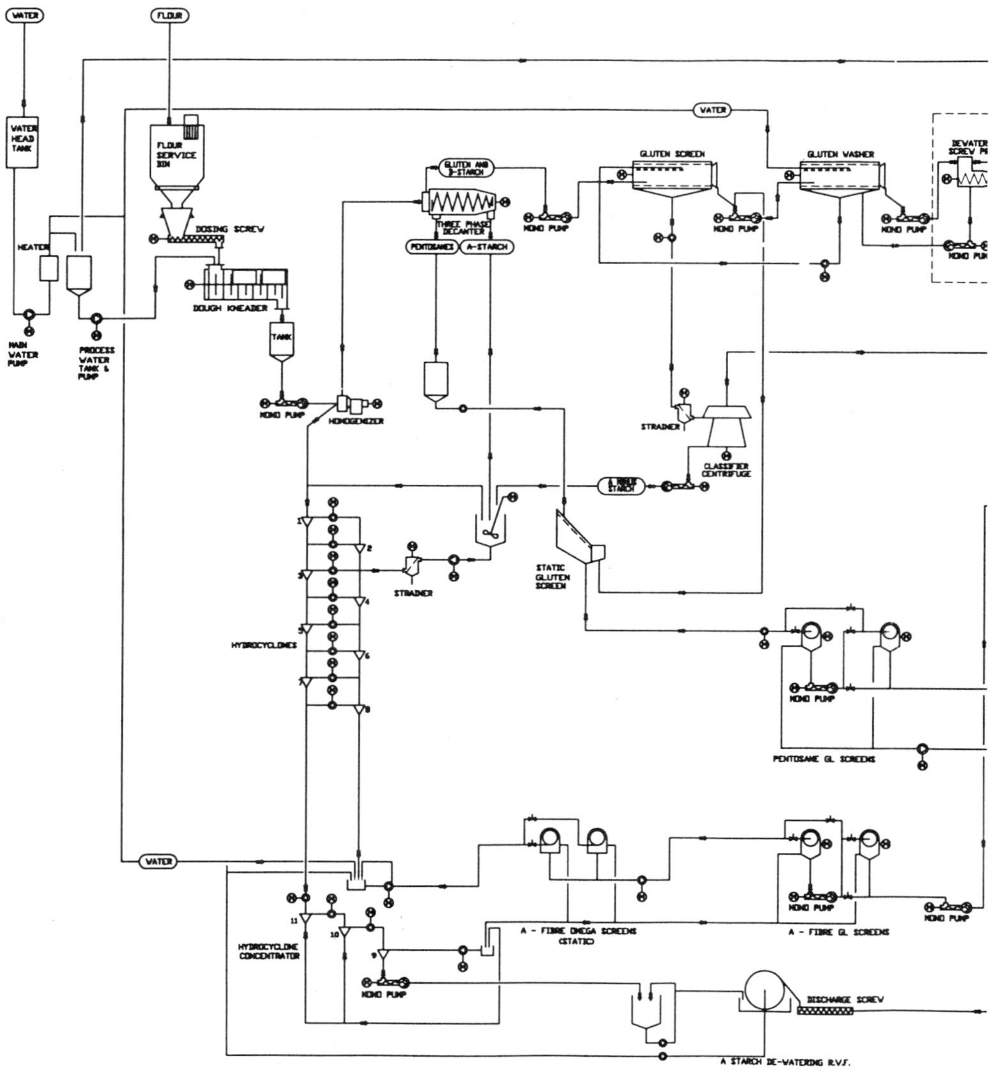

Figure 1. Flow diagram of Barr and Murphy wheat process.

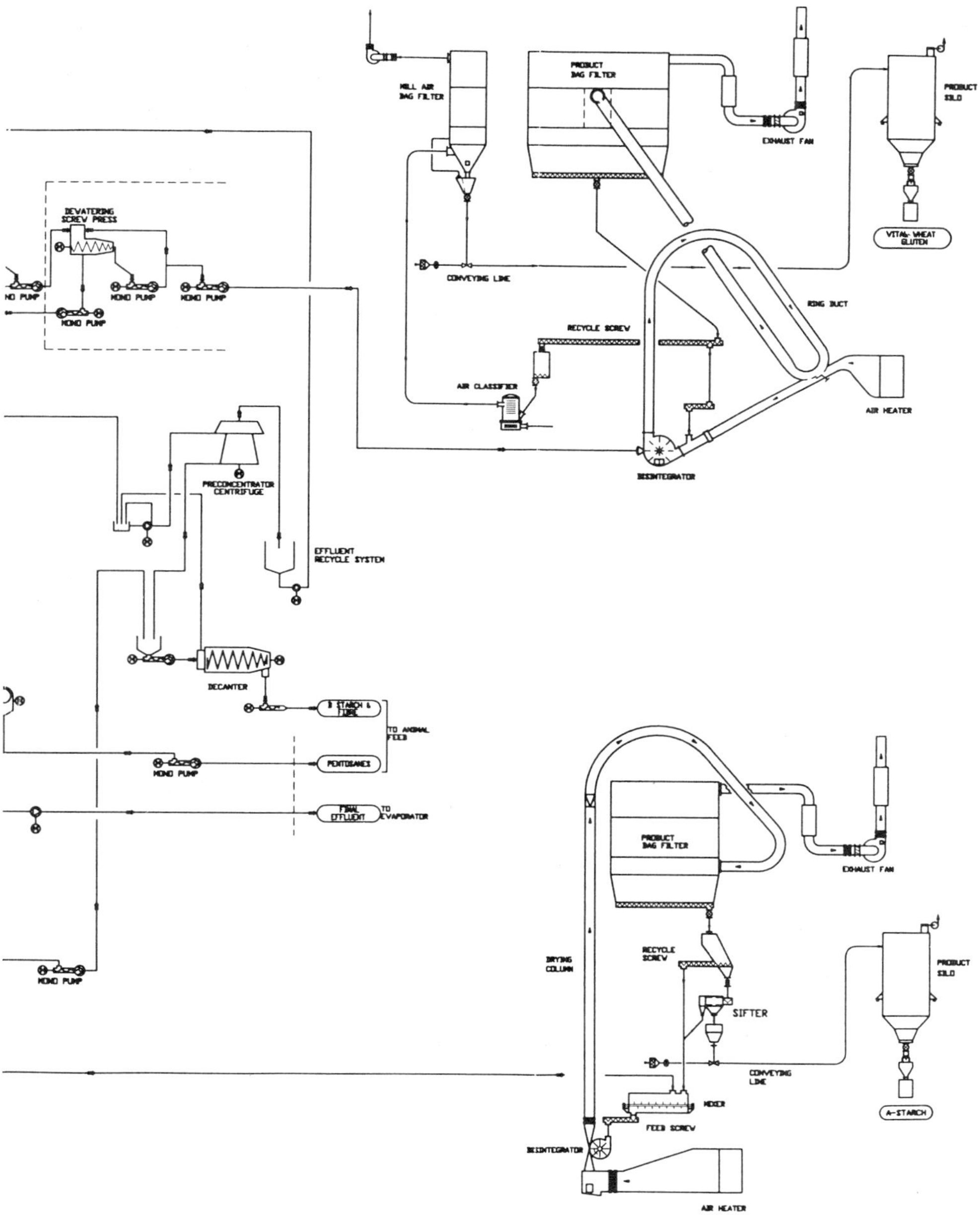

Figure 1. *(Continued)*

also means that the washing and separation can be effectively achieved at higher concentrations; for this reason the water consumption of the process is much reduced, with water requirements of less than 2.2 tonnes of water per tonne of flour achievable.

Following the wet separation process the Vital Wheat Gluten and the 'A' Starch are dried in a Ring Dryer and a Pneumatic Conveying Dryer respectively.

2. PROCESS DESCRIPTION

Figure 1 shows a schematic of the process.

2.1. Dough Preparation and Homogenisation

The process begins with a 20 tonne service bin which receives the flour from the central flour storage system. The flour is discharged by means of a high capacity discharger into a metering weigher which controls the quantity of flour fed into the process. Flour is fed into the process at a controlled rate and mixed with water in a dough mixer. The flour and water are mixed into a dough and the formation of the vital gluten is initiated. The dough is immediately pumped into a high intensity homogeniser where shear forces effectively break up the matrix and form an emulsion which will separate into 3 phases. The mixture of 3 distinct phases from the homogeniser is fed immediately into the special Three Phase Decanter.

2.2. Three Phase Decanter Separation Process

The Three Phase Decanters separate the slurry into 3 distinct phases.

The first phase is the 'A' starch which is almost pure and contains less than 1% protein. This phase is taken from the Three Phase Decanter immediately to a hydrocyclone refining system where it can be refined to a protein content in the range of 0.25%.

The second phase is the heavy phase, containing 'B' starch and vital gluten which are separated by screening out the gluten.

The third phase is the light phase, containing the pentosans and solubles and is effectively the effluent stream of the plant. By removing the pentosans from the 'A' starch and the 'B' starch the viscosity of these streams is reduced and hence the refining of the starch and gluten is greatly enhanced. Because of this the fresh water quantities required for starch refining are kept to a minimum and the waste water stream is dramatically reduced.

2.3. 'A' Starch Washing, Screening and Concentration

The solid phase from the Three Phase Decanter containing 'A' starch is fed to the hydrocyclone feed tank where it is diluted and then pumped into the hydrocyclone refining unit.

The hydrocyclone separation system washes and concentrates the 'A' starch in an 11 stage hydrocyclone battery, reducing its protein content to as low as 0.2 % protein. Feed is pumped into stage 3, while stages 1 and 2 collect starch carryover from the overflow.

'A' starch is counter-currently washed in the first 8 stages. The heavy 'A' starch fraction is screened to remove residual fibre, using a 3 stage screening system consisting of 'Omega' static Screens and Rotary Screens.

In the process of fibre separation and washing, the 'A' starch milk is diluted to approximately 9.4° Baume. The 'A' starch milk is therefore finally concentrated and purified in the last 3 hydrocyclone stages to 21° Baume. The purified starch milk leaves the system from the underflow of the final stage, and discharges to the 'A' starch holding tanks.

2.4. Gluten Handling

The second or "heavy" phase from the Three Phase Decanter contains wheat gluten and 'B' starch, with some fibre.

Gluten is recovered from this stream in Rotary Gluten screens and collected in a hopper, feeding it into a mono type pump and pumped to the rotary gluten washer. Here the gluten is washed with water collected and pumped by a mono type pump to the gluten dryer.

2.5. 'A' Starch Stripping and Cleaning

The filtrate from the gluten screen and washer contains some 'A minus' starch, 'B' starch and fine fibre. The 'A minus' starch is separated from the 'B' starch and the fibre by a disc bowl separator and returned to the hydrocyclone feed tank to achieve maximum 'A' starch yield.

2.6. 'B' Starch Separation and Concentration

The fine fibre is separate from the 'B' starch using a 2 stage screening system consisting of Rotary Cone screens. The 'B' starch is then pre-concentrated by a nozzle separator and further dewatered in a 'B' starch decanter. The decanter overflow is recycled to the nozzle separator.

The concentrated 'B' starch discharges to the 'B' starch holding tanks. The overflow from the separator is effluent, which is largely recycled back into the process, to minimise fresh water consumption.

2.7. Pentosan Screening and Cleaning (Light Phase)

The third phase from the Three Phase Decanter contains pentosans/gums and maybe some fine gluten. Any fine gluten is recovered from this stream in a Gluten Screen and collected in the hopper before the gluten washer.

The filtrate with pentosans is clarified using a 2 stage screening system consisting of Rotary Cone screens. The remaining effluent with only solubles discharges to the final effluent tanks, which feed the evaporator; the fibre and pentosans slurries from the three screening stations discharges to the fibre/gums holding tanks.

2.8. Vital Wheat Gluten Drying

The gluten is dried in a special Gluten Ring Dryer, which operates as follows:

The wet gluten from the gluten screens arrives in a special dewatering screw press where free water is removed. The screw press has a conical screw with a perforated screen plate at the bottom of the inlet hopper. Discharge of the screw press transfers the gluten into the feed pump hopper.

The gluten feed system is designed with two mono type pumps specially arranged to ensure a very steady feed. The circulating pump maintains the wet gluten circulating in a ring main pipe system, from this the feed pump draws sufficient material to feed the dryer at a steady rate. The feed pump is the second mono pump which operates at a controllable rate to feed the gluten into the dryer. Gluten is pumped directly into the disintegrator via a special fishtail feed nozzle where a thin ribbon of wet gluten meets a recirculating stream of partially dried product and is subsequently carried through the drying system. The ring shaped drying duct incorporates a centrifugal classifier, the manifold, which returns partially dried material to the disintegrator, while the dried product leaves the drying system with the spent drying air and is collected in a bag filter.

The dryer system is operated under partial vacuum due to the induced draught fan, drawing the drying air through the system. Prior to entering the system the drying air passes through an air filter and is subsequently heated by a box air heater battery using dry saturated steam or gas.

The dried vital gluten is discharged through a screw conveyor and rotary valve into a discharge screw, from which some is recycled, while the rest is fed to the Air Classifier Mill to ensure the required product particle size.

Milled product is cooled and conveyed to the mill air bag filter, by another induced draught fan drawing air through the milling system.

For environmental protection, the dryer includes a final guard filter, so that, in the event of a bag failure on the main bag filter, no gluten dust is discharged to atmosphere. The dryer is protected with explosion relief doors to V.D.I. standard.

2.9. Gluten Conveying, Storage, and Packing

The finished gluten discharges into a hopper above a rotary valve which in turn feeds a positive pressure pneumatic conveying system. Gluten is conveyed to a storage silo. When the conveying air and product reach their destination, they are separated by a high efficiency reverse jet filter and fan set.

The storage silo is mounted on weight cells with the outlet being fitted with a vibratory bin discharger. From the storage silo, gluten is discharged to a valve sack filling machine.

For environmental protection, a dust extraction system is included at the valve sack filling machine, and a bag failure (dust detector) alarm system is fitted on the vent filter exhaust.

2.10. Effluent Evaporator

The evaporator is a mechanical vapour recompression single effect type with an integrated thermo-compressor finisher.

Effluent (wheat solubles) is pumped from the final effluent holding tanks through the preheater to the 1st stage circulation pumps. These pumps maintain a constant circulation through the evaporator, and evaporation occurs as the liquor descends the calandria tubes.

Vapour passes from the calandria into the separator where liquor carry-over is removed. Partially concentrated liquor is transferred to the 2nd stage, where further evaporation occurs. Vapour from this stage is recovered in the 2nd stage separator drum.

Vapour from the separators is recompressed and forms the heating medium for the 1st stage evaporation. A portion of the vapours undergoes a further temperature increase via the thermo-compressor (steam injection) and is used to heat the 2nd stage evaporator.

Spent vapour from Stage 1 passes to the water cooled condenser. Condensate from here and from the Stage 2 evaporator is discharged via condensate pumps. A vacuum is maintained on the condenser by the vacuum pump.

Final product is discharged from the 2nd stage circulation system by the product discharge pump, and is transferred to the syrup holding tanks.

2.11. 'A' Starch Dryer

'A' starch is dried in a flash dryer which operates as follows:

The 'A' starch is fed to a Rotary Vacuum Filter where the slurry is concentrated to approximately 58% dry solids and transferred to the double shaft mixer on the 'A' starch dryer via a conveying screw.

The filtrate from the vacuum filter is recycled for use as wash water in the starch plant 'A' starch washing process.

The double shaft mixer conditions the raw feed material by back-mixing with dried product. Thus treated, the feed enters the dryer disintegrator via a variable speed feed screw.

In the disintegrator, a fixed beater hammer mill, agglomerates are broken up and the wet starch feed is dispersed into the hot drying airstream. The high speed venturi accelerates the air stream and the wet particles, which are subsequently carried through the drying column and ducting to a bag filter collector, where the dried product is separated from the drying airstream.

The entire system is operated under partial vacuum due to the induced draught fan which draws the drying air through the system. Prior to entering the system the drying air passes through an air filter and is subsequently heated by an air heater battery which can use either gas or steam.

The dried starch is discharged through a screw conveyor and rotary valve in the recycle hopper, from which some is recycled whilst the rest overflows to the pin mill to ensure the required particle size.

Milled product is cooled and conveyed to the mill air bag filter, by another induced draught fan drawing air through the milling system.

2.12. 'A' Starch Conveying Storage and Packing

The finished 'A' starch discharges into a hopper above a rotary valve, which in turn feeds a positive pressure pneumatic conveying system. 'A' starch is pneumatically conveyed to storage silos. When the conveying air and product reach their destination, they are separated by high efficiency reverse jet filters.

The silos are mounted on weigh cells with the outlets being fitted with vibratory bin dischargers.

For environmental protection, a dust extraction system is included at the valve sack filling machine and the vehicle outloading point, and a bag failure (dust detector) alarm system is fitted on the vent filter exhaust.

23

BIOETHANOL FROM CEREAL CROPS IN EUROPE

Chris Wroe

BP Chemicals Ltd
Britannic Tower
Moor Lane, London, EC2Y 9BU
United Kingdom

1. INTRODUCTION

The use of agriculturally derived ethyl alcohol (bioethanol) in motor fuels represents one of the largest potential non-food outlets for cereal crops. As such, its development has attracted considerable debate in Europe over recent years, especially since 1992 when the European Commission proposed tax incentives to encourage the production of non-food crops on land set-aside from food production.

Bioethanol is more likely to be used as a blending component rather than a total substitute for petrol in Europe. Because of its volatility and high water solubility the preferred route is by reacting bioethanol (45%) with isobutylene (55%) to form ethyl tertiary butyl ether (ETBE) which is a close substitute for MTBE, an established petroleum/natural gas based oxygenated blending component. In the unlikely event that Commission plans for 5% substitution of petrol by bioethanol were to be fully achieved, a market of more than six million tonnes of bioethanol would develop, consuming more than twenty million tonnes of wheat.

In the United States, almost four million tonnes of bioethanol made from corn (maize) is blended directly into petrol whilst some nine million tonnes of sugar-cane bioethanol is used in Brazil, the majority of which is as a petrol substitute in specially modified vehicles.

It is interesting to note that these overseas projects have their origins in concerns about security of fuel supplies following the OPEC-led oil shocks of the 1970's, while the current impetus in the European Union follows changes in agricultural policy aimed at reducing food surpluses.

2. ECONOMICS

The value of bioethanol (as ETBE) in Europe is closely related to fossil fuel prices, taking account of its octane value in petrol (or its value related to methanol where MTBE is substituted by ETBE).

Cereals: Novel Uses and Processes, edited by Campbell *et al.*
Plenum Press, New York, 1997

Currently bioethanol costs between 3 - 5 times the wholesale price of petrol (Anon, 1994; Hutcheson, 1995), or alternatively, bioethanol is only cheaper than petrol if crude oil prices exceed \$50 per barrel (compared with about \$18 today).

The sharp changes in crude oil price during the 1970's and 80's triggered intense research into energy conservation, and also into improved oil extraction techniques which have sharply reduced the cost of finding and producing crude oil. As a result, although short-term price volatility can still occur, sustained high oil prices are not expected for many years (Figure 1).

Thus if bioethanol is to develop in Europe, it can only do so if it is heavily subsidised for public policy reasons. Oil prices are not expected, to return to peak levels for many years.

Bioethanol production costs are government supported in both US and Brazil. However, cheaper feedstock costs and scale economies lead to lower costs in US compared with Europe, and Brazilian economics reflect cheap land and labour, together with a contribution from cane-burning to electricity generation in the dry season.

3. PUBLIC POLICY OBJECTIVES

The idea of diverting substantial funds into biofuels production was originally triggered by concerns over security of supply of fossil fuels; more recently objectives have changed towards environmental issues (vehicle emissions and global warming) and concerns about maintaining the rural economy.

Each of these objectives is examined below and the value of bioethanol programmes compared with alternative ways of achieving the policy goals.

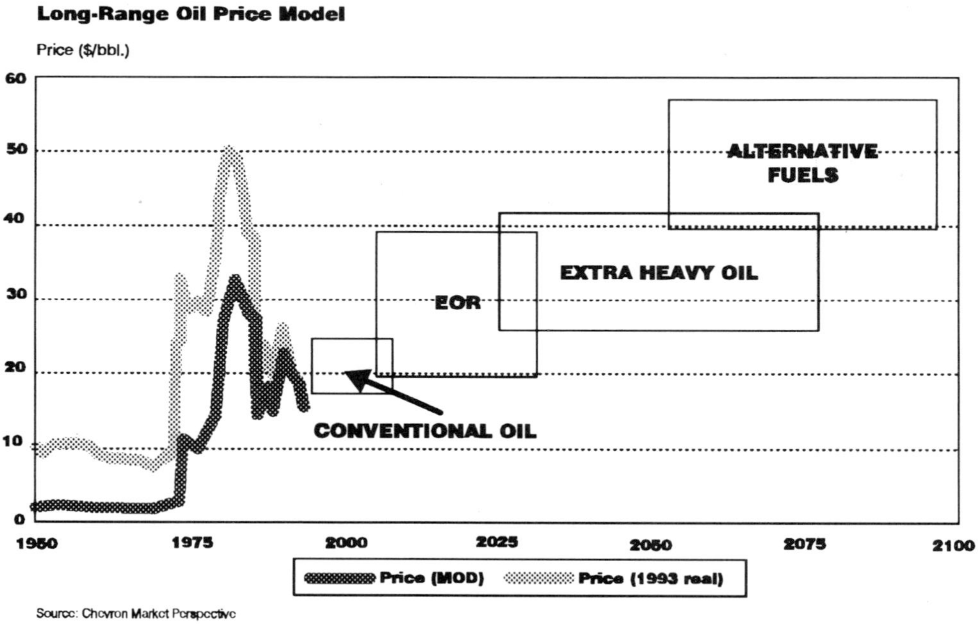

Figure 1. Long range oil price model, compared with actual oil prices. (EOR = Enhanced oil recovery)

3.1. Security of Energy Supplies

As mentioned above, the changes in oil exploration technology have greatly increased the options for producing crude oil and widened the production base outside the politically unstable Middle East. Bioethanol crops are planted annually and the processing technology is relatively simple. Therefore large-scale bioethanol production could be quickly established if a drastic change in energy markets lifted the long-term price of oil above $50/barrel and made biofuel production economic. (A full 5% substitution of transport fuels would reduce EU import dependence on fossil fuels by less than 2%.)

It is likely that measures other than growing fuel crops would provide a more effective response to sustained high fuel prices.

3.2. Environmental Benefits

The effect of adding oxygenates to gasoline on total emissions from vehicles is controversial as some emissions tend to be reduced (CO + HC) while others increase (NOx and aldehydes). Even allowing for some benefit, there is no advantage obtained from agriculturally-based fuels (Anon, 1996) and similar effects could be obtained more cheaply from fossil-based oxygenates (MTBE or synthetic ethanol) or by engine modifications.

Concern over global warming has led to pressure to reduce carbon dioxide emissions which has in turn focused attention on biofuels which absorb CO_2 as they grow. However the benefit is far smaller than expected due to the high fossil fuel consumption in modern mechanised intensive farming. Current bioethanol production is likely to have an energy output of less than 120% of the fossil fuel input, making it misleading to describe bioethanol as a renewable fuel. Thus the unit cost of reducing CO_2 emissions is very high (Figure 2).

Environmental groups express concern about likely heavy fertiliser/insecticide dosages to maximise yield in response to the poor economics of bioethanol production.

3.3. Rural Economy

The final justification, and most potent force currently driving biofuels in Europe, is the retention of infrastructure and employment within the rural economy. This is bound up with the future of the Common Agricultural Policy and particularly how much land will be surplus to food production in the future.

A recent study (Colley, 1994) has suggested that for the majority of EU cereal farmers, income from bioethanol production would be barely economic even with maximum tax reliefs and set-aside payments (Figure 3).

Therefore it appears that public funds would be better applied to other means of supporting the rural economy, rather than growing bioethanol fuel crops.

3.4. Research and Development

Evidence from Brazil suggests that bioethanol production costs have been falling at around 2% per year over many years (albeit from a high base). In Europe agricultural yields continue to improve with genetically modified crops offering attractive long-term prospects. A major breakthrough in processing bioethanol would occur if the lignocellulose within the crop could be fermented as well as the starches, and significant research effort is being applied in this area.

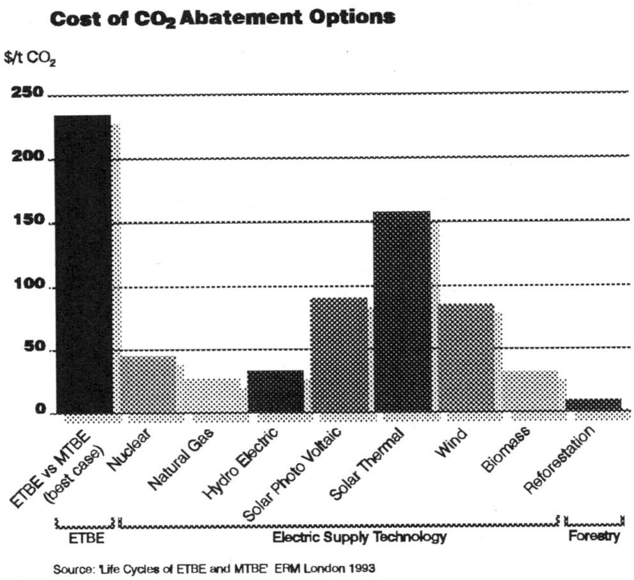

Figure 2. Comparison of the cost of CO_2 abatement options, showing Bioethanol (as ETBE) to be an expensive way to reduce global warming.

4. CONCLUSIONS

- Bioethanol production in Europe is currently highly uneconomic without subsidy and is expected to remain so for many years, while fossil fuels are abundantly available. It would require a huge step-change in energy prices to allow biofuels to compete with gasoline.
- The three policy objectives commonly quoted to justify support for bioethanol (energy security, environmental benefit and the maintenance of the rural economy) can all be met much more economically by alternative policies.

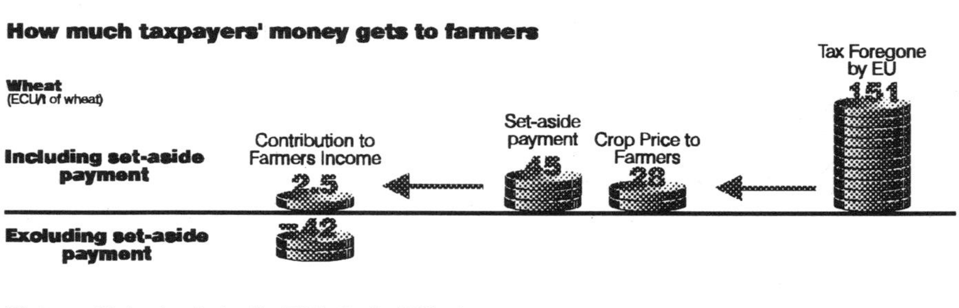

Figure 3. Tax subsidies to wheat farmers.

- Large-scale bioethanol production could be quickly implemented if crude oil supply became scarce or expensive.
- Therefore, while research and development will continue to reduce the cost of bioethanol production, there is no justification for spending public funds on supporting large-scale bioethanol production within the European Union today or in the foreseeable future.

REFERENCES

Anonymous (1994) "Biofuels", International Energy Agency, Paris, p 9

Hutcheson RC (1995) "Alternative Fuels in the Automotive Market", CONCAWE, Brussells, pp 31, 32, 55, 56

Anonymous (1996) "Investigation into Effects of MTBE and ETBE on Exhaust Emissions", BP Technology Centre, Sunbury

Colley RC (1994) "Effectiveness of Tax Incentives: Bioethanol (as ETBE) for Gasoline", ERM London, for BP Chemicals Ltd.

24

DETERMINING THE PROFITABILITY OF A WHOLECROP BIOREFINERY

Eric Audsley and Janet E. Sells

Silsoe Research Institute
Wrest Park, Silsoe, Beds MK45 4HS, United Kingdom

1. INTRODUCTION

The wholecrop biorefinery is founded on the concept that, given a (future) shortage of fossil fuels and the difficulty of harvesting the sun's energy, it is wasteful to throw away parts of the crop that have been grown, particularly when they have a considerable value (see also Chapters 20, 21 and 27). Thus straw internodes are just as good as wood chips, if there were no nodes and leaves. Even with grain it is wasteful to use the starch and discard the remainder. Therefore, in the same way as an oil refinery separates oil into components for different uses, one could insert a bio-refinery between the grower and the user to which the farmer delivers his crop and which separates the crop into components for the different users.

There are different ways of implementing the biorefinery concept. At one extreme, the biorefinery could simply take in wheat grain and fractionate it into different products (starch, gluten, fibre, etc.) for users, which is simply a super-flour mill. At the other extreme the biorefinery takes in the wholecrop (grain and straw) of wheat, rape, linseed, miscanthus, reed canary grass, etc. and produces products for many different markets. This work was carried out as part of an ECLAIR project led by the Danish BIORAF Foundation which concentrates on the wholecrop concept of the biorefinery and was also funded by MAFF to determine the profitability of the biorefinery concept (Audsley et al, 1995). In addition to our modelling work, there were experimental projects on plant breeding, storage of wholecrop and fractionation of wheat, straw and rape.

The objective of the systems analysis was to develop a comprehensive model to enable the performance of full-scale biorefinery systems to be assessed. The model determines the optimum long term profitability of a specific biorefinery system to give the optimum timing and types of harvesting, storage, processes and products. Determining the optimum ensures that different systems are compared on a like basis. The model can analyse the concept for a wide range of crops, for alternative methods of processing those crops, considering the location of the biorefinery in any country, and considering different economic scenarios. The generality of the design allows one to answer a wide range of questions. For example (i) what is the best size of the biorefinery for a given area of crops?; (ii) what is the best area of crops for a given size of biorefinery?; (iii) what is the best area of crops and size of biorefinery for a given demand for some products of the biorefinery?; (iv) what is the best system of crop storage?; etc.

The analysis is founded on the data and relationships between factors determined through experiment by partners in this project. The results of experiments have been generalised to allow predictions for methods not covered by experiments and to allow the tool to be used for the analysis of hypothetical processes or situations, which may occur in the future or which are not considered in the present situation analysed.

The model has been applied to determine the profitability of a wheat and oilseed rape whole crop biorefinery. The wheat biorefinery after grain/straw separation produces baking flour, gluten and starch flour and animal feed (bran and starch residue). A second system (Figure 1) produces baking flour, starch flour, bakery syrup, dietary fibre and animal feed (bran only). An alternative uses feed wheat to produce starch. The rape biorefinery produces oil, protein and animal feed (hulls, rape syrup). Both have a straw milling line producing internode chips/fibre and straw meal.

2. METHOD

The analysis procedure consists of two optimisation models:

a. the effect on the profitability of the farms around the location of the biorefinery due to selling the crop to a biorefinery, on the assumption that the biorefinery contracts to harvest the crop with a wholecrop forage harvester. This will have an effect on the farms, which currently have the optimal labour and machinery for the system the farm employs.
b. the profitability of the biorefinery system compared to traditional processing such as flour milling. The biorefinery system is defined as including long-term storage of the crop (although the storage is assumed to take place at the farms), the transport of crop to the biorefinery, and the processing to produce the various products.

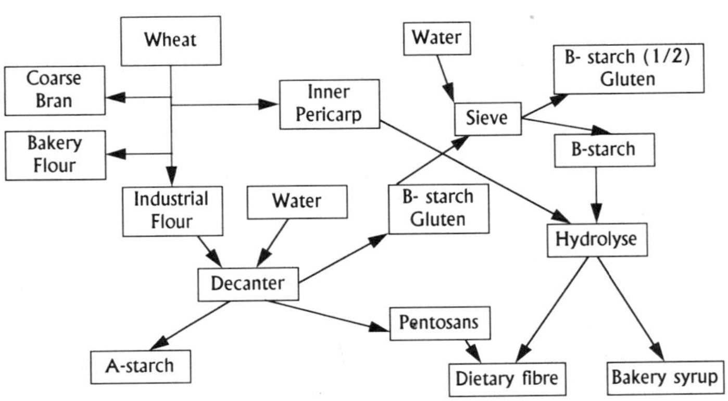

Figure 1. Possible processes for baking quality wheat.

The two parts together determine the profitability of the whole biorefinery system. Although the farm and biorefinery parts of the system are independent as far as decision making, the farmers' income, which is the price paid by the biorefinery for the crop, is exactly equal to the cost of the crop to the biorefinery. As both parts (i.e. farmer and biorefinery) are equally interested in making a profit, the whole system must make an increase in profit for it to be possible to find a crop price at which both sides will participate.

2.1. The Farm Model

The farm model determines, for a given cropping, the optimum farm system with and without supplying to the biorefinery. The change to the farm system for supplying the biorefinery involves a contractor harvesting the appropriate crop(s) and the farm labour providing the transport. The reduction in area of crops to be combined will mean the farmer should reduce the size of combine or even contract out the remainder of his combinable crops. A location is defined by its farming area, the number of farms partitioned by size, labour type, cropping and available working hours. Within a particular size range of farm the proportion of crops grown and the average farm size are known from survey statistics. The savings in labour and machinery costs are calculated for the full range of different sizes and types of farm found in a region, assuming they contracted with the biorefinery. It is assumed there are two labour types:

1. family farms where there is no cost advantage to saving labour and thus the labour is zero costed;
2. full labour cost farms where all labour is paid or hobby farms where labour is at a premium.

For each farm type described in terms of size, zero cost labour and cropping, the optimum labour and machinery requirements are calculated for an average farm using linear programming (Audsley, 1993). The linear programme consists of a cost function to be minimised and three types of constraints: 1) available working hours; 2) operation sequencing; 3) crop rotation. Each crop is described by a set of operations. An operation is defined in terms of timing and machinery requirements. The timing includes timeliness penalties, for example, delaying the drilling of wheat will incur a cost in terms of loss of potential yield. Labour and machinery requirements are determined from the available working hours which are dependent upon factors such as soil type, average rainfall and type of operation. Thus the change in cost of the average farms before and after they contribute to the biorefinery is calculated.

2.2. The Biorefinery Model

The biorefinery model is a linear programme model which determines the optimum system given a set of processing possibilities. The constraints are the material flows between processes and the available working hours of the machinery. Starting from a crop, the biorefinery consists of a sequence of processes in which a 'process' produces 'products' which become inputs to other processes or are sold. Processes can be harvesting, drying, storage or fractionation. Every process step is defined by an input product (wholecrop, grain, straw, etc.) output products (grain+straw, flour+bran, fibre+meal) and a process table which interrelates input and outputs. Each product consists of fractions (grain = endosperm + germ + hull) which have contents (moisture, energy, protein, ash). A process can have input restrictions (e.g. mc $< 14\%$) and output products have restrictions (e.g. ash

< 1%) on either fractions or contents. Fractionation processes aim to extract a maximum amount of a desirable physical fraction with the maximum permissible amount of undesirable fractions. A simple fractionation has two outputs, but complex systems and systems which combine several fractionations can have many outputs. Storage and drying are also processes, but only have a single output.

The model is constructed by determining all the possible processing routes through the biorefinery, including timing of harvest and duration of storage. A route through a biorefinery consists of a sequence of processes which extract from an input product, one or more output products with specified quality parameters such as levels of impurities, plus a residue product. Given the machinery available and hence hours of processing possible, the model determines the optimum utilisation of the machinery and products to sell which maximises overall profit, in terms of the optimum choice of processing routes.

A route for 1 hectare of a crop consists of a harvest date, harvest method, system of storage, duration of storage and fractionation (including drying where necessary) sequence, as shown in Table 1.

A biorefinery system could have two wholecrop fractionation lines with maximum input moisture contents of 25% and 16%, two storage methods (urea with input moisture content of 25% and ambient air with input moisture content of 18%), storage for 60, 90, ..., 300 days, and two alternative methods of dry milling, one producing a flour for baking and one producing an industrial flour for further processing, plus two alternative straw fractionation processes. Drying processes may be necessary before each process depending upon the restrictions on input moisture content. Thus the route shown in Table 1 could branch at P2 to wet separation, at P5 to an alternative dry milling and at P7 to an alternative straw milling.

The progress of each crop along any route is calculated from the performance data specified for each harvesting, drying, storage and fractionation process. This determines the quantity of the outputs from each process and the sales value of the products. At each stage the product can be sold, if this is possible and more profitable than further processing. Constraints are formulated describing the possible progress of a crop along the route. Labour and machinery are associated with each hectare of crop processed at a particular time along a particular route, so that for each machine in each period there is a constraint equation that the total hours required must be less than the hours available.

One of the concepts of the biorefinery is that small batches of different crops could be processed, in other words the same machinery could be used for different crops e.g.

Table 1. An example of a processing route through the Biorefinery

	Process: ijksm=12100	Product Input	Product Output	
P1	Wholecrop harvest	0	1	
P2	BAE-drying	1	2	Input m.c. restriction on P3
P3	Dry separation	2	3+4	May sell products 3 & 4
P4	Drum drying *for sale*	4	5	
P5	Dry milling	3	6+7	Must sell product 7
P6	Drum drying	4	8	Input m.c. restriction on P7
P7	Straw milling	8	9+10	Must sell products 9 &10
P8	Wet milling	6	11+12	Must sell product 12
P9	Flash drying *for sale*	11	13	Must sell product 13

milling wheat and rape. This requires a downtime for cleaning between batches and was considered impractical by some. This option was not considered in the model.

2.4. Data

2.4.1. Fractionation Process Table. The processes are defined generally so that the effect of changes automatically propagate through the system to affect the quantity and quality of the output products. Thus for example, an increase in the protein content of the wheat grain input to a line producing starch, propagates to a reduced yield of starch and increased yield of starch residue with a higher protein content (and hence value).

There are many different ways of fractionating an input into two (or more) outputs, such as sieves, shaking tables, blowing air, and for liquids, separators or decanters with high speed spinning plates. Thus separating dry grain and straw is a combination of a sieve and air, as found in a combine harvester. Separating flour and bran uses dry milling followed by shaking tables, separating oil from a solution of hulls and protein uses a separator. The basic rule of fractionation is that the greater the proportion of the target fraction extracted in the target output, the greater the proportion of undesirable fractions in the target output will be (see Figure 1). If the throughput is increased using the same machinery, the quality of fractionation will be lower and, of course, *vice-versa*. It is possible to specify, by experimentation, a process table that lists for a series of throughputs and proportions of target extracted, the proportion of undesirable fraction that will be obtained. Then given any input, if the proportion of the undesirable fraction in the target product is limited to a maximum, for the available machinery and throughput, the proportion of the target fraction extracted can be calculated.

Consider for example the sub-set of a fractionation table shown in Table 3, where the throughput of the machine is specified as a proportion of the design or standard throughput.

If the throughput ratio is 1, and 90% of the endosperm is extracted, there will be 1% of the non-endosperm fraction in the output and hence the ash content of the output can be calculated, as shown in Table 4.

Thus the ash content would be 0.40% dry matter (DM). If the maximum ash content is 0.5%DM, more endosperm can thus be extracted, with consequently more hull and germ. If the extraction is increased to 92%, the amount of the hull and germ can be determined by interpolation (4.6% by linear interpolation as there are only 2 points) and the output product thus has 0.47%DM ash. The actual amount which gives 0.5% is found by successive iteration.

As a simplification, in this analysis, all fractionation processes are considered as simply dividing one input into two outputs. In practice, some can achieve three outputs in one pass or have two inputs, etc. Where there are three output products, the process is modelled as consisting of extracting the first product from the input and then extracting the second product from its residue.

Figure 2 shows the ash content of the flour *versus* various percentages of flour extracted, for a number of different extraction methods. The theoretically calculated best possible is where first endosperm and then other fractions are progressively removed. Note that no point can be lower than the ash content of the endosperm and the lower asymptote of the ash curve must equal the ash content of the endosperm, which is 0.38% in this case. With the system used in the experimental biorefinery (BIORAF), it is known that it is generally possible to extract 40% of the flour with a maximum of 0.65% ash, and a further 40% with a maximum of 1% ash. These data suggest that the ash content of the

Table 2. An example of a section of the linear programming model matrix, showing for the route R the fractionation products F and the products sold S, the process flows and machinery constraints. Z = maximum profit objective function

	R210	F2101	S2103	F2103	S2104	F2104	F2107	F2241	F2243	S2245	F2245	S2246	F2246	F2249	S2251	F2253	...	mach0	...		
Land	1									1					1					≤	A
Prod12	-12.1	1						1												=	0
Prod42		-0.49	1	1																=	0
Prod52		-0.51			1	1														=	0
Prod72				-0.41			1													=	0
Prod36								-0.98	1											=	0
Prod66									-0.49	1	1									=	0
Prod76									-0.51			1	1							=	00
Prod96											-0.41			1						=	
Prod37															-0.96	1				=	0
Prod67																-0.49				=	0
Prod77																-0.51				=	00
Prod97																				=	
mach02		0.16		0.22		0.59	0.84											-0.029		≤	0
mach12	1.04																			≤	0
mach22		0.28																		≤	0

Table 2. (*Continued*)

	R210	F2101	S2103	F2103	S2104	F2104	F2107	F2241	F2243	S2245	F2245	S2246	F2246	F2249	S2251	F2253	...	mach0	...		
mach32		0.05																		≤	0
mach42				0.33																≤	0
mach52					0.01	0.01														≤	0
mach62						0.59														≤	0
mach72							0.67													≤	0
mach82							1.00													≤	0
Z	-0.85	-1.34	13.90	11.58	1.36	3.37	32.19	-1.21	-0.26	13.90	11.58	1.36	3.37	32.19	-1.21	-0.26		-1.80			

Table 3. A subset of a fractionation table, showing the level of impurities for a given target extraction rate

Throughput ratio	% extracted	% impurities
1	90	1
1	95	10

Table 4. Procedure for calculation of the ash content in the output from the fractionation process

	Ash content of input %DM	Tonnes per tonne input product	Percentage in output product	
Endosperm	0.38	0.83	90	0.7470
Hull	10.25	0.14	1	0.0014
Germ	5.00	0.03	1	0.0003
Ash content, %DM	of input product = 1.90		of output product = 0.40	

endosperm may not be 0.38% as in the figure and other data sources do indeed suggest a range of higher values up to 0.55%. This was one of many cases where the fundamental data were suspect or incompatible with data from other sources, so that calculating performance was challenging. Using the data in Figure 2, derived values for p (extracting endosperm) and g (impurities) can be calculated, as shown in Table 5.

Similar tables were derived for each fractionation process. Table 6 is extracted from a process flow sheet from the biorefinery which gives typical quantities of outputs in each fraction. The flow chart shows that 2000 kg grain is fractionated into baking flour and industrial flour. The industrial flour is then fractionated into 360 kg gluten at 25%DM containing 70% protein. The remainder from this is then fractionated into 329 kg starch flour

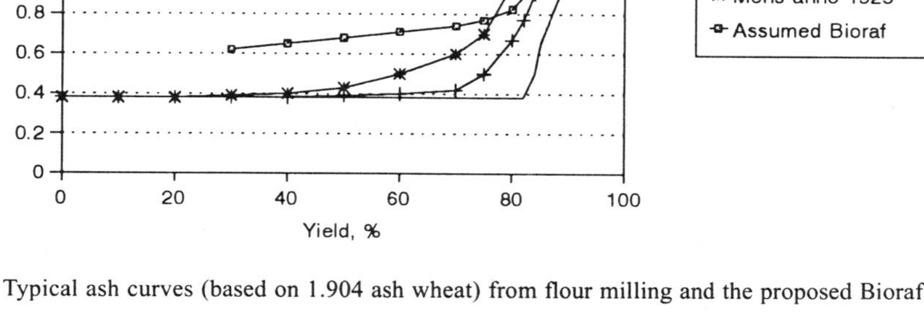

Figure 2. Typical ash curves (based on 1.904 ash wheat) from flour milling and the proposed Bioraf dry milling curve.

Table 5. Derivation of fractionation table for Bioraf dry milling process

Ash content, %	Extracted, %	p, %	g, %
0.62	30	35.6	4.4
0.65	40	47.3	6.6
0.68	50	59.0	9.1
0.71	60	70.5	12.0
0.74	70	82.0	15.3
0.77	75	87.6	17.8
0.79	77.5	90.3	19.3
0.825	80	92.8	21.6
1.08	85	95.7	36.1
1.35	90	98.1	53.0

at 97%DM containing 97% starch. Hence p = 83, g = 4.3 for protein and p = 69, g = 5.8 for starch extraction.

2.4.2. Storage Processes. A number of different methods were examined in the project for storing undried wholecrop. In the end, the only methods useable were urea, which makes the products unsuitable for human consumption, and ambient air drying.

The moisture content and duration of storage determine the effect on the contents, which can be a loss (such as energy value), a change towards a limit (to equilibrium moisture content) or an additive change (addition of urea). In storage the moisture content of the various fractions is assumed to equilibrate. Finished products have a storage cost until they are required to meet demand.

2.4.3. Transport Costs. The area surrounding the biorefinery is divided into concentric rings of equal area. Each ring contains a crop area which can be harvested with progressively higher transport costs. Transport costs are calculated by reference to data published by HGCA (1994) on contract haulage rates for different distances. These can be converted to a cost per load and hence, given a transport vehicle capacity and the bulk density of the product to be transported, the cost in £/load of transporting any product can

Table 6. Example of the calculations to derive a process table for starch and protein extraction from industrial flour from a process flow chart and product contents. Note that the endosperm column is the sum of the starch, protein and non-starch

	DM	Endosperm	Starch	Protein	Non-starch	Germ	Hull
Wheat	1720	1428	1014	164	250	52	241
Flour	683	663	471	76	116	4	17
Rem	1037	765	543	88	134	48	224
IFlour	706	663	471	76	116	8	36
Bran	331	102	73	1218	1840	40	188
Protein flour	90	88	20	63	5	.3	2
Rem	616	575	450	13	111	7	34
Starch flour	319	317	310	1	6	.4	2
Rem	297	258	141	13	105	7	33

be calculated: 84.75 + 0.6625 d (d = distance in miles). The model determines the optimum number of rings from which to harvest.

2.4.4. Market Values. Since products are specified by qualities, either proportions of fractions or contents, prices can be specified as a function of these. The price functions are determined by a market survey (Audsley et al, 1995). Residue products are assumed to be suitable for animal feed and a price is calculated from the energy and protein content. Typically these products are dry (10–14% mcwb) but the process of hydrolysis is carried out typically at 22%DM and the separated products contain a lot of water. Typically the hulls fraction will have a 34%DM content, the protein 17%DM and the syrup residue 4%DM. These can be dried to 95%, 96% and 65%DM respectively before sale, but the hulls and protein can be sold wet (without drying) with appropriate adjustment to the price.

2.4.5. Crop Data. The progress of the crop commences with the harvest. Harvesting loss, yield and moisture content of each of the fractions of a crop are defined as a function of time. Thus in addition to yield and baking quality, the leaf content of straw and the protein content of grain reduce with time over the harvest period.

2.4.6. Machine Data. Machinery is defined by its size - tonnes/hour throughput, tonnes/hour moisture removed, tonnes storage capacity. Data are needed on capital and repair costs, fuel, electricity or power used and costs of inputs to the process such as enzymes, urea or water.

3. RESULTS

Table 7 gives results for UK wheat only biorefineries where wheat is 37% of the croppable area around the biorefinery. The net profitability takes account of the value of the grain and straw that the farmer would have achieved under the conventional farming system. The biorefinery model optimum profit is in parentheses for interest. Values of system profitability greater than zero indicate a system which, from the data, adds value to grain and straw compared with present systems. Thus wholecrop scenarios including straw milling and/or enzymatic hydrolysis are profitable compared with present practice. The most profitable systems involve harvesting and processing grain only, leaving the straw at the farm for incorporation. Use of straw mill A (a combination fine grinder plus disc mill to produce higher value internode fibres) provides the best alternative in the wholecrop scenarios. However the cost of this is £5/ha for feed wheat and £42/ha or £18/ha for baking wheat with or without enzymatic hydrolysis respectively. Thus the value of the straw products are not sufficient to cover the additional costs of forage harvesting, storing and transporting wholecrop plus subsequent straw processing (as opposed to combining, storing and transporting grain). Dry separation is much more profitable than wet.

Table 8 shows how the system of storage influences the profitability of the wholecrop biorefinery system. Where wet separation is used, storing the wholecrop prior to separation increases the profitability by £39/ha where there is no straw processing. Storage after wet separation is more profitable, by about £80/ha, when the straw is processed. For dry separation, scenarios making use of dutch barns with and without ventilation are more profitable than just using one or the other. Storage with ventilation is more profitable than without since ventilation provides some drying. However, if the wholecrop is

Table 7. Profitability of wheat biorefinery for the UK, £/ha of crop

Dutch barn - vent	Dutch barn + vent	Dry separator	Wet separator	Grain store	Dry mill	Wet mill	Straw mill A	Straw mill B	Feed wheat grain: £719/ha	Baking wheat grain: £819/ha	
											+ Enzymatic hydrolyser
x	x	x			x	x			-74(611)	-2(783)	322(1107)
x	x	x			x	x	x		95(780)	160(945)	484(1269)
x	x	x			x	x		x	49(734)	116(901)	440(1225)
x	x		x		x	x			-100(585)	-44(741)	252(1037)
x	x		x		x	x	x		48(733)	98(883)	394(1179)
x	x		x		x	x		x	3(688)	55(840)	351(1136)
				Combine harvesting							
				x	x	x			100(819)	178(997)	526(1345)

Table 8. Comparison of storage methods for baking wheat biorefineries, £/ha of crop

		Straw milling		
Storage method	Sepn	None	Mill A	Mill B
Dutch barn +vent and Dutch barn -vent	Dry	322	484	440
Dutch barn +vent only	Dry	312	474	430
Dutch barn -vent only	Dry	297	451	409
Dry to 15%mc and Dutch barn -vent	Dry	358	522	477
Dutch barn +vent and Dutch barn -vent	Wet	252	394	351
Grain store and straw store	Wet	213	475	428

dried to its equilibrium moisture content (15%) prior to unventilated storage profitability is about £37/ha more then using dutch barns.

Table 9 gives the system profitability of UK rape-only biorefineries where rape is 8% of the croppable area. A rape biorefinery is highly unprofitable. The rape biorefineries giving the least loss are where wholecrop rape is harvested and processed to produce seed and straw products. These scenarios are at least £55/ha better than combining and processing just rape seed, unlike the wheat biorefinery scenarios. Exploration of the data using the second scenario in Table 9 which produces wet by-products, indicates a break-even oil price of about £800/tonne. Valuing oil at £1000/tonne gives a system profit of £147/ha of rape.

Table 10 shows sensitivity analyses for a baking wheat biorefinery of the straw : grain ratio and protein content of grain. The straw : grain ratio is changed from 1.0 to 0.6. The protein content is changed from a fixed 13% DM to a formula varying with harvest date. Decreasing the straw : grain ratio increases the profitability. This is unsurprising since the results clearly show that a wheat biorefinery in the UK where only grain is harvested and processed is more profitable then the wholecrop scenarios. The differences reflect the value of straw for each scenario. The change in the combine harvest scenario profit is due to the change in performance of the straw grain separation. Thus a lower straw : grain ratio means higher profit. Reducing protein during harvest time affects the system profitability by only £1–2/ha. This reflects the slightly lower feed value of bran, the price of which is a function of its crude protein content.

Table 9. System profitability for oilseed rape biorefineries in the UK, £/ha of crop

						Oilseed Rape Seed: £484/ha Straw: £77/ha	
Dutch barn - vent	Dutch barn + vent	Grain store	Dry separator	Enzymatic extractor	Strawmill A	Wet by-products	Dry by-products
x	x		x	x		-556(-82)	-621(-147)
x	x		x	x	x	-384(90)	-451(23)
x			x	x		-551(-77)	-625(-151)
x			x	x	x	-356(118)	-431(43)
			Combine harvesting				
			x		x	-438(46)	-509(-25)

Table 10. Sensitivity analysis of baking wheat biorefinery profit to straw: grain ratio and protein content of grain. Additional system profit, £/ha of crop

Separation		Straw mill		Original scenario system profit	Straw:grain ratio 0.6:1	Protein content = 13 - 0.2day
Dry	Wet	A	B			
x				322	+ 57	- 1
x		x		484	- 8	- 2
x			x	440	+ 10	- 1
	x			252	+ 84	- 2
	x	x		394	+ 29	- 1
	x		x	351	+ 46	- 2
Combine harvest				526	+ 24	- 1

4. CONCLUSIONS

A general method for economic analysis of sequences of fractionation has been developed. This method has been used to determine the optimum annual profit of a central biorefinery system, i.e. where wholecrops are harvested and processed (including storage, drying and fractionation) throughout the year. Use of the model for UK conditions, with appropriate data sought from other tasks in the Whole Crop Biorefinery Project and external sources, has produced an assessment of a biorefinery's competitiveness with present systems, with the following conclusions:

1. A UK wholecrop wheat biorefinery situated in East Anglia is profitable when straw milling and/or enzymatic extraction (for baking wheat only) is used.
2. UK rape biorefineries are highly unprofitable. The oil price needs to be ~£800/tonne to compare with the farmer selling the crop conventionally.
3. Combine harvesting with grain processing in a biorefinery is more profitable than wholecrop harvesting. The value of the straw products do not offset the additional costs of forage harvesting, transporting and storing wholecrop (as opposed to combine harvesting, transporting and storing grain) plus subsequent straw processing.
4. Straw milling wheat straw using a combination fine grinder plus disc mill to produce the higher valued internode fibres is £44/ha more profitable than the simpler disc milling system producing internode chips.
5. Enzymatic hydrolysis of baking wheat grain is valued at £324/ha.
6. The best storage system is drying the wholecrop to its equilibrium moisture content prior to unventilated storage in a Dutch Barn.

REFERENCES

Audsley E (1993) "Labour, machinery and cropping planning." Paper presented at XXV CIOSTA CIGR V Congress, 1993, Wageningen, The Netherlands

Audsley E, Sells JE, Boon A (1995) "Economic Assessment of the Whole Crop Biorefinery System." Final Report of Whole Crop Biorefinery Project Task 6, The Bioraf Denmark Foundation, September 1995

HGCA (1994) Marketing Note Vol. No. 29, Issue No. 21(11), 21 November 1994

25

DEVELOPMENT OF A GENERIC FERMENTATION FEEDSTOCK FROM WHOLE WHEAT FLOUR

Colin Webb and RuoHang Wang

Satake Centre for Grain Process Engineering
Department of Chemical Engineering
UMIST
PO Box 88, Manchester M60 1QD, United Kingdom

1. INTRODUCTION

There is an inevitability that one day much of the finite resource we currently use as feedstock for the production of chemicals and energy will have been consumed (or will no longer be available) and industry will be obliged to turn to renewable resources as replacement raw materials. Of the alternatives available, cereals offer amongst the best potential, being energy intense and environmentally benign. However, whatever other factors influence the choice, the real driving force for the adoption of new raw materials is the relative economics associated with obtaining and processing those materials compared with traditional feedstocks.

With the move to renewable feedstocks, which will necessarily be biological materials, will come a move to alternative processing routes, many of which will also be biologically based. Of these the most exciting prospects for the future lie with fermentation (see Chapters 2, 8, 26 and 27). Inextricably linked to the biotechnology revolution, the fermentation industry is poised to become the supplier of an almost limitless range of products. These will include both entirely new products and replacements for existing products which are currently based on non-renewable raw materials and environmentally harmful processes. Unlike chemical processes, fermentations benefit from the use of complex, natural, raw materials. In essence, fermentation can be regarded as the chemical industry equivalent of farming. What defines the final product is not the raw materials but the organism which is cultivated during the fermentation. The production vessel is supplied with nutrients in the form of carbon, nitrogen and other materials such as minerals and vitamins (and occasionally light). It is then *seeded* with the chosen species, sub-species or strain of micro-organism, which grows and produces the desired product. With the number of possible micro-organisms ranging into the hundreds of thousands (and being added to at increasing rates) the variety of potential products is enormous. Moreover, these products

Cereals: Novel Uses and Processes, edited by Campbell *et al.*
Plenum Press, New York, 1997

cover a similar range to that produced from petroleum and can, in principle, be produced from a single raw material. There is therefore an analogy between fermentation processing and petrochemical processing.

The most common source of both energy and carbon for fermentation is sugar (usually as glucose or sucrose). This is often supplied as molasses, a low priced by-product of the sugar industry. However, as the fermentation industry grows and the sugar industry declines (in part due to competition from cereal-derived glucose and fructose syrups) there is a tendency towards alternative sources of carbon. Of the alternatives, cereal crops are widely considered to offer the greatest potential. Being much lower in moisture, cereals are more energy intensive than sugar crops and have the advantage of being more easily storable and transportable. The sugars are in the form of starch, which is not readily fermentable, but can relatively easily be hydrolysed to glucose when required. In addition, cereal grains contain virtually all the nutrients required to support the majority of microorganisms and so, in principle, require little supplementation of nitrogen, phosphorus, etc.

It is therefore possible to foresee a future in which some industrial products will be extracted directly from grain (e.g. oil, starch) and others will be produced by fermentation of grain flour. Indeed, grain processors are already leading the field in the development of products from renewable feedstocks (Frost, 1996). At the time of writing, within the past month AE Staley Manufacturing Co. have announced plans to build a "multi-million-lb per year" cereal-based lactic acid plant (Milling & Baking News, May 7, 1996). In the longer term, a *total processing* concept can be envisaged in which wheat, say, is milled primarily to produce flour for food use but with lower value streams being taken off for processing to non-food products rather than being blended into higher value streams. Figure 1 illustrates this concept (Figure 1 in Chapter 27 illustrates the same idea using a combination of cereals and the green juice from agricultural residues; Chapters 17 and 21 also describe some similar integrated systems, while Chapter 20 describes the move towards

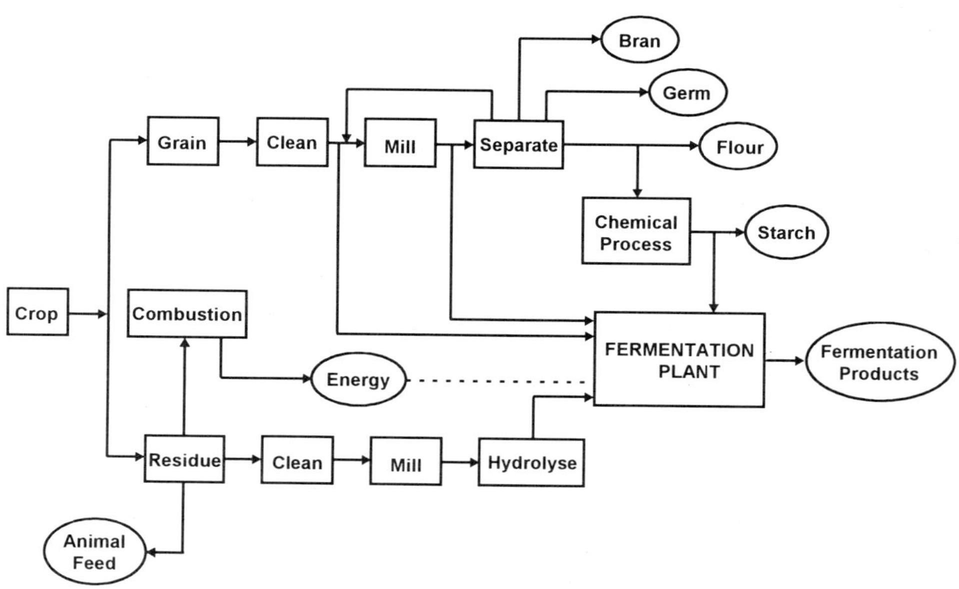

Figure 1. A total processing concept for cereals.

extracting a range of both food and non-food products from immature wet crops). Central to the non-food process would, of course, be a fermentation plant. A key to the success of this approach would be the production of a generic fermentation medium from which a whole range of fermentation products could be produced.

It is current practice (and therefore economic) in some fermentation industries to buy glucose which has been produced by hydrolysis of starch extracted from cereal grains, for use as a carbon/energy source. The starch hydrolysis is carried out using enzymes which have themselves been produced by fermentation! To the glucose must be added a nitrogen source, often in the form of corn steep liquor (a by-product of the starch extraction process). The mix is then supplemented with minerals and other nutrients (which may well have been present in the original grain) to provide a complete fermentation medium. Thus, the total cost of this medium includes costs associated with starch extraction, enzyme production, starch hydrolysis, and nutrient supplementation. All to arrive at a medium which is equivalent to whole grain flour!

In the Satake Centre for Grain Process Engineering at UMIST, research is being conducted into the production of a generic fermentation medium based on whole wheat flour. The process minimises the number of extraction/conversion steps, avoiding unnecessary separation and recombination and preventing loss of nutrients. This chapter describes results obtained to-date in the development of the process, which produces separate glucose-rich and nitrogen-rich streams suitable for use in subsequent fermentations.

2. MATERIALS AND METHODS

2.1. Wholemeal Wheat Flour

Wholemeal wheat flour, applied as the only nutrient throughout this project, was obtained from the Wellesbourne Watermill, Stafford. It is stone ground from a local species of *Mercia*, a soft wheat with a reported protein content of 11%.

2.2. Micro-Organism

A strain of *Aspergillus awamori*, a sequential mutant of *Aspergillus niger* NRRL 3312, was employed in this study. The strain was stored dry in the form of spores in sand at 4°C. Before use it was purified by plating onto a solid medium of wholemeal wheat flour agar.

2.3. Gluten and Bran Separation from Wholemeal Flour

Where gluten-free flour was required, the conventional Martin process was employed for the separation of gluten from the original flour. The flour was mixed with tap water at the ratio of 75 mL water per 100 g flour to form a stiff dough which was aged at about 25°C for two hours before being washed with tap water on a sieve of 500 μm aperture under kneading by hand. The amount of washing water was decided by the desired concentration of flour suspension. Where bran-free flour was required the bran was separated by sifting flour suspensions from gluten separations with a sieve of 125 μm aperture.

2.4. Analyses

Glucose concentration in samples from various sources was analysed using a glucose analyser (Beckman, USA), while free amino nitrogen (FAN) concentration was determined by the ninhydrin colorimetric method of the European Brewery Convention. Dry weight measurements were made, in order to estimate biomass concentration, after first removing suspended starch solids by hydrolysis. To determine the starch content of the original flour, both enzymatic and acid hydrolyses were employed. For acid hydrolysis, three samples of flour, about 1 g each, weighed to 0.0001 g were transferred into 30 mL distilled water. After pH was adjusted to 1.5 using 1 M sulphuric acid, the suspensions were made up to exactly 50 mL. The process of hydrolysis was carried out in an autoclave at 135°C for 15 minutes. The liquids were thereafter cooled to room temperature by running water before glucose concentration was measured. For enzymatic hydrolysis, an enzyme solution with an amyloglucosidase activity of 130 U mL^{-1} at 60°C and pH 4.5, was utilised to convert starch in flour samples into glucose. Three flour samples of about 1 g each, weighed to 0.0001 g, were transferred into three 50 mL volumetric bottles with 20 mL distilled water to form suspensions, which were gelatinised for 20 minutes in an 85°C water bath. After being cooled to room temperature, exactly 20 mL of enzyme solution was introduced into each bottle. The suspensions were diluted with distilled water and placed in a 60°C water bath for 24 hours before glucose concentration was measured.

3. PROCESS DEVELOPMENT

The main aim of the research was to produce a generic fermentation feedstock from whole wheat flour by hydrolysing available starch into glucose and converting protein into free amino nitrogen. The basis for the process was a continuous fermentation using the filamentous fungus *Aspergillus awamori* which produces the necessary amylolytic enzymes to degrade the starch into glucose. The glucose-rich effluent from the fermentation would be filtered to remove cells, in order to minimise glucose consumption by the fungus. The solid residue from the filtration stage, containing both fungal cells and undigested bran would then be autolysed to provide a nitrogen-rich stream which could be blended with the glucose-rich stream to create media to suit a range of subsequent fermentations. It was considered potentially advantageous to extract gluten as a by-product rather than to downgrade it to nitrogen, so tests were carried out with both whole and gluten-free flours.

Initial experiments involved batch fermentations to determine the suitability of whole wheat flour as a fermentation medium for *A. awamori.* These were also used to indicate the optimum residence time for a continuous fermentation to minimise consumption of glucose produced via the enzymes excreted by the fungus. The effects of the presence of bran and gluten were also studied through batch fermentations.

Batch fermentations were operated in a two litre stirred tank fermenter with a working volume of 1.8 litres, fitted with two Rushton turbine impellers. Flour suspensions were cooked at 70 – 80°C for 20 minutes by sparging 1 bar gauge live steam in order to gelatinise completely the starch content. The gelatinised flour mash was then diluted to the desired concentration before 0.002% (v/v) silicone antifoam was added. After being agitated until homogeneous, the liquid was sterilised at 121°C for 120 minutes. During fermentations temperature was controlled at 30°C. In the early stage, i.e. before the completion of spore germination, broth pH was uncontrolled from the initial value of about 5.5. After spore germination, at around 20 hours after inoculation, when the broth pH had decreased to 4.5, pH was controlled at 4.5. During the course of the fermentations, agitation speed was co-ordinated with aeration level to maintain the dissolved oxygen level above 25% of saturation, which is the reported critical oxygen level for the growth of *A. awamori* (Kostka and Kaczkowski, 1989). More specifically, agitation speed was maintained between 200 rpm and 300 rpm during the first 48 hours and was increased to a maximum of 600 rpm in the later stages, whereas aeration rate was increased from 0.1 vvm to 1 vvm.

Continuous fermentations were carried out in a 10 litre stirred tank bioreactor with a working volume of 5 litres. Operations were started batchwise and switched to continuous mode at about 40 hours, after glucose concentration had increased to its maximum value. Fermentation pH was controlled at 4.5 in early experiments but was later reduced to 3 to discourage the establishment of bacterial contaminants. Temperature was controlled at 30°C, agitation speed at 500 rpm and aeration rate at 1 vvm throughout.

Effluent from the continuous fermentation was concentrated by centrifugation at 4000 rpm for 5 minutes. The concentrated slurry was sealed in bottles which were stored in a water bath to initiate autolysis reaction. The supernatant liquor was investigated as a source of enzymes for further glucose production by blending it with flour suspensions from the gluten washing stage. Both gelatinised and ungelatinised flour were tested.

4. RESULTS AND DISCUSSION

4.1. Batch Fermentations

In addition to following the profiles of glucose concentration and total dry weight, microscopic studies on bran destruction during fermentation were also performed. Preliminary fermentations on various flour concentrations revealed that a concentration of 8% original flour was most suitable at laboratory scale. Higher concentrations usually caused system blockage.

Figure 2 presents the results of a batch fermentation using 8% wholemeal wheat flour. Glucose formation started at about 12 hours after inoculation and led to a visible decrease in broth viscosity. After reaching a maximum of 40.2 g L^{-1} at approximately 32 hours, glucose concentration started to decline following starch exhaustion. The dry weight measurements presented in Figure 2 represent all of the solids present in the fermentation broth, as no method was found to separate the fungal hyphae from wheat bran residues. During the experiment, it was found that the aleurone layer, accounting for more than 50% of the wheat bran or more than 7% of the whole wheat kernel on a dry basis (Shetlar et al, 1947; Hinton, 1959), was digested. It is generally believed that the utilisation of aleurone cells demands a satisfactory means of disrupting the thick and indigestible cell wall and discharging the contents (Bradbury et al, 1956; Fulcher et al, 1972; see also Chapter 29). In this study, the sterilisation procedure was originally considered to constitute such a satisfactory means, but later experiments showed that the broth filtrate also possessed the ability to hydrolyse unsterilised aleurone cells.

At about 10 hours after inoculation cell growth led to a sharp increase in total dry weight. Microscopic observation revealed that the subsequent decline in dry weight was primarily due to the destruction of very small aleurone specks. The next increase in dry weight suggested completion of these tiny specks. Beyond about 100 hours, the decrease in dry weight was attributed to the combined effects of the various hydrolytic enzymes produced by the fungus on the undissolved components, being in excess of the contribu-

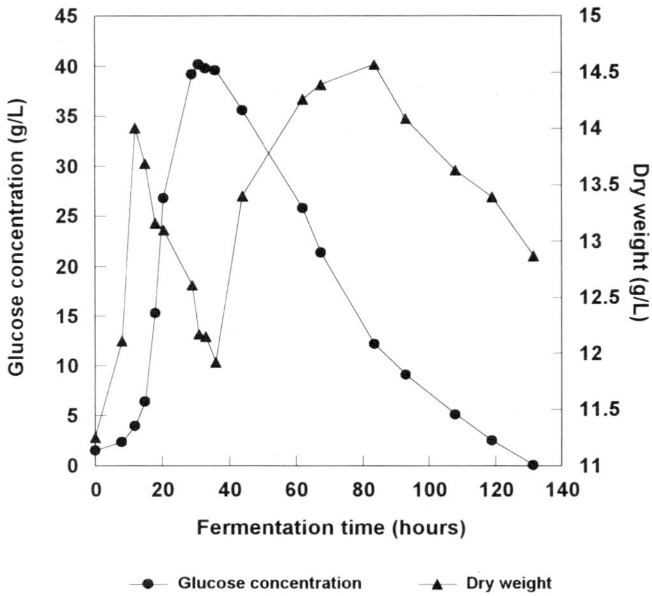

Figure 2. Results of a batch fermentation on 8% original wholemeal flour.

tion of cell growth to total dry weight. A further experiment using more finely ground flour showed that undissolved solids were digested in a shorter period of time (see Table 1).

Batch fermentations involving gluten-free flour showed similar glucose production but with much reduced glucose consumption, presumably because of nitrogen limitation on cell growth. It was therefore decided that gluten extraction could be carried out without adverse effects on starch hydrolysis. When gluten and bran were left out of the flour suspension, fermentation still took place though the final amounts of glucose produced were lower. The results of these batch fermentations are summarised in Table 1.

4.2. Continuous Fermentations

The maximum specific growth rate of *A. awamori* was estimated from the batch fermentations to be 0.034 h^{-1}. Accordingly, the first continuous fermentation was started with a dilution rate corresponding to this value (i.e. a residence time of 29.4 hours). Glucose

Table 1. Comparison of maximum glucose production on various flour-based media

Medium composition	Maximum glucose concentration (g L^{-1})	Starch conversion ratio (%)
original wholemeal flour	40.2	74.2
further ground wholemeal flour	44.8	82.7
further ground gluten free flour	42.8	79.0
bran free flour	38.6	71.3
gluten and bran free flour	37.6	69.4

concentration, at steady state, reached an average value of 38.3 g L^{-1}, corresponding to a starch conversion ratio of 70.7%. Following an increase in dilution rate to 0.04 h^{-1}, glucose concentration increased to an average of 41.7 g L^{-1}, equal to a starch conversion ratio of 77%. Thorough microscopic analysis found neither aleurone layer nor freely suspended aleurone cells in the effluent under both dilution rates. Early studies on the chemical constituents of wheat showed that the aleurone layer was rich not only in protein but also in phosphorus and different vitamins, namely thiamine, niacin and riboflavin. The release of these components after the destruction of the aleurone layer and aleurone cells apparently favoured cell growth, resulting in a higher apparent maximum specific growth rate than that obtained in batch fermentations, so that it looked possible to operate continuous fermentation under a higher dilution rate.

A dilution rate of 0.05 h^{-1} was chosen for the next continuous fermentation. After an operating period of more than six residence times this was raised to 0.054 h^{-1} and then was further increased to 0.06 h^{-1} after another six residence times. Figure 3 shows the time course for this second continuous fermentation. As had been expected, glucose concentration in the effluent increased slightly following the increase in dilution rate, and approached an average of 44.2 g L^{-1}, corresponding to a starch conversion ratio of 81.6%. No obvious increase in glucose concentration was detected after further hydrolysis of samples from the effluent at 60°C for 24 hours, confirming that the residence time of 16.67 hours was still long enough for complete conversion of the starch present in the medium. Microscopic studies, however, found some freely suspended aleurone cells in the samples from the operating period at a dilution rate 0.06 h^{-1}. The fact that the entire destruction of aleurone cells required longer residence time than starch conversion raised a conflict between the preservation of glucose and the other nutrients in the effluent. With the presence of sufficient enzyme activities in the broth, continuous fermentation with a shorter resi-

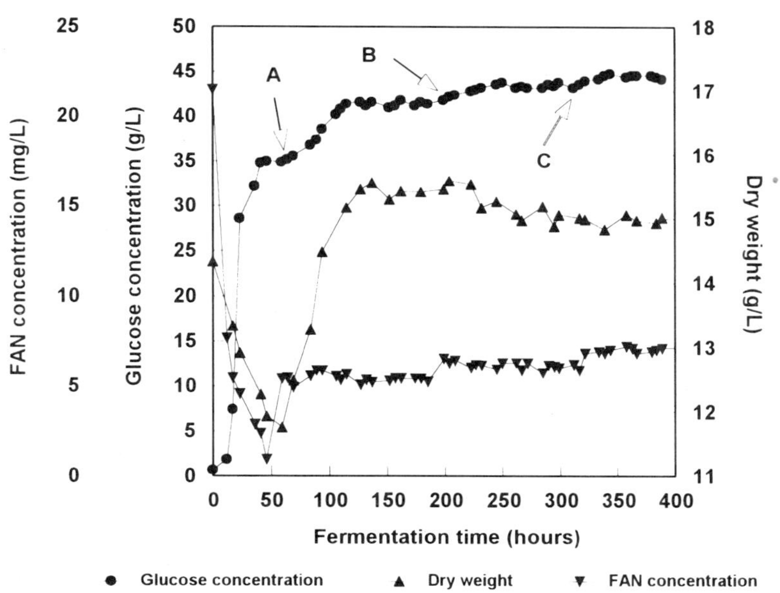

Figure 3. Time course of continuous fermentation on 8% gluten-free flour. (A) - Start of continuous operation at a dilution rate of 0.05 h^{-1}; (B) - increase to 0.054 h^{-1}; (C) - increase to 0.06 h^{-1}.

dence time undoubtedly benefited glucose production due to lower consumption. On the other hand, it maintained some raw materials intact. Among these materials, there possibly existed what would be potential nutrients for secondary fermentation after proper conversion. Referring to the results obtained in batch fermentations, further milling, or in other words, smaller particle size, effectively accelerated solid destruction. Considering the overall project aims it seemed unnecessary to apply any of the traditional wheat milling process to the wheat which would be used in this project since the flour quality produced by the complex milling strategy is not important to the conversion by micro-organisms, as long as the particle size is small enough. A simple industrial mill, like the one employed in the dry-milling of corn, is most likely sufficient.

The effects of gluten content on effluent composition was studied in a further continuous fermentation using flour suspensions of varying gluten content. Dilution rate was fixed at 0.05 h^{-1}, and the fermentation was carried out long enough to establish steady state for each feed. Table 2 summarises the average outputs during each steady state. The decreases in glucose concentration due to the removal of gluten from the original flour verified that for *A. awamori* gluten not only provides a nitrogen source but also takes precedence over starch as carbon source, so that the reduction of gluten content in the medium led to a decline in FAN output and at the same time resulted in higher starch consumption.

4.3. Glucose Enhancement

The maximum concentration of medium fed to continuous fermentations was limited to 8% original flour due primarily to the very high viscosity of the gelatinised flour mash. Consequently, glucose concentration in the effluent barely reached a maximum average of 51.4 g L^{-1}, which is much lower than the desired concentration for many commercial fermentations. To increase this figure, experiments on glucose enhancement were carried out using filtered effluent as a source of amyloglucosidase to hydrolyse flour suspensions. Both gelatinised and ungelatinised flour suspensions were used, and in all experiments temperature was controlled at 60°C, the optimum for amyloglucosidase activity. The results are shown in Figure 4.

In the case of gelatinised flour, glucose concentration increased very rapidly within the first 30 hours, reaching a value of 180 g L^{-1} in 48 hours, corresponding to a starch conversion ratio of 94.8%. The apparent glucose production rate reached a maximum value of 9.42 g L^{-1} h^{-1}. For ungelatinised flour, the apparent rate of glucose production was considerably lower at 4.72 g L^{-1} h^{-1}, though conversion to the same level was achieved. It can therefore be concluded that gelatinisation (which, at industrial scale, will involve liquid

Table 2. Average outputs of glucose and FAN for different levels in the feed

	Glucose (g/L)	FAN (mg/L)	Starch conversion ratio (%)	FAN productivity (mg FAN/g flour)
Wholemeal flour	51.4	98.8	94.9	1.24
25% gluten removed	48.0	75.5	88.8	0.94
50% gluten removed	45.5	55.9	84.0	0.70
75% gluten removed	42.5	33.6	78.5	0.42
gluten free flour	39.8	8.05	73.5	0.10

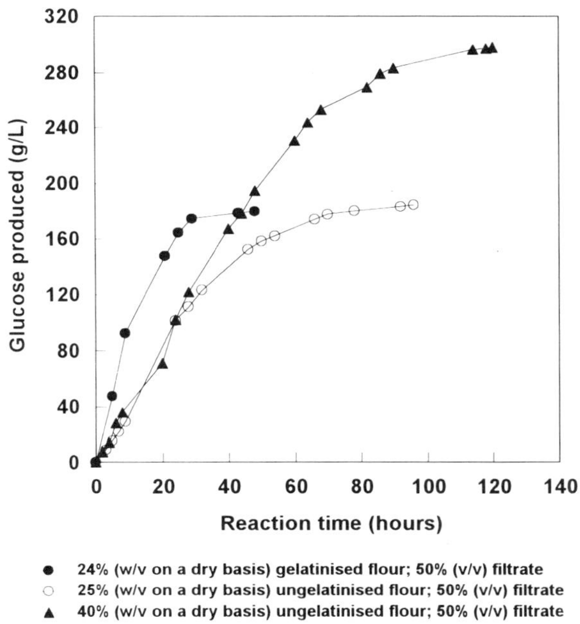

Figure 4. Comparison of glucose enhancement between experiments on gelatinised and ungelatinised gluten-free flour.

transportation equipment as well as heating and cooling systems - for a flour mash of very high viscosity) is not strictly necessary. A further test on ungelatinised flour at a higher concentration, unattainable for gelatinised flour, showed satisfactory conversion to around 300 g L^{-1} over a similar period.

4.4. Cell Autolysis

Continuous fermentation on 8% wholemeal flour produced an effluent containing 51.4 g L^{-1} glucose and 98.8 mg L^{-1} FAN on average. After glucose concentration had been increased to desired levels by the glucose enhancement operation, the comparatively low concentration of assimilable nitrogen source became the main challenge to the production of a suitable generic fermentation feedstock. Considering that a large number of micro-nutrients and macro-nutrients (including nitrogen) present in the original flour will have been incorporated into the fungal cells during continuous fermentation, an economical method of converting these into assimilable forms would be of great benefit. Studies were therefore carried out to investigate the possibility of achieving autolysis of cells separated from the fermentation effluent.

The conversion of microbial cells into digestible materials for other micro-organisms requires not only the degradation of the intracellular materials but also the disruption of the cell wall. Theoretically, the cell wall, as the defence of a living system, is more resistant to physical and chemical attack than the cell membrane and cytoplasm. Cell wall disruption, therefore, basically involves the application of intense physical, chemical or enzymatic means. From the viewpoint of industrial applications, physical methods usually result in high capital and operating costs (Chisti and Moo-Young, 1986) whereas the principal problems associated with chemical treatments are the recovery of the chemicals used, potential cost and toxicity. Enzymatic approaches may be generally divided into two categories, namely, the application of foreign enzymes capable of promoting cell lysis, and cell autolysis. A wide range of lytic enzymes are known. They are of potential commercial interest due to their gentle and selective nature. However, their current cost and difficulty of reuse limit their application at industrial scale. Cell autolysis, on the other hand, is a natural degradation process which starts after the exhaustion of major nutrients and reserves. The degradation process takes place not only in the cytoplasm (Isaac and Gokhale, 1982) but also in the cell wall (Mitchell and Sabar, 1966). During autolysis, nitrogen, in the form of free amino acids and ammonia; carbon, mainly in the form of glucose; and minerals, including phosphate, magnesium and potassium, are released from the cells into the surrounding liquid phase.

In this study, the possibility of using cell autolysis as an approach for the conversion of *A. awamori* cells or cell components into digestible materials was investigated. Experiments showed that pH in the range 3 - 7 had little effect on either FAN formation or dry weight reduction, while temperatures in the range 25 - 55°C had considerable influence, with higher temperatures leading to higher reaction rates. The results of an experiment carried out at 45°C are shown in Figure 5. Using this approach, solutions containing up to 500 mg L^{-1} FAN were produced.

4.5. Evaluation of the Feedstock

In order to evaluate the product as a potential fermentation feedstock, a number of fermentations have been carried out. These include yeast growth, ethanol production, lactic acid production and glycerol production. It was apparent, from these trials, that the ratio of autolysate in the feedstock controlled glucose consumption and cell growth,

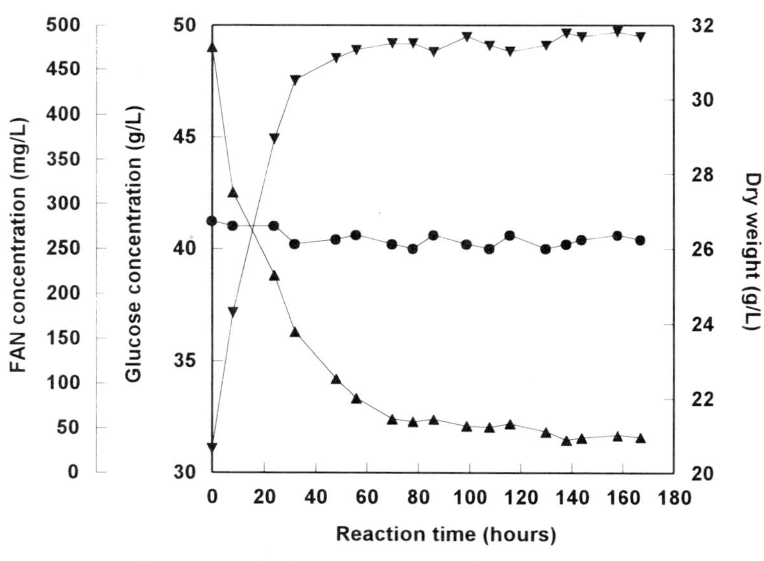

Figure 5. Results of typical operation of cell autolysis at 45°C.

suggesting that the fermentations were essentially nitrogen limited. Figure 6 illustrates the dependence of glucose consumption (as % autolysate in the medium) for a typical yeast fermentation. The exhaustion of glucose in the medium containing 40% autolysate suggests that the cell autolysis operation released not only nitrogen but also the other essential micronutrients for normal yeast growth. In ethanol fermentations, lower amounts of autolysate were required due to the lower cell growth. Ethanol yields in these fermentations were of the order of 96% of the theoretical maximum.

5. THE PROCESS

Based on the results outlined above, a detailed process scheme has been put together for the production of a generic fermentation feedstock from whole wheat flour. The process (Figure 7) includes a gluten extraction stage followed by continuous fermentation of an 8% flour suspension to produce the necessary amylolytic enzymes for hydrolysis of the starch to glucose. The liquid fraction of the effluent from this fermentation is then used to convert further flour suspensions into a glucose-rich stream, while the solids pass to an autolysis stage, where amino nitrogen and other cell based nutrients are released. Only one tonne of flour in 12 is required to pass through the continuous fermentation and autolysis stages, the other 11 tonnes can be treated directly through gluten separation followed by glucose enhancement. The nitrogen-rich stream from the autolysis operation is blended as required with the glucose-rich stream to produce the complete feedstock.

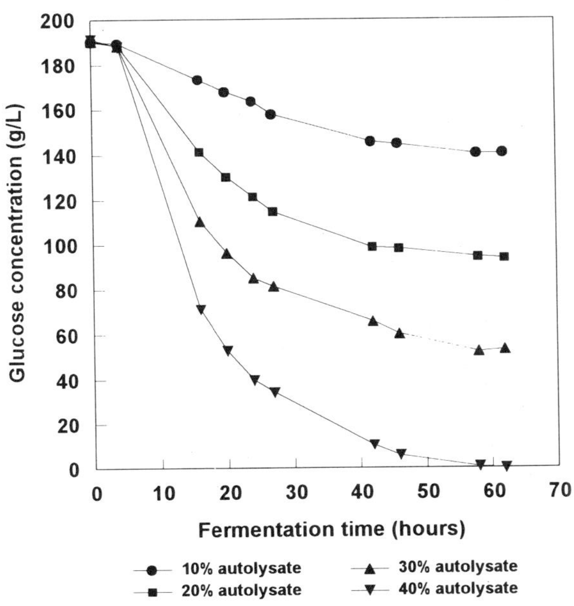

Figure 6. Effects of FAN concentration on glucose consumption during incubation of *Saccharomyces cerevisiae.*

Table 3. Costs of various carbon sources for the production of 1 m^3 feedstock

Materials	Market price (£/T)	Expenditure (£)	Sales (£)	Net cost (£)
wheat	125	46.73		
gluten	800		30.16	16.57
wheat starch	370	97.07		97.07
glucose	775	193.15		193.15
glucose syrup (80.4% DS)	400	123.99		123.99

5.1. Economic Considerations

The relative costs of a variety of potential fermentation feedstocks are compared in Table 3, which shows wheat to be significantly less expensive than the alternatives. In addition, the feedstock produced by the proposed process contains all the necessary nutrients for many different fermentations. All the others, by contrast, must be combined with an appropriate nitrogen source and other minor nutrients, making the financial advantage of the whole wheat raw material even greater. There are, of course, substantial processing costs associated with the use of raw wheat grain. At this stage, it is difficult to estimate the operating costs for the whole process, but the enormous difference in raw material costs between this process and traditional alternatives is likely to be sufficient to cover operating costs. Furthermore, this process also provides environmental advantages over conventional technology because it consumes no extra materials beyond the wheat flour and produces no waste products.

6. CONCLUSIONS

The feasibility of producing a generic fermentation feedstock from whole wheat flour has been demonstrated by the results presented in this work. Batch fermentations with *A. awamori* showed that satisfactory growth and glucose production could be achieved using whole flour, gluten-free flour and bran-free flour. Based on preliminary kinetic analysis, continuous fermentations with residence times between 18 and 20 hours were carried out, which successfully produced an effluent with constant concentrations of glucose and FAN. A glucose enhancement operation was developed to increase glucose

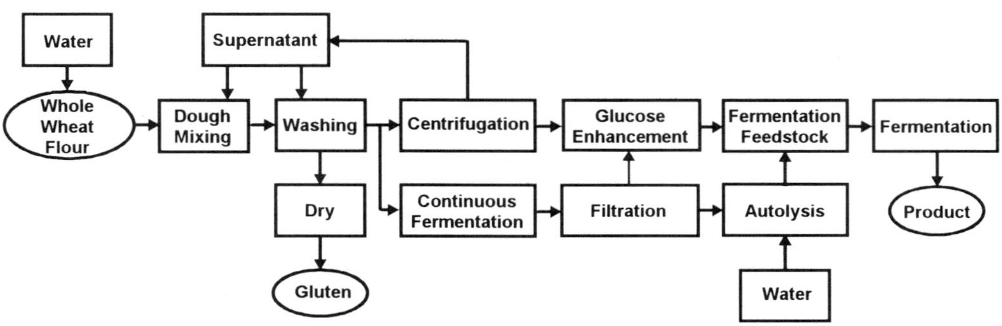

Figure 7. Proposed process for the production of a generic fermentation feedstock from whole wheat flour.

concentration in the effluent filtrate to levels of up to 320 g L^{-1}, by the further utilisation of the amyloglucosidase produced during the fermentation stage. Autolysis of the solids from the continuous effluent provided a potential approach for the production of a nitrogen-rich nutrient stream by effectively releasing all the essential nutrients from the fungal cells.

A. awamori demonstrated not only the ability to hydrolyse wheat starch into glucose and wheat gluten into FAN but also the capability of disintegrating the aleurone layer in wheat bran. The complete destruction of aleurone cells provided an opportunity for the assimilation of the nutrient components of these cells, resulting in less starch consumption for growth of the fungus. Consequently, the conversion ratio from starch to glucose in continuous fermentations was almost 95%. In addition, the release of protein and vitamins from the aleurone cells obviously accelerated the fungal growth so that the residence time for continuous fermentation was considerably reduced from the original estimates. Comparison between batch fermentations on more finely ground flour and on the original stone ground flour suggested that small particle size positively benefited the disintegration rate of the aleurone layer. This suggests that a simple dry milling operation, which provides smaller bran specks, rather than the traditional wheat milling strategy could be used in this process and could lead to a further reduction in the residence time of the continuous fermentation.

Without the addition of any extra nutrient, the feedstock generated by the proposed process was successfully used for several subsequent fermentations, thus confirming its applicability. In general the nutrient requirement of each particular industrial fermentation is different, so various mixtures of the autolysate and the glucose-rich liquor would be used to produce individually suitable feedstocks. Taking into account that the maximum glucose concentration from the glucose enhancement operation was 320 g L^{-1}, a medium composed of 40% autolysate would result in a medium with a glucose concentration of about 194 g L^{-1}, and FAN concentration of about 145 mg L^{-1}. Any requirement to increase glucose or nitrogen concentrations further would necessitate improving the unit operations of glucose enhancement and/or cell autolysis, or carrying out further operations such as evaporation. For fermentations requiring less nitrogen, feedstock composition might easily be modified by adjusting the ratio of autolysate to glucose solution. It would also be possible to use the autolysate as a nitrogen source in fed-batch fermentations, in order to control nitrogen concentration at limiting levels for the production of various products, such as glutamic acid, active dried yeast and so on.

Finally, preliminary mass balance calculations based on the results obtained from *S. cerevisiae* fermentations on the feedstock showed that the FAN released by continuous fermentation and then autolysis, from one tonne of wholemeal flour, is more than the amount of FAN required for the consumption of all the glucose produced from 11 tonnes of wheat. In other words, 92% of the gluten from the original raw material might be separated as a by-product, thereby alleviating some of the operating costs of the process. Even without this added advantage, and the fact that it is environmentally clean, the relative costs of raw materials for this and traditional processes strongly favour the proposed process.

REFERENCES

Bradbury D, MacMasters MM and Cull I (1956) "Structure of the mature wheat kernel. III. Microscopic structure of the endosperm of hard red winter wheat." Cereal Chemistry 33, 361–373

Chisti Y and Moo-Young M (1986) "Disruption of microbial cells for intracellular products." Enzyme Microb. Technol. 8, 194–204

Frost J (1996) "Renewable feedstocks." The Chemical Engineer, 16 May, 1996, 32–35

Fulcher RG, O'Brien TP and Lee JW (1972) "Studies on the aleurone layer. I. Conventional and fluorescence microscopy of the cell wall with emphasis on phenol-carbohydrate complexes in wheat." Aust. J. Biol. Sci. 25, 23–29

Hinton JJC (1959) "The distribution of ash in the wheat kernel." Cereal Chemistry 36, 19–31

Kostka G and Kaczkowski J (1989) "Oxygen requirements of *Aspergillus awamori* fungi in the process of glucoamylase biosysnthesis." Acta. Biotechnol. 9(3), 227–231

Isaac S and Gokhale AV (1982) "Autolysis, a tool for protoplast production from *Aspergillus nidulans*." Trans. Br. Mycol. Soc. 78(3), 389–394

Mitchell R and Naama Sabar (1966) "Autolytic enzymes in fungal cell walls." J. Gen. Microbiol. 42, 39–42

Shetlar MR, Rankin GT, Lyman JF and France WG (1947) "Investigation of the proximate chemical composition of the separate bran layers of wheat." Cereal Chemistry 24, 111–122

26

THE EFFECT OF NUTRIENTS AND α-AMYLASE INACTIVATION ON THE FERMENTATIVE LACTIC ACID PRODUCTION IN WHOLE WHEAT FLOUR HYDROLYSATE BY *LACTOCOCCUS LACTIS* SSP. *LACTIS* ATCC 19435

Karin Hofvendahl and Barbel Hahn-Hagerdal

Department of Applied Microbiology
Lund Institute of Technology/Lund University
PO Box 124, S-221 00 Lund, Sweden

1. INTRODUCTION

Lactic acid is a widely used chemical in the pharmaceutical and food industries (Vickroy, 1985). It is produced either by fermentation or by chemical synthesis (Vickroy, 1985). For the fermentative production, lactic acid bacteria can use various sugars such as glucose, lactose, maltose and sucrose, from whey, molasses and starch wastes (Vickroy, 1985; Atkinson and Mavituna, 1991). The synthetic production demands precursors such as lactonitrile. Lactic acid naturally occurs in two optical isomers, D-(-)- and L-(+)-forms. The D-isomer is harmful to humans (Expert Committee on Food Additives, 1967), and therefore the L-form is more useful. The synthetic production results in a racemic mixture of the two isomers, while fermentative production can yield either form alone or a racemace, depending on the organism used.

Poly lactic acid, PLA, is a biodegradable polyester produced by the polymerisation of lactic acid. It is used for medical applications such as self-degradable prosthetic devices or clips and sutures for wound closure (Kharas et al, 1994). Depending on the optical purity of the lactic acid used for polymerisation, PLA with different properties can be produced. Pure D- and L-polymers are crystalline and more stable than amorphous, racemic polymers (Lipinsky and Sinclair, 1986; Kharas et al, 1994), and they therefore find applications in different areas. PLA can be further diversified if other substances, e.g. glycolyc acid, are copolymerised with lactic acid.

Lactic acid bacteria have complex growth factor requirements because of limited synthesis abilities, and require nitrogen and B-vitamins in large amounts. In industrial processes nutrients are added such as corn steep liquor or malt sprouts (Vickroy, 1985; Atkinson and Mavituna, 1991), but yeast extract, milk and hydrolysed whey proteins have

Cereals: Novel Uses and Processes, edited by Campbell *et al.*
Plenum Press, New York, 1997

also been used (Vickroy, 1985; Lund et al, 1992). In order to upgrade starch fractions lacking technically useful polymer qualities, hydrolysed whole-wheat flour has been used as a substrate for L-lactic acid production by several lactic acid bacteria (Hofvendahl and Hahn-Hägerdal, 1996). For *Lactococcus lactis* spp. *lactis* ATCC 19435 the flour, containing both gluten and bran, was found to contain all necessary nutrients (see Chapter 25 for a similar conclusion regarding the merits of whole wheat flour for fermentations). The yield and productivity of lactic acid were almost the same with and without yeast extract added, 88%, 3.3 g/Lh and 76%, 3.0 g/Lh, respectively. Only the L-form, with no byproducts, was produced in both cases.

In the development of an industrial process, a simulation model of the fermentation is useful, such as that described in Chapter 8. Kinetic parameters, e.g. saturation constant, maximum production rate and yield factor, have to be determined. For that purpose a fermentation medium which permits the determination of cell mass is needed. It is also important that the sugar concentrations remain constant. The whole wheat flour contains large particles, which make it impossible to measure the cell mass, and the hydrolytic enzymes continuously produce more sugars during the fermentation. On the other hand, a medium for obtaining kinetic parameters should to the largest possible extent resemble the hydrolysate to be used in an industrial process. Therefore a synthetic medium would not be preferred. In order to design a suitable medium, cultivations were performed in whole wheat flour hydrolysates (WFH) and in supernatants after centrifugation of WFH (SUP), with the initial glucose concentration adjusted to 40 - 60 g/L and at constant pH of 6.0. The enzyme was either left active, or was inactivated with heat or acid. Yeast extract was added to investigate whether the flour was limited in any nutrient. Cultivations of unhydrolysed flour (UF) and a reference medium (RM) containing all required nutrients were also carried out.

2. MATERIALS AND METHODS

Whole wheat flour, with the coarsest fraction of the bran sieved off, was suspended to a concentration of 0.32 kg/L in water at 50°C. The enzyme α-amylase Termamyl 120 L (Novo Nordisk, Bagsvaed, Denmark) was added to a concentration of 62.5 μL/L (Hofvendahl and Hahn-Hägerdal, 1996). The suspension was heated to 95°C over a 30 minute period, and held at this temperature for 20 minutes. Following this, the hydrolysate (WFH) was cooled and diluted with water to a concentration of 0.22 kg/L, and the pH was adjusted to 6.5. In some experiments the α-amylase was inactivated before the hydrolysate was diluted. The hydrolysate was either heated to 120°C in an autoclave for 5 minutes and then cooled and diluted, before it was centrifuged, or the pH was decreased to 4.0 and kept there for 1 hour during stirring, after which the pH was adjusted to 6.0 and the hydrolysate diluted. In the latter case no centrifugation took place. In some of the experiments the hydrolysate was centrifuged at 6370 rpm (heat-inactivated enzyme) or 17700 rpm (active enzyme) for 30 minutes at 4°C, and the supernatant (SUP) was used for the fermentations. In all cultivations glucose (BDH, Poole, England) was added to a final concentration of 40 - 60 g/L. In some experiments yeast extract, YE, (Difco, Detroit, MI, USA) was added to a concentration of 5 g/L.

In two experiments the flour was only suspended in water (UF) to a concentration of 0.32 kg/L. Glucose, and in one case YE, was added as stated above. The reference medium (RM) had the following composition per litre: yeast extract 5 g (Difco, Detroit, MI, USA), bacto tryptone 5 g (Difco, Detroit, MI, USA), casamino acids 1 g (Difco, Detroit,

MI, USA), K_2HPO_4 2.5 g (Merck, Darmstadt, Germany), KH_2PO_4 2.5 g (Merck, Darmstadt, Germany), $MgSO_4 \cdot 7H_2O$ 0.5 g (Merck, Darmstadt, Germany), and glucose 40 g (BDH, Poole, England).

The microorganism *Lactococcus lactis* ssp. *lactis* ATCC 19435 (Shattock and Mattick, 1943), a type strain, was obtained from the American Type Culture Collection (Rockville, MD, USA). The inoculum was prepared in three steps of 24 hour cultivation at 30°C each. Bacteria stored at -80°C were transferred to a RM-agar plate. In the next step, one single colony was transferred to 5 ml of a medium of the same composition that was going to be used in the cultivation. The cells were harvested by centrifugation for 2 min at 5700 rpm (Wifug, Doctor, Stockholm, Sweden) and resuspended in 30 mL fresh medium. The final inoculum was treated in the same way. The batch fermentations were carried out in one litre double-walled thermostated vessels, sealed with a rubber stopper with ports for pH electrode (Schott Geräte H63, Germany), sampling device, base addition and ventilation. The working volume was 500 ml and the temperature 30°C. The broth was kept mixed by stirring at 150 rpm using a magnetic stirrer and a magnetic rod. The pH was kept constant at pH 6.0 by the addition of 20 % (w/v) NaOH (Eka Nobel, Bohus, Sweden), using a pH meter and titrator (pHM61 and TTT80, Radiometer, Copenhagen, Denmark).

Sterile samples were taken regularly. The number of cells were analysed as viable counts, plating series of diluted samples on RM-agar plates, and propagating them for 48 hours at 30°C. In the SUP and RM fermentations the optical density was also measured, using water as a reference. Double analyses of the dry weight were carried out in the RM fermentation only. The concentrations of glucose, maltose, lactate, formate, acetate and ethanol were analysed by HPLC (Glison, Middletown, WI, USA), using an ion-exchange column, Aminex HPX 87-H (BioRad Laboratories, Richmond, CA, USA), at 65°C and a refractive index detector, RID-6A (Schimadzu, Kyoto, Japan). Sulphuric acid 5 mM was used as mobile phase, and the flow-rate was 0.6 ml/min. All samples were injected twice. The samples containing flour were centrifuged at 13000 rpm for 1–2 minutes, and the supernatant was used for the HPLC analyses. All samples were diluted 10 or 20 times, and then filtered through 0.2 mm cellulose acetate filters (Sartorius, Göttingen, Germany). Double injected standard samples were used to make a calibration curve, integrating the area under each peak, using the software UniPoint (Glison, Middletown, WI, USA).

3. RESULTS

In order to find a medium suitable for the study of kinetic parameters, i.e. a medium allowing the cell mass to be determined and in which no hydrolysis occurred, eleven cultivations were carried out (Table 1). The bacteria were cultivated in hydrolysed whole wheat flour (WFH) and in the supernatant after centrifugation of WFH (SUP), with and without the addition of extra nutrients in the form of yeast extract (YE+ and YE- respectively). Fermentations in unhydrolysed whole wheat flour suspended in water (UF), and in a reference medium (RM) containing all required nutrients were also carried out. The hydrolysates were either used with the enzyme α-amylase still active (enz+), or after inactivation (enz-) with either heat or acid. The initial glucose concentration was adjusted to 40 - 60 g/L in all experiments. The hydrolysates contained 40 - 60 g/L of maltose in addition to glucose, which the RM did not. The variations in maltose concentrations are due to the fact that each hydrolysis is individual, resulting in slightly different distribution patterns of the starch break-down products. The UF contained a small amount of maltose, 4 - 6 g/L.

Table 1. Results of the fermentations with L*actococcus lactis* ssp. *lactis* ATCC 19435

conditions	time[a] (h)				glu (g/l)		mal (g/l)		lac (g/l)		$Y_{L/G}$[c]		$Y_{L/GM}$[c]	v c (CFU/ml)		Byprod[d]	Figure 1
	lag	glu	mal	final	initial	used[b]	initial	used[b]	glu	final	glu	final	final	initial	final		
RM	9	22	-	33	39	39	1	1	32	32	0.90	0.87	0.80	7E+5	5E+9	-	A
UF, YE+	4	17	-	42	50	50	4	3	28	32	0.64	0.68	0.64	5E+7	3E+11	A, E, F	-
WFH, YE+, enz+	<9	16	-	39	51	50	54	1	40	42	0.89	0.93	0.83	2E+7	9E+9	-	-
WFH, YE-, enz+	10	70	105	161	41	41	61	58	32	53	0.83	1.5	0.54	6E+5	3E+7	A,E,F	-
WFH, YE+, enz-	<10	22	-	156	38	38	58	13	43	58	1.3	1.7	1.1	6E+5	2E+9	A	B
WFH, YE-, enz-	63	138	-	204	43	43	57	0	38	43	1.0	1.0	1.0	4E+5	1E+9	A	C
SUP, YE+, enz+	<10	22	153	161	38	38	60	57	37	64	1.1	2.0	0.67	4E+3	1E+8	A,E,F	D
SUP, YE-, enz+	9	~70	-	44	36	14	61	0	-	8	-	0.57	0.57	9E+6	4E+9	-	-
SUP, YE+, enz-	18	62	-	109	42	42	43	23	35	44	1.0	1.2	0.68	1E+3	3E+9	-	-
SUP, YE-, enz-	25	~250	-	109	35	11	38	0	-	11	-	1.0	1.0	5E+3	3E+8	-	-

[a] lag = lag phase, when glucose consumption or lactate production starts
glu, mal = when consumption stoppes
final = total time of experiment

[b] concentrations adjusted for dilution caused by base addition, not the actual concentrations

[c] yields in g lactate per g glucose or g glucose and maltose utilised
calculated using concentrations adjusted for dilution caused by base addition
glu = when glucose consumption ends

[d] byproducts: A = acetate, E = ethanol, F = formate

In Table 1 the concentrations of glucose, maltose and lactate, yields and cell numbers are summarised, in addition to information about byproducts formed. In two fermentations, the time needed for total glucose utilisation had to be approximated by extrapolation (indicated by ~). In the RM fermentation all glucose was consumed in 22 hours, and 32 g/L of lactate was produced (Figure 1 A). The increase in viable counts was about 10^4 CFU/ml. Similar results were obtained in the UF fermentations. However, the yields were lower and byproducts such as formate, acetate and ethanol were also produced.

The WFH was centrifuged to obtain a medium in which measurements of optical density and the dry weight of the cells were possible. In the WFH fermentations all glucose was consumed regardless of the presence of YE, in contrast to the SUP fermentations. When YE was added, all glucose was consumed also in the SUP fermentations. YE decreased the time needed for total glucose consumption to 15–20 hours in all studied media, the exception being the (SUP, YE+, enz-) fermentation, where it took 60 hours. The fermentations of hydrolysates without YE needed at least 70 hours for complete glucose consumption. Thus, YE compensates for the removed flour in the SUP, making total glucose consumption possible.

The α-amylase used for hydrolysing the flour was inactivated (enz-) by acid or heat, to obtain a medium with stable concentrations of maltose and glucose, suitable for kinetic studies. The enzyme inactivations were not totally effective: the glucose and maltose concentrations increased initially when no YE was added to the medium, Figure 1 C, but not when YE was present, Figure 1 B. This could be due to the rapid initial lactate production

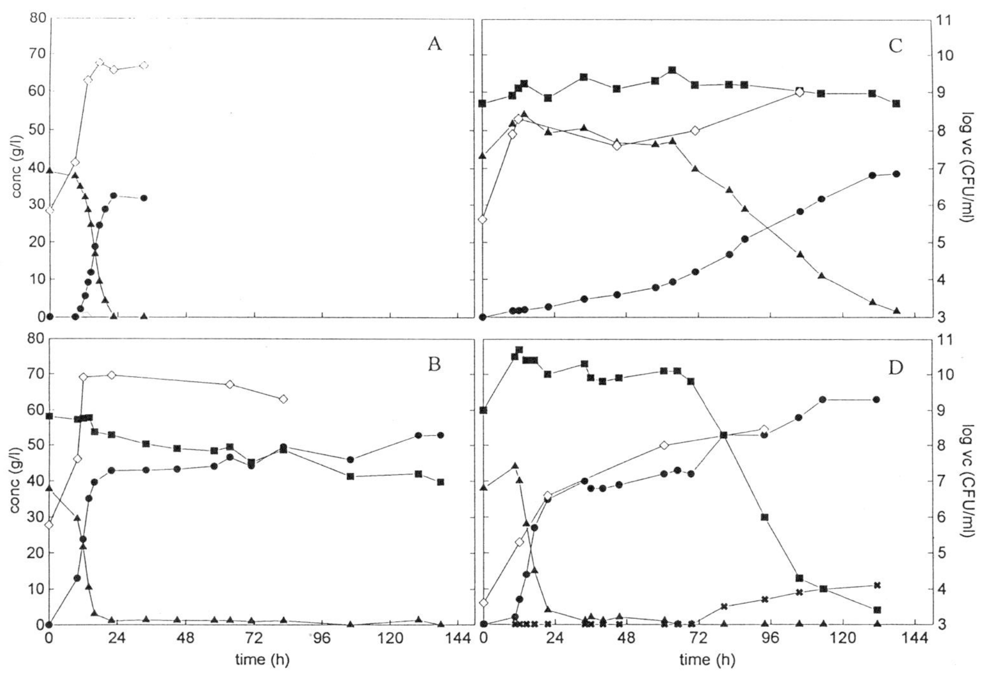

Figure 1. Maltose and glucose utilisation, lactate, and formate production, and viable count in RM **A**, (WFH, YE+, enz-) **B**, (WFHm YE-, enz-) **C**, and (SUP, YE+, enz+) **D**. Maltose (-■-), glucose (-▲-), lactate (-●-), formate (-⌘-), and viable count (-◇-).

which could mask the increase with equal usage. Presence of active enzyme (enz+) decreased the lag phase and the time needed for total glucose consumption.

In the (SUP, YE+, enz+) fermentation, glucose was first totally consumed, with subsequent use of maltose, Figure 1 D, and formate was produced in considerable amount, 11 g/L. Small amounts (< 5 g/L) of ethanol and acetate were also formed. The production of formate, acetate and ethanol, and the utilisation of maltose, started simultaneously. During maltose consumption, approximately equimolar amounts of formate and lactate were produced. Similar results were obtained in the (WFH, YE-, enz+) fermentation. The same three byproducts were found in the two UF fermentations, but in smaller amounts, < 2 g/L. Acetate alone was produced in the two cultivations on WFH with inactivated enzyme (enz-), again in small amounts, 2 g/L.

The cell mass was determined by three different methods: viable counts, optical density and dry weight. Dry weight was only measurable in the RM, and viable count was the only possible method in the WFH and UF. For modelling purposes, dry weight is the preferred method. It correlated well with the lactate concentration (Figure 2 A), whereas the log viable counts initially increased faster than the dry weight (Figure 2 B). The log viable counts and optical density correlated well (Figure 2 C).

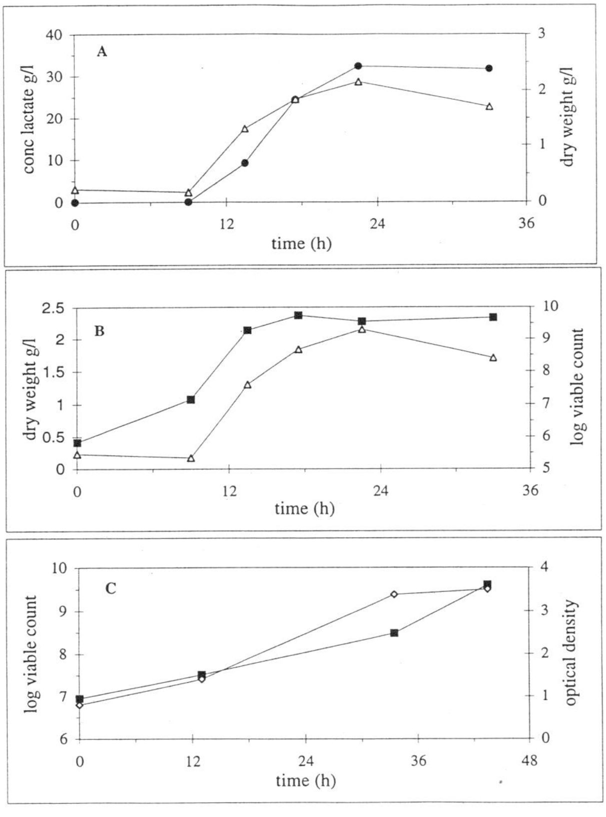

Figure 2. Correlation between lactate concentration (-●-) and dry weight (-▵-) in the RM fermentation **A**, dry weight (-▵-) and log viable counts (-■-) in the RM fermentation **B**, and between log viable counts (-■-) and optical density (-◇-) in the (SUP, YE-, enz+) fermentation **C**.

4. DISCUSSION

The enzyme inactivation increased the time required for complete glucose consumption, and centrifugation had the same effect. This suggests that the nutrients in the flour were denatured or precipitated by the enzyme inactivating treatments heat and acid. This is further supported by the short times obtained in the unhydrolysed flour. Similarly, nutrients appear to be bound to large particles in the flour. In concordance with this, YE made complete glucose consumption possible also in the SUP. Hydrolysed barley flour has been reported to contain enough nutrients for lactic acid fermentation (Javanainen and Linko, 1995).

Cell growth measured as viable counts was higher when YE was present than in its absence. Viable count measurement is a crude method, but the only one possible in the WFH and UF fermentations. Neither optical density nor dry weight measurements were possible due to the high particle content. Dry weight measurements were not possible in the SUP fermentations, due to microscopic particles clogging the filter. However, optical density was measured and correlated rather well with the viable counts. In the reference medium all three analyses were carried out, but dry weights and viable counts did not correlate. Dry weight measurements are most suitable for modelling, but any method that correlates well with the dry weight can be used. The dry weight correlated well with the lactate concentration in the RM fermentation. For the kinetic studies a medium with maximum resemblance to the one to be used in the process, WFH or UF, should be used, and the cell number could be measured as lactate concentration correlated to dry weight.

For the kinetic studies stable and known sugar concentrations are a prerequisite. Therefore, various methods to inactivate α-amylase were investigated. None of the enzyme inactivations were fully effective. An initial increase in maltose concentration was observed in all cases, except in the (SUP, YE+, enz-) fermentation, in which case the maltose concentration decreased slowly throughout the entire fermentation. This indicates enzyme activity, since *Lactococci* do not normally use glucose and maltose simultaneously, when present in large amounts (Qian et al, submitted). Normally glucose is used first, and then maltose, as seen in the (SUP, YE+, enz+) fermentation. A comparison of the two fermentations of SUP with YE further supports the notion that the decrease in maltose concentration, seen in the experiment with inactivated enzyme, is due to hydrolysis rather than to fermentative action by the bacteria. The maltose concentration in the (WFH, YE-, enz-) fermentation decreased to the initial level, therefore no maltose consumption is seen. In the (SUP, YE-, enz-) fermentation, the glucose concentration also increased initially.

The stability and activity of Termamyl are dependent on several factors, e.g. temperature, pH, Ca^{2+} concentration, amount of dry substance and conductivity (Rosendal et al, 1979). Decreasing pH and increasing temperature have negative effects on the activity and stability of the enzyme, but the effect is also time dependent. Probably, treatment at 120°C for 5 minutes was not enough for total inactivation of the enzyme. Time was kept to a minimum due to the risk of toxic Maillard product formation. Apparently, decreasing pH to 4.0 for 1 hour was not totally effective either. An alternative solution is to exclude the enzymatic hydrolysis, as in the UF fermentations. All sugar concentrations are kept stable, and almost no maltose is present. However, the UF medium could not be heated, since this would result in so-called retrograded starch when cooled, which is a non-viscous gel and precipitate (Swinkles, 1985). Treating starch with hot water results in the unfolding and dispersion of the starch molecules from the granules, and in this aspect, the UF medium was less similar to the one to be used in the process. The process conditions most closely resemble the (WFH, YE-, enz+) fermentation, where fermentation and hydrolysis take place simultaneously (Hofvendahl and Hahn-Hägerdal, 1996).

Compared with the yields obtained when subsequent hydrolysis and fermentation occurred, 0.88 and 0.76 with and without YE, respectively (Hofvendahl and Hahn-Hägerdal, 1996), the yield of the RM fermentation, 0.85, seems rather low. However, these yields were based on a theoretical amount of sugar in the medium, while the yield in the RM fermentation was based on actual measurements. The yields in the present study were based on the difference between the initial concentration and that at the end of glucose utilisation, or the final concentration. Therefore, maltose or glucose produced through hydrolysis were not taken into account, and the yields could be overestimated. Thus, some yields are even higher than the theoretical. In the (UF, YE+), (UF, YE-), (WFH, YE-, enz+) and (SUP, YE+, enz+) fermentations, where formate, acetate and ethanol were formed, the yields were lower than those obtained in the RM fermentation, since some of the sugar was utilised for byproduct formation. In the UF fermentations, byproduct formation could not account for the entire sugar consumption. A relatively high cell growth was observed in the (UF, YE+) fermentation.

Formate, acetate and ethanol were formed in four cultivations. In the (WFH, YE-, enz+) and (SUP, YE+, enz+) fermentations, glucose was first entirely consumed and subsequently maltose was used. The molar amounts of byproducts produced during maltose utilisation were 1 lactate : 1 formate : 0.5 ethanol, with acetate being produced to an even lesser extent than ethanol. Byproduct formation started simultaneously with maltose utilisation. In the UF fermentations, the initial amount of maltose was much lower, only 4 and 6 g/L compared to 60 g/L in the others. The utilisations of maltose and glucose took place simultaneously. The molar amounts of byproducts were much smaller than the amount of lactate. The probable pathway used is the mixed acid fermentation, which *Lactococci* are known to use for the fermentation of maltose (Lohmeier-Vogel et al, 1983; Qian et al, 1994). In this pathway pyruvate is both converted into lactate by lactate dehydrogenase (LDH), and into formate, ethanol and acetate by the action of pyruvate formate lyase (PFL), in the molar ratios of 1 formate : 0.5 ethanol : 0.5 acetate. An alternative route is the conversion into lactate by LDH, and into acetate by pyruvate dehydrogenase (PDH), leading to an additional production of NADH and CO_2 (Fordyce et al, 1984). Physiologically, the cell conserves additional energy, since one ATP is produced together with each acetate. When the bacteria switches from growing on abundant glucose to maltose, transport over the cell membrane and subsequent conversion into glucose 6-phosphate, the starting point of the Embden-Meyerhof pathway, requires more energy (Sjöberg et al, 1995). LDH is activated in times of surplus glucose by fructose-1,6-biphosphate, and PFL is simultaneously inhibited by trios-phosphates (Thomas et al, 1980), resulting in predominant lactate production. When glucose is no longer in surplus, the regulations are less expressed, and thus pyruvate meets its alternative fates. PDH is probably used in favour of PFL when reoxidation of NADH to NAD is not needed through ethanol or lactate production, e.g. if O_2 is present, but ATP is still required, and consequently homofermentation does not take place. The gene encoding the enzyme β-phosphoglucomutase, which catalyses one of the steps in the conversion of maltose to glucose and glucose-6-phosphate, has been cloned (Qian et al, submitted). This opens ways to create genetically engineered lactic acid bacteria, capable of more efficient maltose consumption, and thus with a higher intracellular concentration of fructose-1,6-biphosphate. This would probably lead to decreased or ceased byproduct formation, if the regulation is switched over in favour of LDH. Another advantage could be that the sugar present in the wheat flour is fermented faster, since it would not have to be converted into glucose first.

ACKNOWLEDGMENTS

This work was supported by the Swedish National Board for Industrial and Technical Development. We thank Liana Sandström for technical assistance.

REFERENCES

Atkinson B and Mavituna F (1991) "Industrial Microbial Processes Biochemical engineering and biotechnology handbook." New York, Macmillan Publishers Ltd, 1181–1183

Expert Committee on Food Additives (1967) "Lactic acid." WHO/Food Add 29, 144–148

Fordyce AM, Crow VL and Thomas TD (1984) "Regulation of product formation during glucose or lactose limitation in nongrowing cells of *Streptococcus lactis*." Applied and Enviromental Microbiology 48(2), 332–337

Hofvendahl K and Hahn-Hägerdal B (1996) "L-lactic acid production from whole-wheat starch hydrolysate using strains of *Lactobacilli* and *Lactococci*." Enzyme and Microbial Technology, accepted for publication.

Javanainen P and Linko Y-Y (1995) "Lactic acid fermentation on barley flour without additional nutrients." Biotechnology Techniques 9(8), 543–548

Kharas GB, Sanchez-Riera F and Severson DK (1994) "Polymers of lactic acid, Plastics from microbes, Microbal synthesis of polymers and polymer precursors.", DP Mobley, Munic, Vienna, New York, Hanser Publishers, 93–137

Lipinsky ES and Sinclair RG (1986) "Is lactic acid a commodity chemical?" Chemical Engineering Progress 82, 26–32

Lohmeier-Vogel E, Häggström M, Wittgren H-B and Hahn-Hägerdal B (1983) "Levels of metabolic intermediates in *Streptococcus lactis* grown on different carbon sources and the effect on product formation." Acta Chemica Scandinavica 37(8), 751–753

Lund B, Norddahl B and Ahring B (1992) "Production of lactic acid from whey using hydrolysed whey protein as nitrogen source." Biotechnology Letters 14(9), 851–856

Qian N, Stanley GA, Bunte A and Rådström P (submitted) "Carbohydrate metabolism in *Lactococcus lactis*, Cloning and characterization of a novel phosphoglucomutase gene"

Qian N, Stanley GA, Hahn-Hägerdal B and Rådström P (1994) "Purification and characterisaton of two phosphoglucomutases from *Lactococcus lactis* subsp. *lactis* and their regulation in maltose- and glucose-utilizing cells." Journal of Bacteriology 176(17), 5304–5311

Rosendal, P, Nielsen BH and Lange NK (1979) "Stability of bacterial alpha-amylase in the starch liquefaction process." Starch/Stärke 31(11), 368–372

Shattock PMF. and Mattick ATR (1943) "The sereological grouping of *Streptococcus lactis* (group N) and its relationship to *Streptococcus faecalis*." Journal of Hygiene, Cambridge 43, 173–188

Sjöberg A, Persson I, Quednau M and Hahn-Hägerdal B (1995) "The influence of limiting and non-limiting growth conditions on glucose and maltose metabolism in *Lactococcus lactis* ssp. *lactis* strains." Applied Microbiology and Biotechnology 42, 931–938

Swinkles JJM (1985) "Sources of starch, its chemistry and physics" In "Starch conversion technology" GMA. v Beynum and JA Roels. New York, Marcel Dekker, Inc. 14, 15–46

Thomas TD, Turner KW and Crow VL (1980) "Galactose fermentaton by *Streptococcus cremoris*, pathways, products, and regulation." Journal of Bacteriology 144(2), 672–682

Vickroy TB (1985) "Lactic acid: The practice of biotechnology, commodity products" HW Blanch, S Drew and DIC Wang, Elmsford, NY, Pergamon Press. 3, 761–776

27

AGRICULTURAL RESIDUES AND CEREALS AS FERMENTATION MEDIA

Margrethe Andersen and Pauli Kiel

Institute of Biomass Utilization and Biorefinery
South Jutland University Centre
Industrivej 11, DK-6870 Ølgod, Denmark

1. INTRODUCTION

In many traditional industrial processes based on biomass, the production only takes place in the harvest season and the products are main products such as starch, sugar, green pellets, potato chips etc. plus a residue used as fertilizer or animal feed stuff. From economical and environmental points of view such processes are not very rational. A year-round, integrated production with more biotechnologically-based products would seem more reasonable.

In order to work towards a more sustainable process system where all parts of the crops are used for production purposes, we have looked at agricultural residues as microbiological substrates. Using fermentation technology it is possible to produce all kinds of organic chemicals, such as organic acids, amino acids, enzymes, peptides, antibiotics etc. In this way it will be possible to plan new integrated process strategies combining the traditional production of, for example, starch from grain with fermentation technologies where new or traditional organic compounds are produced (Andersen et al, 1993).

In order to combine a green crop drying plant with production of cereal starch and different fermentation products we have studied the residue: green juice as a fermentation medium (Andersen and Kiel, 1991, 1995). The green juice is well suited for fermentation. It has a dry matter content of 3 - 12 per cent and contains all necessary nutrients. To form a good fermentation medium the green juice needs supplementation with carbohydrates. A potential process for production of fermentation medium from green juice and grain is presented in Figure 1; this process clearly has similarities with that proposed in Chapter 25. The green crops are wet separated into juice and press cake. The press cake is dried and used for animal feed. The juice is pretreated with a primary lactic acid fermentation and separated into leaf protein and the supernatant which, in combination with carbohydrates from cereal starch, forms the final fermentation feedstock.

After harvest the green crops are stored for a short period of time in the field and at the drying plant. Then the crops are pressed in order to remove some of the water content

Cereals: Novel Uses and Processes, edited by Campbell *et al.*
Plenum Press, New York, 1997

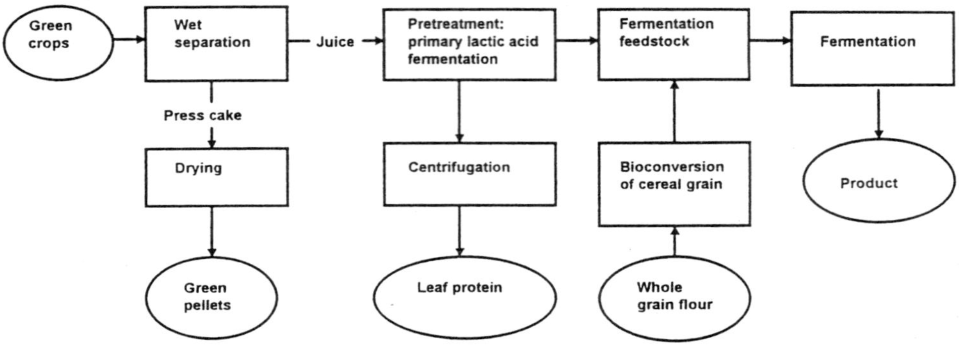

Figure 1. A potential process for production of a fermentation medium from green juice and grain.

before drum drying. The quality of the juice depends on the type of crops, the weather conditions, yearly season (cut 1, 2, 3 or 4), the pressing equipment and the storing time.

2. MATERIALS AND METHODS

The green juice was obtained as the residue from an industrial production of green pellets (Dangroent Products Ltd, Oelgod, Denmark). Concentrations of L-threonine and L-methionine were estimated using microbiological testing methods.

3. RESULTS

The green juice was evaluated as a fermentation medium for L-lysine production using *Corynebacterium glutamicum* ATCC 21253. This strain has experimentally determined cell mass yields of 0.0341 g cells per mg L-threonine and 0.127 g cells per mg L-methionine (Kiss and Stephanopoulos, 1992).

To obtain an ideal cell mass of 20 g cells per litre of medium for L-lysine fermentation, the medium must contain 586 mg L-threonine/litre and 157 mg L-methionine/litre.

In order to get as high a content of these amino acids in the juice as possible and to obtain a stable product, the juice was pretreated in different ways, as follows:

1. sterilization in an autoclave (121°C in 30 minutes)
2. lactic acid fermentation with a selected strain of *Lactobacillus salivarius* BT1001
3. acid hydrolysis in order to obtain the total amino acid content
4. centrifugation (10.000 g for 10 minutes) for separation in leaf protein pellet and supernatant
5. combinations of the above treatments.

Table 1 shows the results of pretreatment of Italian rye grass and alfalfa from two normal industrial produtions of green pellets. For comparison, the results are calculated on the basis of a dry matter content of 10 per cent. The results show that sterilization by heat does not change the content of bio-available amino acids in the juice. As sterilisation is a normal part of an industrial fermentation process, this fact is important. On the other hand,

Table 1. Content of bio-available L-threonine and L-methionine in raw and pretreated clover/grass and alfalfa juice, calculated on the basis of a dry matter content in the juice of 10 per cent

Crops	Alfalfa L-threonine mg/L	Alfalfa L-methionine mg/L	Clover/Grass L-threonine mg/L	Clover/Grass L-methionine mg/L
Raw juice	363	110	617	99
Sterilized (120°C in 30 min.)	334	101	592	72
Sterilized and lactic acid fermented	372	79	517	66
Acid hydrolysis, total	987	328	770	166
Supernatant from raw, lactic acid fermented juice	854	257	798	188
Supernatant, total hydrolysed	1010	228	-	-
Sediment from raw, lactic acid fermented juice	922	278	886	220
Sediment, total hydrolysed	1033	326	1133	388

Table 1 shows that sterilisation prior to lactic acid fermentation prevents the liberation of more bio-available amino acids.

Acidification using lactic acid bacteria resulted in sedimentation of some of the proteins and in partial hydrolysis of the other part of the proteins and peptides. In the case of alfalfa (Table 1) the content of bio- available L-threonine rose from 363 mg/L to 854 mg/L after lactic acid fermentation. The content of bio-available L-methionine rose by the same order of magnitude. The lactic acid fermentation resulted in a stable product. As shown in Table 3, it is possible to store the fermented juice for up to one year at normal outdoor temperatures without any loss in amino acid content.

Comparing the results from alfalfa and clover/grass in Table 1, it is clear that the results of pretreatment vary a lot. The reasons can include type of crop, harvesting time (cut 1, 2, 3 or 4), storage time after harvesting, weather conditions: temperature, wind, sun, rain etc.

Table 2 shows the first results from a more systematic study of parameters influencing the content of bio-available amino acids in green juice. These results show that storage for one day after harvest, prior to pressing, increases the content of available L-threonine

Table 2. Content of bio-available amino acids in juice from freshly harvested, cut and pressed Italian rye grass and in juice from grass stored one day after harvesting. All juices are calculated on the basis of a dry matter content of 10 per cent

	L-threonine mg/L	L-methionine mg/L
Juice from freshly harvested, 2nd cut (24/7)	-	32
cut and pressed grass 3rd cut (13/9)	121	20
4th cut (25/10)	216	45
Juice from grass stored one 2nd cut (24/7)	-	89
day after harvesting and 3rd cut (13/9)	411	118
cutting 4th cut (25/10)	459	123

Table 3. Stability of lactic acid fermented alfalfa juice after storage under anaerobic conditions

Storage after fermentation, (days)	0	70	146	182	210	237	266	382
Content of L-threonine (mg/L)	918	944	775	669	728	640	863	51
Content of L-methionine (mg/L)	276	206	266	192	197	263	248	246

and L-methionine. This is a possible explanation of the difference between alfalfa and clover grass in Table 1.

4. CONCLUSION

From the results of pretreatment and analysis it can be concluded that green juice is well suited as a fermentation medium for L-lysine fermentation. After pretreatment with lactic acid fementation it is possible to obtain high enough contents of bio-available essential amino acids, L-methionine and L-threonine, to reach a final cell mass concentration of 20 g/L and at the same time obtain a stable product that can be stored for months without loss of amino acid content. More investigations are planned in order to find out how the different factors influence the final content af amino acids and other essential components in the juice when used as a fermentation medium.

REFERENCES

Andersen M, Madsen AG and Kiel P (1993) "Biotechnological conversion of agricultural residues from agro-industries(Danish)" Danish Ministry of the Environment

Andersen M and Kiel P (1991) "Method for obtaining a culture medium from plant sap." PTC, No. WO 92,19716

Andersen M and Kiel P (1995) "Green Biomass for Fermentation in the Green Biorefinery." Poster, 7th European Congress on Biotechnology, Nice, France

Kiss RD and Stephanopoulos G (1992) "Culture Instability of Auxotrophic Amino Acid Producers." Biotechnol. Bioeng. 40, 75–85

28

FUNCTIONAL FOODS FOR HEALTH

Opportunities for Novel Cereal Processes and Products

Peter J. Wood

Centre for Food and Animal Research
Agriculture and Agri-Food Canada
Ottawa, Ont K1A 0C6, Canada

1. INTRODUCTION

The term "functional food" has been coined to describe a product which, as well as providing the normal attributes expected of food (basic sustenance and pleasing texture and flavour) also confers a specific health benefit or benefits. The terms "designer food" and "nutraceutical" are also used and highlight the fact that a population with a specific health problem might be targeted, much as with a pharmaceutical product. The underlying rationale for producing such foods is that many of the diseases of industrial, developed societies are life-style related, with diet playing a major role. The conditions take many years to develop and consequently prevalence increases with an ageing population. Once established, the illnesses, such as heart disease and cancer, are difficult or impossible to treat successfully. Diet-related disease is therefore an enormous social and financial burden on society and diet modification has become a worthwhile goal for that society. Individuals, however, respond variously to advice on diet and it is notoriously difficult to obtain a response, even when acute health problems such as diabetes are present. Nevertheless, all surveys in North America suggest that nutrition does now play a role in consumer choice (Wrick 1993).

In response, the "healthy choice market" has developed significantly, particularly in the area of calorie-reduced and low fat products, a market in which cereals play a significant role, for example, maltodextrin-based fat replacers. This article will focus, however, on more specific health claim based foods and the potential for cereals in such a market.

2. FUNCTIONAL FOODS AND HEALTH CLAIMS

The regulatory and marketing issues involved are complex, but given that health claims are allowed, the potential advantages of development of functional foods are:

- Reduction in health care costs and improved well-being of the population.
- Improved understanding and public awareness of diet - health relationships.

Cereals: Novel Uses and Processes, edited by Campbell *et al.*
Plenum Press, New York, 1997

- New opportunities for value-added processing of cereals.

Many countries are presently evolving a functional food policy. Japan is well advanced with their Foods for Specified Health Use (FOSHU) policy. In Canada, health claims for foods are not allowed. That is to say, such a claim would require evaluation and marketing of the product as a drug. In the USA, the Nutrition Labelling and Education Act (NLEA), has since 1990 developed certain categories for health claims.

Two of the allowed claims are for fibre-containing foods, in general, as promoting reduced risk of cancer and heart disease:

- fibre-containing grain products, fruits and vegetables and cancer (foods with this claim must be low in fat and contain 4 gram total dietary fibre per reference amount)
- fruits and vegetables and grain products that contain fibre, particularly soluble fibre (0.6 gram soluble fibre per reference amount) and risk of coronary heart disease.

(From the USA "Code of Federal Regulations" Volume 21-Food and Drugs; parts 100–161—published by Office of Federal Registrar—US Government.)

This chapter will deal with the heart disease claim and focus on an offshoot of the soluble fibre category, which is the acceptance by the FDA of a petition by the Quaker Oats Company for a specific health food claim for oatmeal and oat bran on the basis that this food may lower serum cholesterol levels, a risk factor for coronary heart disease. This claim is for a particular food and, indeed, for a particular component in that food.

Based on evaluation of 37 clinical trials, this is being referred to as a landmark decision, although at the time of writing a final decision had not been taken. A good many of the studies reviewed are difficult to interpret for a variety of reasons, such as: lack of appropriate control diet, lack of reliable diet records, changes in fat intake or weight of subjects during the study.

The following is an excerpt from what is now up for acceptance (Federal Register 1996, Vol. 61, No. 3, 296–313):

Nature of the claim. A health claim associating diets high in oatmeal or oat bran with reduced risk of coronary heart disease may be made...provided that:

(A) The claim states that oatmeal or oat bran "may" or "might" reduce the risk of heart disease.

(B) In specifying the disease, the claim uses the following terms: "heart disease" or "coronary heart disease".

(C) The claim states that:
(1) Diets high in oatmeal or oat bran may reduce the risk of coronary heart disease: and
(2) The effect of dietary intake of oatmeal or oat bran on the risk of coronary heart disease is particularly evident when these foods are consumed as part of a diet that is low in saturated fat and cholesterol.

(D) The claim does not attribute any degree of risk reduction for coronary heart disease to diets high in oat bran or oatmeal and low in saturated fat and cholesterol.

(E) The claim does not imply that consumption of oat bran or oatmeal is the only recognised means of achieving a reduced risk of coronary heart disease.

Nature of the food

(A) The food shall contain no less than 20 g oatmeal or 13 g oat bran that provides, without fortification, at least 1 g of ß-glucan soluble fibre per reference amount customarily consumed. Beta-glucan will be determined by...

---

3. CLINICAL STUDIES WITH OAT ß-GLUCAN

The key component by which the oat products are to be judged is ß-glucan or (1→3)(1→4)-ß-D-glucan. This is the predominant so-called soluble fibre of oats which most studies had suggested might be the active ingredient of oats. The following studies (Braaten et al, 1994, Wood et al, 1994a) were designed to test the hypothesis. Since there was no commercial supply, it was necessary first to isolate the ß-glucan on the large or pilot plant scale (Wood et al, 1989), to allow suitable animal and clinical trials. The product obtained, oat gum, contained about 80% ß-glucan. This material has been fully characterized (Doublier and Wood 1995, Wood et al, 1991a, Wood et al, 1994b).

The product was administered to hypercholesterolemic subjects in drink form in a placebo-controlled cross-over trial which minimized interference with the daily routine of the subjects. A significant mean reduction in LDL-cholesterol of about 10% in 4 weeks, without change in HDL-cholesterol, was observed (Braaten et al, 1994). The results are summarized in Table 1.

Viscous gums, and other dietary interventions which reduce serum insulin levels, are associated with reduction in serum cholesterol levels (Jenkins et al, 1995). Accordingly, postprandial blood glucose and insulin response were also examined (Wood et al, 1994a).

The data (Table 2) showed that oat ß-glucan attenuated postprandial blood glucose and insulin response to a 50 g oral glucose load, and the attenuation was proportional to the logarithm of apparent viscosity at 30 s^{-1}. The analysis showed that around 90% of changes in response could be accounted for by the flow viscosity of the ß-glucan in the mixture as consumed by the subjects.

Viscosity is dependant, *inter alia*, on concentration of polymer in solution, and, at the molecular level, structure and molecular weight. Thus, for physiological response, not only does the dose, or amount, of ß-glucan delivered have to be considered, but also physicochemical characteristics such as solubility and molecular size.

4. SOURCES OF ß-GLUCAN

ß-Glucan is the main endospermic cell wall polysaccharide of oats and barley but is also present, in lesser amounts, in wheat and rye (Table 3).

Oats and barley are the major sources, and these levels can be increased by milling to produce bran, or coarse, fractions. The potential for ß-glucan enrichment is indicated in Table 4.

Barley is perhaps easier to mill than oats; dry milling and sieving can yield about 20–30% bran containing close to 20% ß-glucan - a three fold enrichment from the starting seed. Milling of oats is made more difficult by the soft nature of the kernel and the lipid

Table 1. Comparison of blood lipid levels (mmol/L) at baseline and at week 4 of oat gum phase

Measurement	Baseline	Week 4	% change
Total Cholesterol	6.77	6.15*	–9.2
HDL-Cholesterol	1.29	1.27	0
LDL-Cholesterol	4.62	4.16*	–10.0

* Significantly different from baseline (p<0.0001)

Table 2. Regression analyses of relationships between blood glucose and insulin variables and $\log_{10}$[viscosity][1]

Variable	a	b	r^2	P<[2]
Peak D glucose[3]	3.00	–0.31	0.90	0.0001
AUC glucose (mmol/min/1)	136	–12.1	0.79	0.0001
Peak D insulin (pmol/l)	533	–71	0.96	0.0001
AUC insulin (pmol/min/1)[4]	29474	–3342	0.84	0.0001

[1]Variable = a + b $\log_{10}$[viscosity]; viscosity units, mPa.s, at 30 s^{-1}
[2]Significance of regression relationship, or equivalently, correlation coefficient
[3]D, difference from baseline
[4]AUC, area under 2 hr curve

distribution throughout. The coarse fractions are referred to as oat bran but oat bran is not an exact parallel to wheat bran. The American Association of Cereal Chemists' definition only requires that it be a coarse fraction that is less than 50% of the groat, contains at least 16.0% dietary fibre of which one third must be soluble, and contains no less than 5.5% ß-glucan.

The anatomical definition of bran is the outer layers of the seed - percarp, seed coat, aleurone layer (See Stenvert, Chapter 29). Even in wheat it is difficult to access and isolate these layers specifically by conventional milling, and the product normally produced contains some endosperm. The processes shown for oats in Table 4 that achieve ß-glucan contents greater than about 10% are wet or organic solvent based or use defatted oats. The separation of the coarse brans results in starch enrichment in the flour (Table 5) but significant starch amounts are also normally found in the bran.

5. PRE-PROCESSED WHEAT

Wheat is not generally thought of as a ß-glucan source; levels are usually less than 1%, commonly about 0.6% (Table 3). However, specific histochemical techniques reveal a distinct localisation in the aleurone and adjacent sub-aleurone region (Fulcher and Wood, 1983). Although this still contrasts with the much greater deposits in oats, new friction/abrasion pre-processing techniques for wheat (Dexter and Wood, 1996; see also Forder, Chapter 32) are able to access this region in a more precise fashion than has been achieved with conventional milling. These new processes, designed primarily to improve mill capacity, increase extraction rate and yield better quality flour, also give a by-product

Table 3. $(1\rightarrow3)(1\rightarrow4)$- ß-D-Glucan content of cereals

Cereal	ß-D-Glucan (% dry weight basis)
Barley	3 – 11
Corn	0.1
Oats	3 – 7
Rice	0.1
Rye	1 – 2
Wheat	< 1

Adapted from Wood, 1992

Table 4. Enrichment of (1→3)(1→4)-ß-D-glucan in oat and barley by laboratory-scale milling processes

Process	ß-Glucan (% dry weight basis) Whole Grain	Flour	Bran	EF[1]	Bran yield (% of whole)	Ref[2]
Barley						
Roller mill, sieve	4.2–11.3	3.7–9.0	4.9–15.4	1.1–1.6	28.4–31.5	1
mean	*6.0*	*5.0*	*8.2*	*1.4*	*30.3*	
Udy mill, sieve	5.1–7.2	1.2–2.0	16.0–19.9	2.3–3.5	18.7–30.1	2
mean	*6.1*	*1.6*	*17.8*	*3.0*	*23.6*	
Air classify	5.8	–	14.7	2.5	13.2	3
	8.0	–	14.6	1.8	27.3	
(Defatted sample)	19.6	–	28.5	1.5	26.0	2
Oats						
Falling number mill, sieve	3.9–6.8	1.6–2.3	5.8–8.9	1.3–1.6	48–58	4
mean	*5.1*	*1.8*	*7.4*	*1.5*	*53*	
Steep, EtOH sieve	3.9–6.8	-	14.7–19.1	2.8–3.5	13–23	5
mean	*5.1*	-	*16.7*	*3.3*	*19*	
Dry sieve, EtOH sieve	4.2–6.7	-	10.8–13.4	1.6–3.2	26–34	6
Defatted, roller mill, sieve	4.7	0.6	21.2	4.5	18	2

[1]Enrichment Factor; ß-glucan in bran/ ß-glucan in original seed
[2]References: 1, Bhatty, 1992; 2, Knuckles et al, 1992; 3, Wu et al, 1994; 4, Wood et al, 1991b; 5, Wood, 1992; 6, Wood et al, 1991c

with possibly useful ß-glucan levels, as shown in Table 6. The process of yielding starch-enriched flour resulted in a by-product with a 5-fold increase in ß-glucan content.

The pre-processed fractions of wheat analysed here are potential novel food ingredients. As found in wheat bran, these layers have been available only together and in the presence of significant proportions of the endosperm. This is strikingly reflected in starch analysis of the friction product, which contained just 4.3% starch compared to about 15% for a typical wheat bran.

There are potential commercial and nutritional advantages, although also some potential problems, associated with the concentration of components within the different fractions obtained by these friction/abrasion processes. The most obvious commercial and nutritionally significant component of the pre-processed fractions is dietary fibre. However, a multitude of other functionally distinct components such as waxes, lignin, phytate, vitamins, minerals, phenolics are also concentrated in these layers and are further separated between the different friction and/or abrasion products. Some of these compounds

Table 5. Starch content of fractions produced by laboratory-scale dry-milling of oats and barley (from Knuckles et al, 1992; Wood et al, 1991b)

Sample	Starch (% dry weight basis) Whole grain	Flour	Bran	Flour yield (% of whole)
Barley (hulless or dehulled)	61.9–70.7	70.9–79.2	47.2–50.7	50.6–73.1
mean	*66.8*	*75.1*	*49.3*	*66.7*
Whole oat groat	54.9–63.6	65.1–73.7	41.1–52.6	42–52
mean	*59.1*	*69.7*	*49.3*	*47*

Table 6. Analysis of commercially produced pre-processed wheat fractions[1]

Sample	Starch (% dry weight)	ß-Glucan (% dry weight)
Starting wheat	59.6	0.5
Pre-processed wheat	67.5	0.6
Finished flour	75.2	0.2
Friction product	4.3	0.4
1st Abrasion product	13.4	2.6
2nd Abrasion product	31.9	1.9

[1]Kindly provided by E Timm and Son Ltd., Goole

are powerful antioxidants, may possess potent pharmacological properties and might therefore become desirable ingredients in the developing market for functional foods for health, or nutraceuticals.

These fractions of wheat have been consumed by people for millennia as part of whole wheat and, lately, wheat bran products. Thus, despite the novelty of the fractions, there should not be concern that some unknown component might render the fractions harmful. Nevertheless, the presence of specific components in much greater concentration than obtained in conventional wheat bran is cause for caution. Mycotoxins, from fungal contamination, and pesticide or herbicide residues, are of particular concern.

With the present state of knowledge, the value of the pre-processed outer layer fractions can be evaluated as a particularly rich source of dietary fibre. The expected physiological function would be those attributed to insoluble fibre, and particularly wheat bran-namely improved colonic function and reduced risk of some diseases of the large bowel, in particular cancer. It is nevertheless of interest that there is also a significant concentration of a variety of potentially active components including ß-glucan, identified by the FDA in the USA and by Braaten et al (1994) as the active ingredient of oat bran.

REFERENCES

Bhatty RS (1992) "ß-Glucan content and viscosities of barleys and their roller-milled flour and bran products." Cereal Chem. 69, 469–471

Braaten JT, Wood PJ, Scott FW, Wolynetz MS, Lowe MK, Bradley-White P and Collins MW (1994) "Oat beta glucan reduces blood cholesterol concentration in hypercholesterolemic subjects." Eur. J. Clin. Nutr. 48, 465–474

Dexter JE and Wood PJ (1996) "Recent applications of debranning of wheat before milling." Trends in Food Sci. Technol. 7, 35–41

Doublier J-L and Wood PJ (1995) "Rheological properties of aqueous solutions of (1→3)(1→4)-ß-D-glucan from oats (*Avena sativa* L.)." Cereal Chem. 72, 335–340

Fulcher RG and Wood PJ (1983) "Identification of cereal carbohydrates by fluorescence microscopy." In "New Frontiers in Food Microstructure" DB Bechtel, ed., pp 111–147, Amer. Assoc. Cereal Chem., St. Paul, MN

Jenkins DJA, Jenkins AL, Wolever TMS, Vuksan V, Rao VA, Thompson LU and Josse RG (1995) "Effect of reduced rate of carbohydrate absorption on carbohydrate and lipid metabolism." Eur. J. Clin. Nutr. 49 , S68-S73

Knuckles BE, Chiu MM and Betdchsrt AA (1992) "ß-Glucan-enriched fractions from laboratory-scale dry milling and sieving of barley and oats." Cereal Chem. 69, 198–202

Wood PJ, Weisz J and Fedec P and Burrows VD (1989) "Large-scale preparation and properties of oat fractions enriched in (1→3)(1→4)-ß-D-glucan." Cereal Chem. 66, 97–103

Wood PJ, Weisz J and Blackwell BA (1991a) "Molecular characterization of cereal ß-D-glucans. Structural analysis of oat ß-D-glucan and rapid structural evaluation of ß-D-glucans from different sources by high-performance liquid chromatography of oligosaccharides released by lichenase." Cereal Chem. 68, 31–39

Wood PJ, Weisz J and Fedec P (1991b) "Potential for ß-glucan enrichment in brans derived from oat (*Avena sativa* L.) cultivars of different (1→3)(1→4)-ß-D-glucan concentrations." Cereal Chem. 68, 48–51

Wood PJ, Weisz J and Mahn W (1991c) "Molecular characterization of cereal ß-glucans. II. Size-exclusion chromatography for comparison of molecular weight." Cereal Chem. 68, 530–536

Wood PJ (1992) "Aspects of the chemistry and nutritional effects of non-starch polysaccharides of cereals." In "Developments in Carbohydrate Chemistry." RJ Alexander and HF Zobel, eds. pp. 293–314, Amer. Assoc. Cereal Chem., St. Paul, MN

Wood PJ, Braaten JT, Scott FW, Riedel KD, Wolynetz MS and Collins MW (1994a) "Effect of dose and modification of viscous properties of oat gum on blood glucose and insulin following an oral glucose load." Brit. J. Nutr. 72, 731–743

Wood PJ, Weisz J and Blackwell BA (1994b) "Structural studies of (1→3),(1→4)-ß-D-glucans by ^{13}C-nuclear magnetic resonance spectroscopy and by rapid analysis of cellulose-like regions using high-performance anion-exchange chromatography of oligosaccharides released by lichenase." Cereal Chemistry 71,301–307

Wrick KL (1993) "Functional foods, cereal products at the food-drug interface." Cereal Foods World, 38, 205–214

Wu YV, Stringfellow AC and Ingett GE (1994) "Protein and ß-glucan enriched fractions from high-protein, high ß-glucan barleys by sieving and air classification." Cereal Chem. 71, 220–223

29

NOVEL NATURAL PRODUCTS FROM GRAIN FRACTIONATION

N. L. Stenvert

Goodman Fielder Milling and Baking Group
PO Box 1, Summer Hill, NSW, 2130
Australia

1. INTRODUCTION

At present we have evidence for a positive role in health of nutrients and many non-nutrients in foods such as whole grains (Clydesdale, 1994). This role is not normally evident in refined flours, indicating that this characteristic resides in the bran and germ components of the grain which are normally separated and discarded.

In the wheat grain the bran and germ represent ~15% and ~2.5% respectively of the grain. The wheat bran consists of numerous layers, mainly fibrous in nature, as well as the aleurone cell layer. This cell layer is dormant but living, as is the wheat germ, and together these layers represent the essential vitality of the wheat grain, containing the grain's essential nutrients (vitamins, minerals, oils, proteins) required for growth. On germination these cells become the site of intense synthetic activity so as to mobilise the energy reserves of the grain and transfer them to the newly growing plant. To enable this to occur the aleurone cell layer and wheat germ are rich in minerals required for enzyme functions, including magnesium, zinc, iron and phosphorus; B-group vitamins, including thiamine, niacin and folic acid; protein which is of the high quality required for dynamic growth - the essential amino acids are present in proportions close to ideal for human nutrition; lipids which are contained in small droplets called spherosomes and provide the first energy needs of the germinating grain.

The contents of the aleurone cells are enclosed in very thick cellulosic cell walls which are relatively indigestible in human nutrition. Hence the highly nutritious contents of these cells are normally trapped inside the fibrous layers of the grain, resulting in wheat bran being unpalatable with unavailable nutrients, even when present in normal wholemeal bread. For this reason bran is normally separated from the flour during milling and sold as a residual product for animal feeds.

Cereals: Novel Uses and Processes, edited by Campbell *et al.*
Plenum Press, New York, 1997

2. RECOVERY OF ALEURONE CELL CONTENTS AND WHEAT GERM

After many years of research and development to find alternative uses for the bran layer of wheat, the Goodman Fielder Milling and Baking Group perfected a novel milling system for capturing the nutritious components of the mill by-product. This is achieved by cutting through the cell walls of the aleurone cells to unlock the nutrient rich contents of these cells and then separating the contents from the fibrous grain layers.

At the same time the wheat germ is comminuted and recovered with the aleurone cell contents. The recovery process is applicable to all wheat types and grades. The resultant product is totally natural and rich in nutrients; its name is being registered as "Nature's Gold".

3. NATURE'S GOLD NUTRITIONAL PROFILE

It is possible to modify the relative nutrient density/dietary fibre levels of Nature's Gold by adjustment of the recovery process. However, the following proximate analyses are typical of overall composition (Table 1).

3.1. Protein

Whole grains and many legumes are lacking in some of the essential amino acids. The proteins in aleurone cells and wheat germ are, however, of high quality since they are required for dynamic growth. It is not surprising therefore that Nature's Gold has protein quality close to ideal for school age children, when compared to recommended FAO/WHO/UNU levels (Table 2).

Thus not only is the protein of high quality, but the high protein level results in a major contribution to % daily value. Vegetarians (particularly vegans) could benefit from such a protein source.

3.2. Dietary Fibre

Nature's Gold is an excellent source of dietary fibre. The fibre is derived from wheat bran, hence is predominantly insoluble and will contribute to regularity and bowel health. Numerous studies have outlined the role of wheat bran fibre in anti-tumour action, with a possible explanation being the continuous production of short chain fatty acids along the

Table 1. Proximate analysis of Nature's Gold (per 100 g)

Protein	22g
Moisture	11.8g
Ash	4.9g
Dietary fibre	20.2g
Fat	4.4g
Cholesterol	0g
Carbohydrate	56.9g

Table 2. Amino acid composition of Nature's Gold

Essential amino acid	Nature's Gold (mg/g crude protein)	FAO/WHOMinimum Requirements10–12 year olds (mg/g crude protein)
Histidine	28	19
Isoleucine	32	28
Leucine	61	44
Lysine	49	44
Methionine + Cystine	37	22
Phenylalamine + Tyrosine	73	22
Threonine	37	28
Tryptophan	8.2	9
Valine	45	25

length of the large bowel (McIntyre et al, 1991). There is also evidence that inclusion of bran components, including the aleurone cell contents, can promote bacterial fermentation and growth (Hansen and Hansen, 1994), providing a possible role for Nature's Gold as a prebiotic.

3.3. Carbohydrate

Due to the high level of other nutrients in Nature's Gold, the overall carbohydrate level is reduced compared to whole grains. However, the metabolism of this carbohydrate is of considerable interest. When Nature's Gold is incorporated into the bread Farrer's Gold, the release of glucose during metabolism is markedly reduced. When compared to white bread (Glycaemic Index: 100) at equicarbohydrate consumption levels (after also correcting for dietary fibre) a Glycaemic Index of 80 is obtained. Reasons for this are not apparent since wholemeal flour or intact wheat bran does not produce this effect. However, it is potentially of significant benefit to diabetics.

3.4. Vitamins and Minerals

The mineral and vitamin profile of Nature's Gold is shown in Table 3. Since many vitamins and minerals are concentrated in the aleurone cell layer, Nature's Gold has vita-

Table 3. Mineral and vitamin content of Nature's Gold

Vitamin	Level (per 100g)	Mineral	Level (per 100g)
Thiamine	1.57mg	Iron	22.7mg
Riboflavin	0.26mg	Magnesium	530mg
Pyridoxine	0.34mg	Potassium	1160mg
Folic Acid	800μg	Phosphorus	1140mg
Vitamin E	2mg	Zinc	8.3mg
Niacin	24mg	Manganese	10.1mg
		Sodium	20mg

min and mineral levels up to 20 fold greater than white flour and up to 10 fold greater than whole grain. Nature's Gold is a good source of thiamine, niacin and folic acid. The latter is of particular interest to women of child bearing age, since a lack of folic acid can lead to birth defects. Nature's Gold is also a good source of the minerals magnesium, phosphorus and iron. The latter is again of relevance to women where iron deficiency is a common problem. Sodium levels are extremely low. The product also contains significant quantities of zinc and manganese which are involved in numerous essential enzyme systems, including enzymes that can detoxify free radicals (superoxide dismutase) and thus have an anti-oxidant role. When reinforced by the presence of other anti-oxidants including vitamin E, phenolics, lignans and phytic acid, Nature's Gold may have a significant role in modulating the effects of free radical damage.

4. NATURE'S GOLD AS A FUNCTIONAL FOOD

Nature's Gold is a totally natural, nutrient dense product derived from the portions of the grain believed to provide health benefits, namely the bran and germ. It is evident that the role of certain minerals and vitamins in Nature's Gold are well understood, and can be duplicated by synthetic compounds; it is equally clear that an incomplete knowledge exists of the role of other nutrients and micronutrients in Nature's Gold and their possible synergistic action in influencing health. Nevertheless a number of postulated benefits can classify Nature's Gold as a functional food (Table 4). (See Chapter 28 for a discussion of issues surrounding functional food claims.)

5. FOOD APPLICATIONS

While Nature's Gold is a natural, nutrient rich grain product, this in itself will not guarantee usage as an ingredient in other foods. To achieve this it is essential that the following criteria apply: acceptable taste profile; acceptable colour characteristics; storage stability; compatibility with other food systems; microbiological acceptability; minimal pesticide residues; and production at an ISO 9000 accredited facility.

Nature's Gold complies with all the above criteria; its first food application was in the highly nutritious bread Farrer's Gold (Stenvert, 1995). Within 4 weeks of launch this bread had grown the total Australian wholemeal bread market by 9.6% and achieved leadership in this category. Other product applications, including high nutrient breads, pasta,

Table 4. Proposed therapeutic effects of Nature's Gold

1.	*Promote bowel health	- Dietary fibre - Prebiotic
2.	*Prevent neural tube defects	- Folate
3.	*Reduce glycaemic index	- ?
4.	*Minimise free radical damage	- Various anti-oxidant functions
5.	*Rectify protein deficiency states	- Quality/quantity of protein
6.	*Rectify mineral/vitamin deficiency states	- High mineral/vitamin levels.

biscuits, breakfast cereals, flours and nutraceuticals, are currently being actively pursued in Europe, USA, South Africa and Australasia.

6. ALTERNATIVE GRAINS

While Nature's Gold is derived from wheat, the process is readily applicable to other grains. This can result in the other nutrients being combined with alternative fibre sources e.g. oat or barley fibre. Alternatively the conceptual approach can be applied to fractionation of a range of legumes and oilseeds to provide nutritionally advantageous products.

7. SUMMARY

An outline has been given of a natural product obtained by grain fractionation. The product is nutrient dense, with many desirable health attributes. Its potential use in other food products has been considered and the applicability of a fractionation process to other grains has been discussed.

REFERENCES

Clydesdale FM (1994) "Optimising the Diet with Whole Grains." Critical Review in Food Science and Nutrition, 34 (5&6), 453–471

Hansen A and Hansen B (1994) "Influence of Wheat Flour Type on the Production of Flavour Compounds in Wheat Sourdoughs." Journal of Cereal Science 19, 185–190

McIntyre A, Young GP, Taranto T, Gibson PR and Ward PB (1991) "Different Fibers have Different Regional Effects on Luminal Contents of Rat Colon." Gastroenterology, 101, 1274–1281

Stenvert NL (1995) "New High Fibre Bread - Farrer's Gold." Food Australia 47 (10), 462–463

30

APPLICATION OF FERMENTED FLOUR TO OPTIMISE PRODUCTION OF PREMIUM CRACKERS AND BREAD

Hans Moonen

Food Science & Technology Centre
Quest International
PO Box 2, 1400 CA Bussum - Holland

1. INTRODUCTION

Yeast fermented bakery products such as premium crackers and bread are characterised by a specific texture and unique flavour. Premium crackers are appreciated by consumers as high quality baked products, which offer good margins for their producers. The key feature in the processing of premium crackers is their extensive sponge fermentation time. This leads to the unique flavour, texture and shelflife of the final cracker. But it is also a complex process, which can lead to many breakdowns, long production runs, high capital investment, need for a relatively high labour input, and high occupied space in factories.

2. MATERIALS AND METHODS

The formulation and processing used for premium soda crackers, comparing sponge and dough with straight dough, is shown in Table 1. For convenience this figure is based on 1000 g of flour. Sponge fermentation took 18 hours at 30°C. Enzymes were added at the dough side, either Biobake BSC, a Quest International product consisting of bacterial amylase and protease, or Biobake BCC consisting of fungal xylanase.

A weak flour was used (650 g), which was mixed with 310 g water and 4 g yeast, and fermented for 18 hours at 30°C. After the fermentation, 350 g of flour, 110 g of fat, 18 g of salt and in standard trials 4.5 g soda was added and the dough remixed and rested for 4 hours at 30°C. Thereafter the dough was laminated, sheeted and stamped, and the crackers baked at 210°C for 8 minutes. Optionally 50 ppm Biobake BSC was added, to improve machinability and lamination, to regulate spread and shrink, and to improve crispiness and colour.

In the all-in straight dough process 1000 g of flour was mixed with 4 g yeast, 320 g water, 110 g fat, 18 g salt, 4.5 g soda and 30 g Hy-Bake PCM. Hy-Bake PCM is produced by Quest International and a patent application for its production and application has been

Cereals: Novel Uses and Processes, edited by Campbell *et al.*
Plenum Press, New York, 1997

Table 1. Formulation for soda crackers

Sponge and Dough		All-in	
Sponge		White flour	1000 g
White flour	650 g	Yeast	4 g
Yeast	4 g	Water	320 g
Water	310 g	Fat	110 g
(18 hours fermentation time)		Salt	18 g
Dough = Sponge plus:		Soda	4.5 g
White flour	350 g	Hy-Bake PCM®	30 g
Salt	18 g	Biobake BSC®	50 ppm
Soda	4.5 g	*optional for higher lift*	
Biobake BSC®	50 ppm	Biobake BCC®	100 ppm
(3 hours resting time)		*(3 hours resting time)*	

filed. The dough was rested for 4 hours at 30°C and laminated, sheeted, stamped and baked at 210°C for 8 minutes. Optionally 100 ppm Biobake BCC was added, to improve lift in the oven.

3. RESULTS

Replacement of the laborious sponge fermentation by one single ingredient might be beneficial for the industry, by streamlining the processing without adversely affecting final product properties such as flavour and texture. It is clear that during the sponge fermentation acid and flavour precursors are produced. Figure 1 shows the characteristic decrease of the sponge pH over time. The pH decreases from 5.8 at the start of the fermentation to 3.8 after 18 hours fermentation. By adding 4.5 g soda to the sponge at the remixing phase, the pH increases to 7.5, which decreases to 7.0 after 4 hours resting time. The acid which affects the gluten structure is neutralised by the soda, and the flavour precursors are transferred into final flavours during the baking process.

The challenge was to develop an innovative ingredient, which contains these acids and precursors in an approximately 100 fold concentration and can be added as a powder

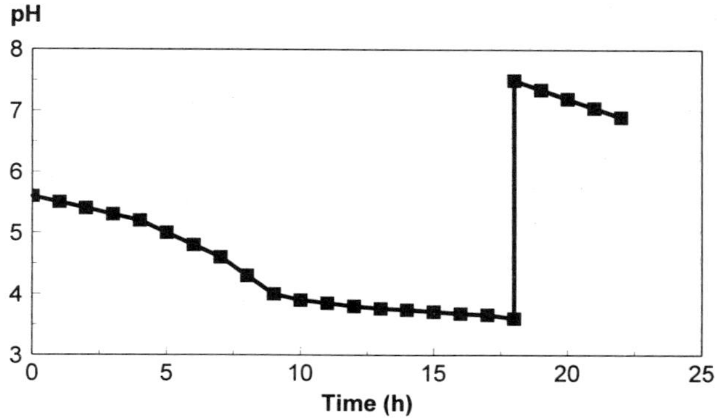

Figure 1. pH changes during the sponge and dough process.

in a straight dough process. This ingredient, Hy-Bake PCM, has been developed in the following way (Figure 2). Specific wheat flour fractions e.g. germs, bran, are extensively hydrolysed by using a combination of enzymes consisting of α-amylase, hemicellulase and protease. The next step is that these hydrolysed wheat fractions are fermented with microorganisms traditionally used in the baking industry, namely:

- lactobacilli such as *Lactobacillus plantarum* or *Lactobacillus brevis*
- yeast such as *Saccharomyces cerevisiae*.

After this fermentation the entire broth is transferred into a powder by e.g. spray drying, freeze-drying or mixing with dry carriers.

The final product, cultured wheat fractions, is a multifunctional ingredient with positive effects on dough rheology and final taste, aroma and texture of the baked product.

The dosage of the Hy-Bake PCM in premium cracker manufacturing is dependent on

- flour pH and "strength"
- design of the sponge process
- proportion of "sponge" to "dough"
- final pH of sponge dough
- desired lift from sodium bicarbonate
- desired final product pH.

In a well defined system, the quantity of Hy-Bake PCM addition is linearly related to the quantity of soda added.

One of the most prominent advantages of the application of Hy-Bake PCM in premium cracker production is the dramatic reduction of processing time required. With the conventional approach, using 18 hours sponge fermentation, the process takes 22.8 hours. With the innovative approach described here it takes only 4.55 hours (Figure 3).

Other advantages of applying the new approach in comparison with the traditional approach might be in the areas of: capital investment, process management, labour, working capital, energy use, factory space, consequence of sponge failure, reproducibility be-

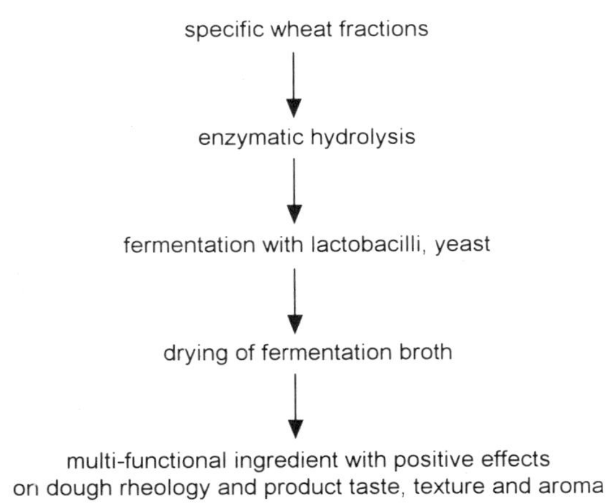

Figure 2. Hy-Bake PCM production.

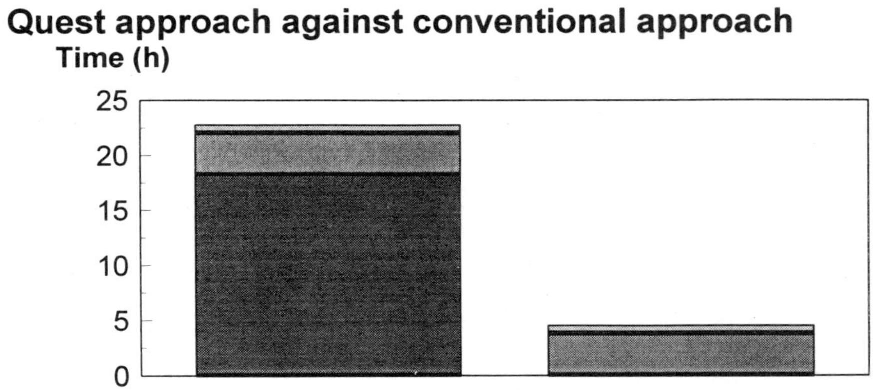

Process	Conventional	Hy-Bake PCM
Mixer	0.25	0.00
Sponge	18.00	0.00
Mixer	0.25	0.25
Dough process	3.50	3.50
Laminating, sheeting, cutting	0.30	0.30
Baking	0.50	0.50

Figure 3. Time saving with the innovative approach to premium cracker production, based on pre-fermented flour fractions.

tween sponges, on-line control, recipe stability, new product development, yield and lack of alcohol emission.

A very promising application of this new technology can be *upgrading* the quality of crackers produced by only chemical leavening in a straight dough process. By inclusion of the Hy-Bake PCM in this recipe, the final obtained cracker will have a texture and flavour characteristic for the sponge and dough process.

This new approach, concentration of specific acids and flavour precursors by a factor of 100, produced by fermentation, in a dry, powdered product has found major applications in bread baking as well. Increasingly it is realised that taste and aroma of bread has been decreased due to different processing methods and new varieties. Typical examples of this are

- Chorleywood breadmaking process
- retarded dough systems
- frozen dough systems
- in-store baking
- high-fibre bread.

The application of concentrated, prefermented flour fractions in these applications has been found beneficial as well.

4. CONCLUSIONS

In conclusion the combination of bakery application skills and flavour expertise, together with cereal chemistry, enzymology and fermentation knowledge, has resulted in a new class of fermented flour products, which help to streamline and upgrade all kinds of fermented and unfermented, speciality baked products in the areas of crackers and bread.

31

NEURONAL AND EXPERIMENTAL METHODOLOGY TO IMPROVE MALT QUALITY

Myriam Fliss,[1,2] Francoise Maurel,[1,3] Jean Louis Delatte,[1] Joseph Boudrant,[2] Marie-Christine Suhner,[3] and Marc Gabriel[3]

[1] Malteries Soufflet, Quai Sarrail
10400 Nogent sur Seine, France
[2] CNRS - LSGC
2 Avenue de la Forêt de Haye, 54500 Vandoeuvre les Nancy, France
[3] CRAM IM
Rue Jean Lamour, 54500 Vandoeuvre les Nancy, France

1. INTRODUCTION

The aim of the malting process is to produce malt, which is homogeneous, by modifying barley as efficiently as possible. It begins with the steeping of barley in water to achieve a moisture level sufficient to activate the metabolism of germination, leading to the development of hydrolytic enzymes. After a period of germination where even modification is achieved, the green malt is kilned to arrest germination and stabilize the malt. Then, colour and aroma of the malt are developed during this phase. The malting process lasts 8 or 9 days. Malt in turn becomes the raw material of the brewery. Brewers' malt specifications and quality requirements are becoming progressively more demanding. The maltster should then be able to predict the influence of the malting process on the quality of malt when selecting barley. However, barley quality is affected by the genotype, the environment under which it is grown and the effectiveness of selection programmes.

The Soufflet Group has developed, through its Agricultural Division, efficient relationships with breeders to select barleys. In addition, Soufflet Maltings has installed equipment and automation systems in its plants. The use of these automation systems has provided increased accuracy and highly automated control of grain handling and other operations in the malthouse. Furthermore, tests in a micromalting unit are made to optimize the process conditions. Despite this, it is difficult for the maltster to achieve a malt which meets the customer's specifications at the first attempt. Apparent consistency can be made by blending of malts having marginally different analyses which when mixed give the required analysis to meet the customer specifications. The maltster has often to modify set points of his process. These process adjustments are based on an iterative method; how-

Cereals: Novel Uses and Processes, edited by Campbell *et al.*
Plenum Press, New York, 1997

ever this is quite complex because of the inter-relationships between the various criteria of the analysis and the individuality of the malting plant. Indeed, the process does not have a maximal reproducibility.

A programme of research was therefore initiated in order to limit the variability of analysis and to progress in the knowledge of the relationships between the malting process and the quality of the malt. This programme is based on the development of a "decision making tool" and of some mathematical relationships between malt quality and the malting conditions, including experimental designs and linear regression analysis. Initial results are presented in this chapter.

2. MATERIALS AND METHODS

When choosing the best conditions of the process, the maltster uses a method of qualification. Trials are performed at three stages: a micromalting scale, the industrial unit and through industrial series of batches.

The micro-malting unit works on 0.6 kg of grains. It is essentially used to assess the malting quality of the newly harvested barleys. Micromalting can give much useful information towards the malting regime to be used in the industrial plant. It is suitable to appreciate, rapidly, the malting behavior of the barley and to establish initial conditions of process. However, scaling up from a kilogram or so in the micromalting plant to hundreds of tons in the industrial plant results in a loss of direct reproducibility. So, an industrial scale trial on 120 tons is realised where the external conditions are taken onto account. The malting process is determined with greater precision. After that, series of six malting batches are made on the industrial scale. A gradual adjustment of the process often occurs in the light of the performance and the analysis of the previous batches. Moreover, the maltster has no suitable information on which to base immediate remedial action on his process.

The way of choosing the process conditions depends on three sources of information: the mathematical tool; the human expertise; and the industrial trials. This information is complementary, but is different in nature and type.

To formalise the maltster's knowledge and perfect it with the information on the performance of previous industrial batches, a tool of distributed artificial intelligence was developed (Pham, 1994). This system can evolve and adapt continuously. It is a decision-making system whose objective is to optimize the process conditions, using the same information sources as are available to the maltster.

The functions of this system are first to help the maltster in the choice of the process conditions. This is the main first function (Fp1). It can also be used to update the maltster's knowledge or to "teach the malting art" to a new maltster; this becomes the main second function (Fp2). Morever, various parameters must be integrated in the system such as (i) the barley characteristics (named Fc1), e.g. the barley variety or the germinative capacity; (ii) the malt specifications (Fc2) i.e. the analytic values of the quality; (iii) the external conditions e.g. humid external temperature (Fc3); and (iv) economic constraints such as the yield and the quality rating (Fc4). The definition of the system was realized with the help of the APTE method (Blaison, 1991), and is illustrated in Figure 1. The choice of the best malting conditions is based on the definition of predictive models between these various parameters. This research of models takes part equally in a method of qualification of the process.

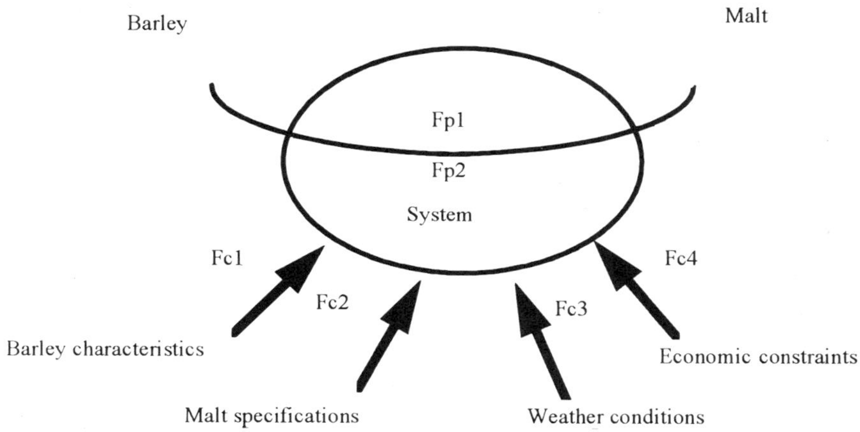

Figure 1. The A.P.T.E. method.

A methodology based on experimental designs and linear regressions was used to establish mathematical models from the process (Fliss et al, 1995a). Experiments were first carried out at the micromalting unit with two factorial designs of 2^6 processings. Then a fractional design of resolution III was applied at a malting pilot plant previously described (Fliss et al, 1995c). This pilot unit of 130 kg barley capacity creates an interface between the laboratory and the industrial maltings. Through its design, it offers greater flexibility and reproducibility of the process conditions. Using this pilot unit, the effect of the malting process on the quality of the malts produced can be studied systematically. More precisely, the following experiments can be performed: grain respiration during steeping, the effect of temperature variations during the germination process on quality and the enzyme potential of the malt. Furthermore, the possibility of recycling the air during kilning can allow the production of special malts.

These two approaches will constitute the decision-making system.

3. RESULTS

The models of knowledge, which were the decision-making tool, and the mathematical model from the experimental designs, have been realised, independently of each other.

3.1. A Tool of Distributed Artificial Intelligence

The development of the distributed artificial intelligence (DAI) tool (or the decision-making tool) depends mainly on two information sources:

- The knowledge of the maltster.
- The data of the industrial production covering the malting processes. In our study, data of 60 industrial productions were collected and combined. These data linked the raw material and the final product, the malt.

From this information, two models of knowledge have been created. The first one is based on the *knowledge of the expert*, and is called the "generic model". This model re-

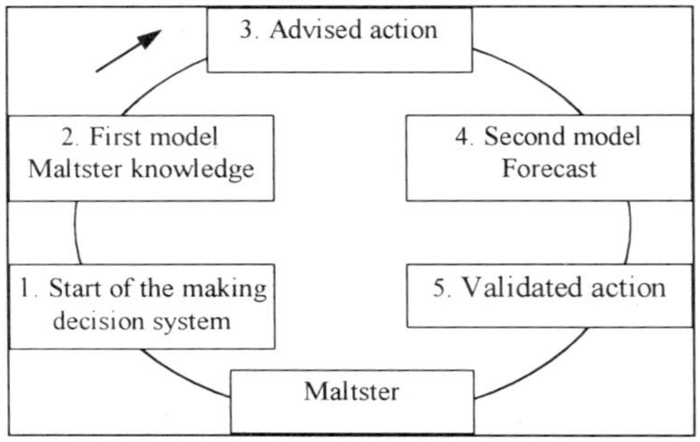

Figure 2. The models of knowledge created.

constitutes partially the reasoning of the expert on a decision of optimization of the process. A method of know-how capitalization was set up (Maurel et al, 1995).

The second model, based on industrial data, is a *forecasting* model. This model depends on the construction of a system of neurons. The establishment of this system of neurons consists in the determination of the input and output parameters that constitute the structure of the network, and the examples that define the content of the network.

Here, the input parameters are the process data (temperature, time of each step, air flow etc.). The output parameters are the results of the malt analysis. The content of the system is obtained by a sorting undertaken on the couple (variety, external temperature). The result of this model is a synthesis of these examples (Maurel et al, 1995) and permits anticipation of the results of the malt analysis.

This estimate unfolds in two stages. A first stage is based on the "generic model" and proposes an action. A second stage, using the second model, validates this action. Once the action is validated, it is proposed to the maltster that takes the decision to modify or not the malting process. Figure 2 illustrates the system.

3.2. Experimental Strategy Tool and Multivariate Analysis

The second tool uses the methodology of experimental designs. This enables the experimentation time to be reduced and the essential factors and interactions of the process to be defined. All the process factors are studied at the same time. The study of each factor involves as much precision as if it were studied alone, as has often been the case (Bettner and Meredith, 1969; Piendl, 1975, 1976; Reeves et al, 1980). Different experimental designs of two levels have been realized (Fliss et al, 1995a). From the resolution of the designs, process factors such as the duration of steeping, the temperature of germination air and the total recycling of air flow during the 5th germination day of the air appeared to be the most influential on the quality of malts. Simple models taking into account these factors as well as the value of their effects have been established.

In the same way, a statistical analysis of industrial data has been initiated. Indeed, for each plant, there is a mass of historical data covering the analysis of each batch of malt produced and the malting process parameters under which each batch was made. More precisely, it concerns a multivariate analysis. This methodology uses principal component analysis (PCA) and linear regression (Fliss et al, 1995b). These analyses allow, respectively, summary of the information contained in the production data, and identification of relationships between variables in the form of polynomial models. Some correlations exist between the different qualitative variables of the malt on the one hand; on the other hand, correlations between these variables and the process variables have been identified. Simple mathematical models explaining these relations have been established. Nevertheless, the established models concerning relationships between conditions of process and the quality of malts, have too weak correlation coefficients to explain correctly the impact of the process on the quality of malts. More thorough research of these models is under way, using a hierarchical classification of the important process variables and of various observations.

4. CONCLUSION

Up till now, the two methodologies have been established independently of each other: on the one hand, the development of a capitalization tool of the expert's know-how; and on the other hand the statistical and experimental approach. The future objectives are the fusion of these two systems to improve the knowledge of the expert, and to allow qualitative malt variations to be anticipated and quantified. The mathematical modelling should allow a more rapid convergence of the decision-making system. Conversely, the latter has to provide new directions for the experimentation of tests and to orientate the statistical analysis towards new variables to be integrated in its model so as to improve its strength.

REFERENCES

Blaison G (1991) "Manuel de la qualité." Rapport interne SECTOR, Les Ullis

Bettner RE and Meredith WOS (1969) "The effect of duration of initial wet steep period on malt quality." ASBC Proceeding, 70–77

Fliss M, Delattre D, Brault R, Baret JL, Kobilinsky A and Boudrant J (1995a) "Les plans d'expériences en malterie." Récents Progrès en Génie des Procédés 9(36), 79–86

Fliss M, Delattre D, Brault R, Baret JL, Kobilinsky A and Boudrant J (1995b) "Etude de l'impact des paramètres physicochimiques du maltage sur la qualité des malts par l'analyse statistique et les plans d'expériences." Récents Progrès en Génie des Procédés 9(39), 85–90

Fliss M, Delatte JL, Delattre D, Brault R, Baret JL and Boudrant J (1995c) "A new pilot unit to study the influence of the conditions of the malting process on the quality of malts." BIOS 255, 109–112

Maurel F, Delatte JL, Kasprzyck P, Criton G, Suhner MC and Gabriel M (1995) "A neuroagent approach in the malting process." Second International Workshop on Learning in Intelligent Manufactoring Systems, Hungary, April

Reeves SG, O'Farrell DD and Wainwright T (1980) "The effect of increased steeping temperature on malt properties." J. Inst. Brew. 86, 226–229

Pham KM (1994) "A neural multi-agent approach for modelling, distributed processing and learning." Ed. S Goonatilake, S Khebbal, J Wiley, London

Piendl A (1975) "Degree of steeping and malt quality. Part I. Effect of steeping on carbohydrates." The Brewers Digest 34–38, 42

Piendl A (1976) "Barley variety and malting technology as influencing factors on the properties of malt - An evaluation by means of the analyse of variance." MBAA Techn. Quarterly 13, 131–141

32

FLOUR MILLING PROCESS FOR THE 21ST CENTURY

D. E. Forder

Satake UK Ltd
PO Box 19, Bird Hall Lane, Cheadle Heath
Stockport, SK3 0RX, UK

1. INTRODUCTION

During the 20th century the major process for converting wheat into flour for human consumption has undergone a period of gradual refinement following the replacement of stones by rollermills towards the end of the last century. With improved engineering, machine capacities have increased significantly and this, along with the introduction of pneumatic conveying, has caused a steady move towards plants of higher output per unit of building volume. This trend has continued to be driven by the commercial necessity to contain costs in a highly competitive industry.

However, the underlying process has essentially remained unaltered. Wheat is conditioned by the addition to it of a small amount of water in order to cause the bran layers to fuse together, while reducing the hardness of the endosperm and attaining the desired flour moisture content. This treated wheat will normally have a moisture content in the range 15.5 - 17.5%. Once the moisture has penetrated the kernels, which conventionally takes many hours, they are subjected to a grinding process in which, as far as is possible, the bran is left intact and the endosperm is scraped from it for subsequent refinement and particle size reduction.

For many years alternative technologies which would enhance the commercial performance of the process have been sought. Higher process yields and/or product quality gains at viable levels of capital investment are required.

One approach which has been attempted on a number of occasions and which is well documented, is that of pearling or 'debranning' the kernels to remove the majority of the bran prior to milling the resulting grains (Dexter and Wood, 1996). These attempts generally failed because the process could not be shown to provide sufficient overall economic benefit over the, by now, highly refined conventional process.

For the last seven years the Satake Corporation has been applying its extensive expertise as the world's leading rice milling engineer to this challenge. Recently rapid progress has been made and the breakthrough has been achieved in developing a commercially viable process, PeriTec (Satake, 1990; McGee, 1995; Goldsmith and Forder, 1994).

2. THE PERITEC PROCESS

PeriTec is a complete system starting at the raw wheat silos and ending at the finished product bins. It has sufficient flexibility to allow it to be tailored to the specific requirements of most millers. The necessary hardware has rapidly evolved as a result of experience from the various commercial plants built by Satake as well as extensive testing in our own research mills.

The heart of the system is the PeriTec process in which the wheat is debranned and milled. The milling section, whilst using conventional rollermilling technology, exploits the very different physical characteristics of debranned kernels and the reduced quantity of bran present to produce a more compact unit with enhanced levels of extraction and product purity.

The tempering and debranning section, shown in schematic form in Figure 1, accepts conditioned wheat at typically 14% moisture content. The conditioning can take place as part of either the intake or cleaning operations. After bran removal the wheat, still at 14% moisture content, passes to the hydrating section where sufficient water is added to take account of moisture loss due to milling and any adjustment for customers' flour moisture specification. Due to the rapid absorption of this water, and hence relatively short time span between hydrating and milling, it is quite feasible to have this system respond to actual flour moisture by means of a feedback loop, to give better control of the final flour moisture.

In conventional milling, wheat is conditioned to a moisture content above 14% and hence, once conditioned, must be milled otherwise it will deteriorate. Typically it would be milled 8–16 hours later. In the PeriTec system wheat does not have to be committed to milling until it actually reaches the debranning section. Typically flour will be produced from it some 1/2–2 hours later.

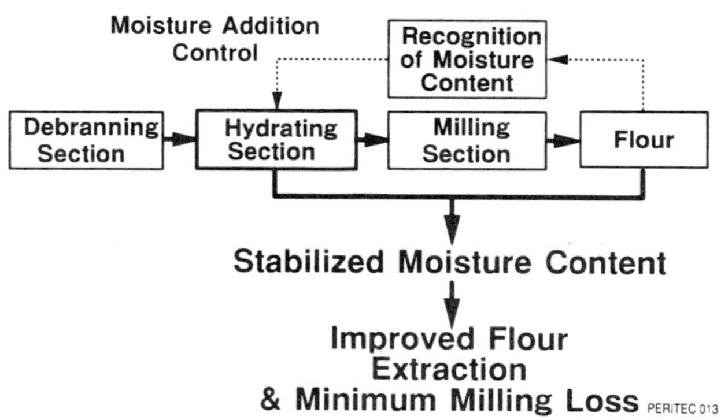

Figure 1. Schematic of the Peritec process.

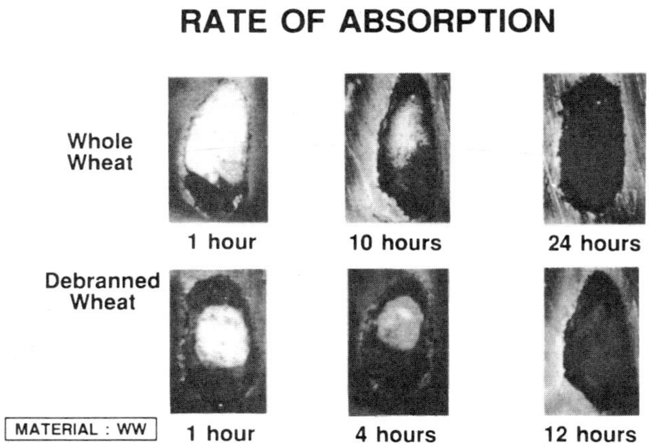

Figure 2. Rate of absorption of Eosin-dyed water into whole and debranned wheat.

3. ASSESSMENT OF THE PERITEC PROCESS

The relatively rapid uptake of water by debranned wheat was confirmed by experiments in which the time dependent penetration was determined by the use of a water soluble Eosin dye. The progress of the dyed water towards the centre of the kernel after various times is shown in Figure 2. As an example for DNS, a relatively hard wheat, the rate of penetration for both whole wheat and debranned wheat was measured by the increase in weight, and hence water content, after total immersion in water. Figure 3 shows the uptake of water to be much more rapid for debranned wheat, continuing for several hours, while for whole wheat the initial rapid uptake falls off after less than one hour. As would be expected this phenomenon is strongly temperature dependent, as illustrated in

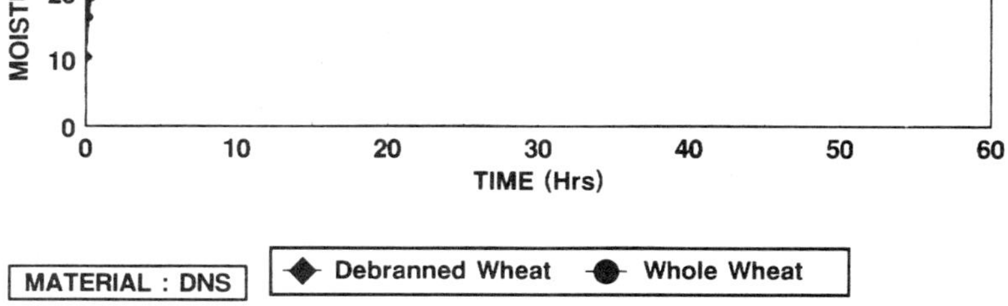

Figure 3. Rate of absorption of water, measured by weight increase, into whole and debranned wheat.

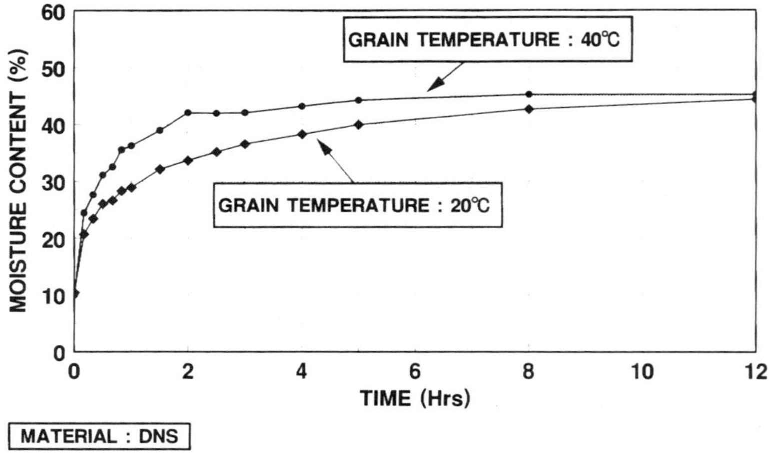

Figure 4. Effect of debranned grain temperature on rate of absorption of water.

Figure 4. Wheat from the debranning section would normally enter the hydration section at 30–38°C.

The hardware used in the PeriTec process has undergone rapid development and is now into its fourth generation. Progress has been spurred by the need to simplify the process, reduce capital and running costs and to give greater control of the work done on the individual wheat kernel.

The VCW series of machines, Figure 5, are single units which complete the whole debranning operation in two internal stages using abrasive processes initially followed by interparticle friction. With this equipment it is possible to control closely the level of bran

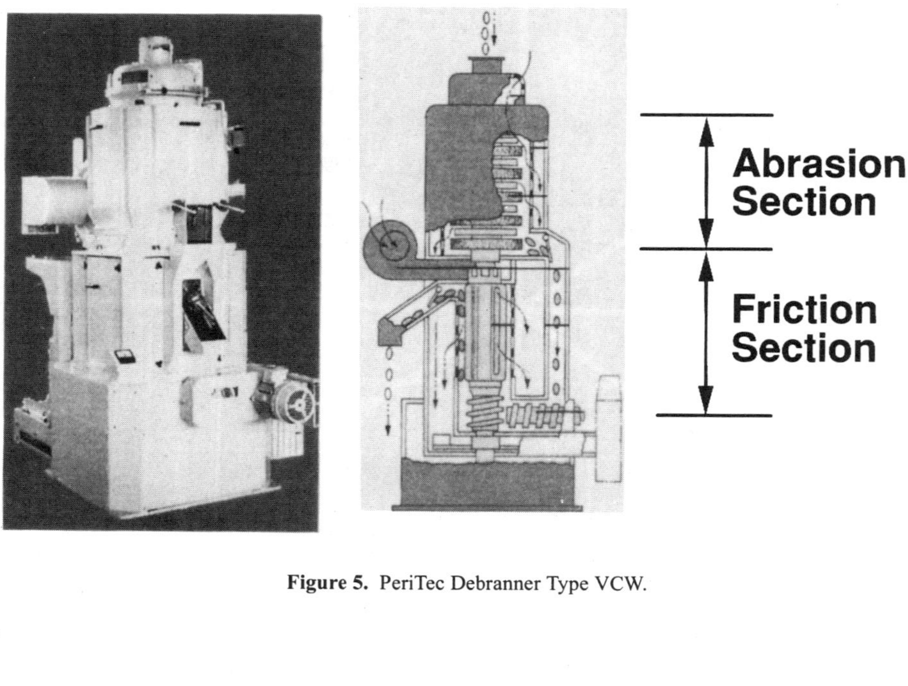

Figure 5. PeriTec Debranner Type VCW.

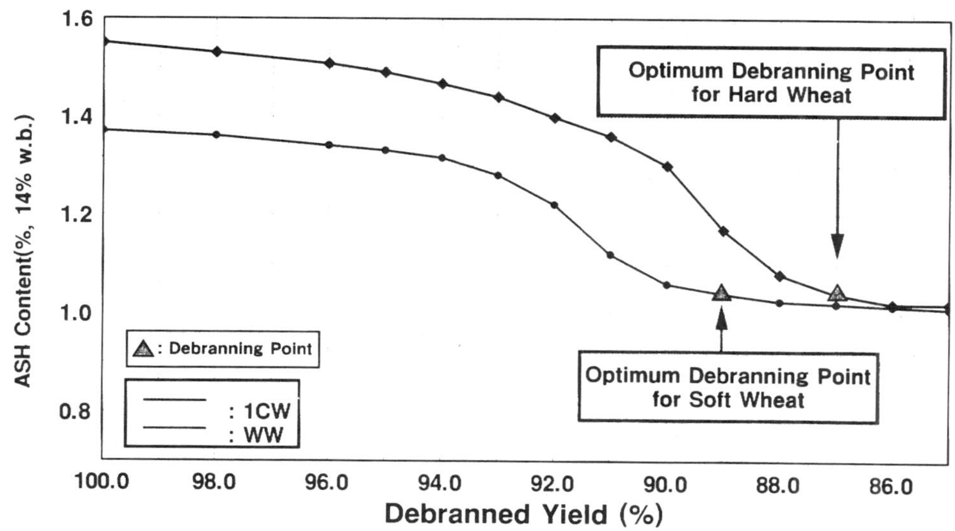

Figure 6. Ash content *versus* Yield as bran layers are removed, showing the optimum debranning points for different wheats.

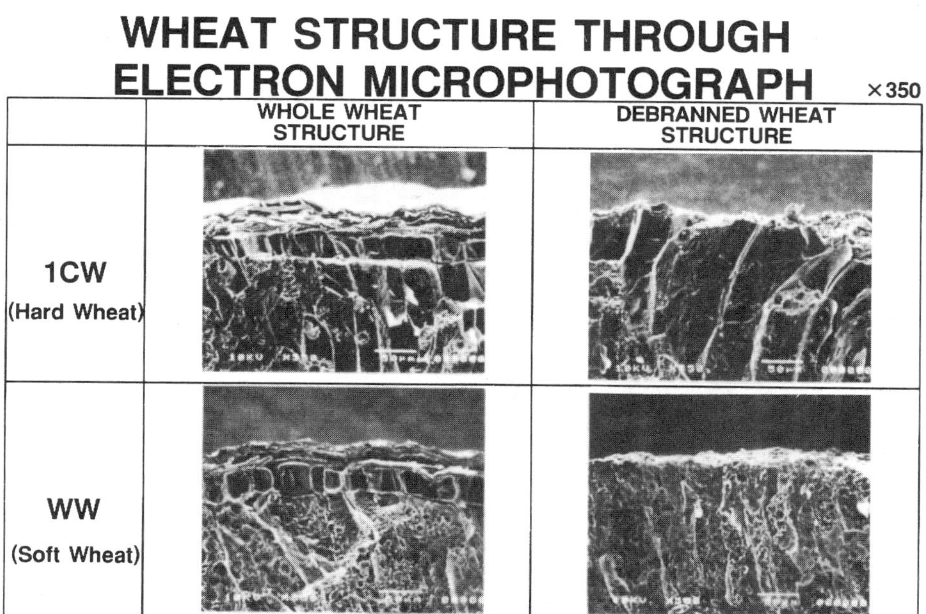

Figure 7. Electron microphotographs of wheat structure before and after debranning.

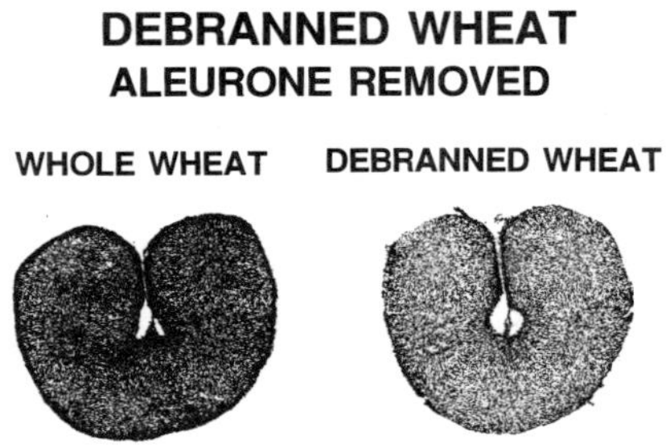

Figure 8. Concentration of aleurone magnesium by X-ray emission microscopy.

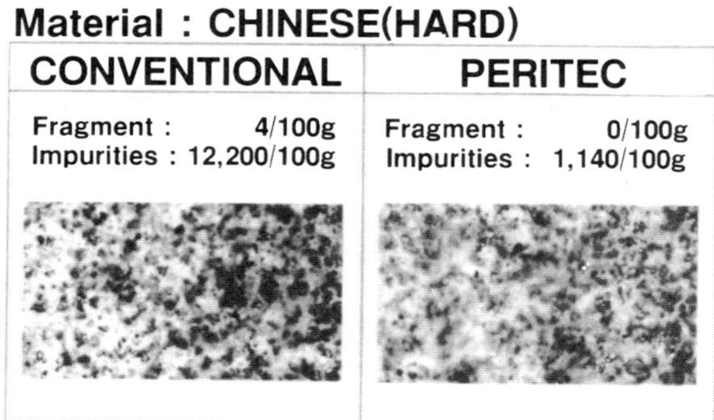

Figure 9. Reduced fragment count in flour from the PeriTec process, compared with conventionally-milled flour.

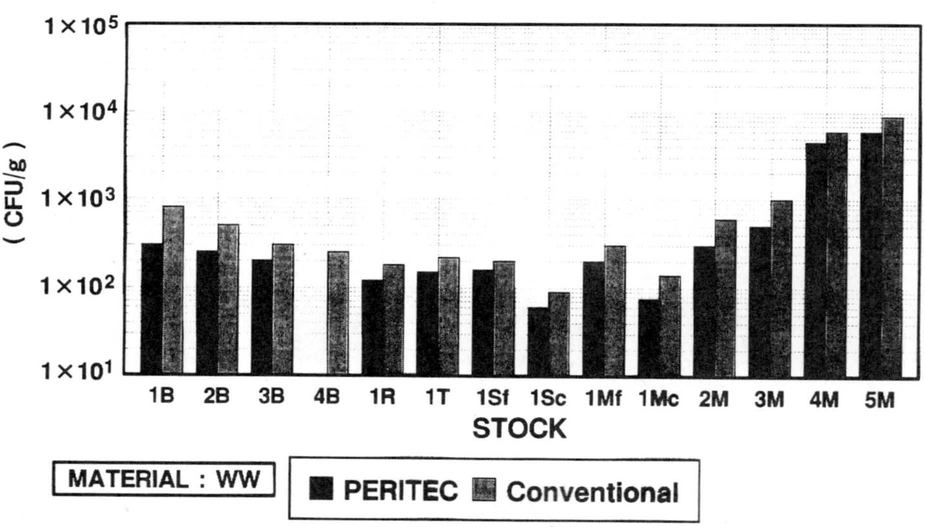

Figure 10. Mould residues in PeriTec and conventionally-milled flour.

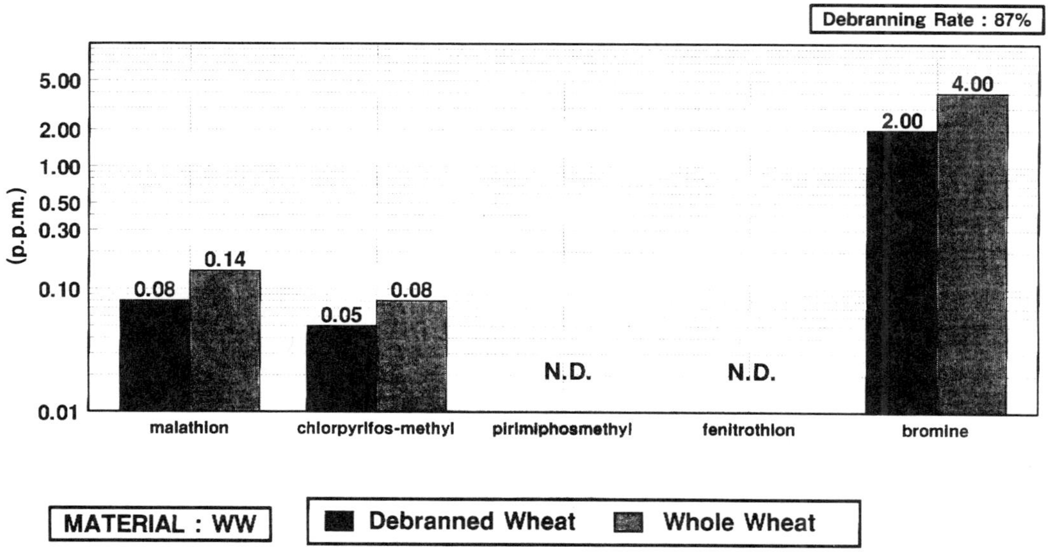

Figure 11. Chemical residues in PeriTec and conventionally-milled flour.

removal to suit the subsequent milling and the end product. Figure 6 illustrates how the wheat ash content decreased in a non-linear fashion as the Bran layers are removed. The steep parts of the curves are thought to be associated with the removal of the high ash aleurone layers.

In both these cases, the end point was chosen so that the majority of the aleurone had been removed; however, the system is sufficiently flexible for other end points to be chosen if desired. Figure 7 presents sections through typical kernels before debranning at the optimum points shown in Figure 6.

It is not always possible to identify readily the degree of debranning from this type of electron micrograph. As an alternative, the concentration of trace elements such as magnesium or phosphorus in the aleurone has been used to produce x-ray emission micrographs, as shown in Figure 8. In this case enhanced magnesium emissions were used.

The level of bran removal of 10–12% cannot be achieved by simple scouring; the purity of the final flour can be directly attributed to this level of removal. As examples of this insect fragment count, impurities (bran and external), moulds and yeasts and chemical residues were determined for PeriTec and conventional flours produced from the same batches of wheat. Figure 9 shows results of the tests for fragment count and impurities in accordance with AACC methods. In this context, impurities were classed as non-white, non-endosperm particles, which in fact were mainly bran particles.

Figure 10 shows the results of the determination of the level of moulds and yeasts in individual flour streams. In all cases, the counts were lower in the PeriTec flour. There is no data for 4Bk on the PeriTec system because the shorter PeriTec break system does not have a 4Bk passage. Figure 11 presents the results of testing for residual chemicals which might be used pre- or post-harvest. Where detected, the levels in the PeriTec flour were about half that for the same wheat conventionally milled.

4. SUMMARY

The current state of the rapidly evolving technology of the PeriTec process has been described. A number of specific areas where scientific endeavours have made significant contributions have been highlighted.

The picture is far from complete. Two major areas not dealt with are flour functionality and novel bran co-products. Research into these areas, with particular emphasis on the effects of specific kernel components, will provide fruitful opportunities for cereal chemists in the future (See Wood, Chapter 30; and Stenvert, Chapter 31).

With the current PeriTec technology, debranning has come of age. Commercial plants are in operation whose performance confirms expectations and point towards PeriTec as the milling system for the 21st Century.

ACKNOWLEDGMENTS

Acknowledgements are due to all my colleagues in Satake on whose behalf this paper has been prepared.

REFERENCES

Dexter JE and Wood PJ (1996) Recent applications of debranning wheat before milling. Trends in Food Science and Technology 7: 35–41

Goldsmith R and Forder DE (1994) Verfaren und seine Anwendung bei der Vernahlung von Durum Weizen. Durum und Teigwaren-Tagung. Detmold. May 1994

McGee BC (1995) The PeriTec Process and its Applications to Durum Wheat Milling. Assoc. Oper. Millers. Bull. (1995) pp 6521–6528

Satake RS (1990) Debranning Process is New Approach is Wheat Milling. World Grain 8(b), 28 30–32

33

SORGHUM PROCESSING TECHNOLOGIES IN SOUTHERN AFRICA

Trust Beta and Kennedy Dzama

University of Zimbabwe
Box MP 167, Mt Pleasant, Harare, Zimbabwe

1. INTRODUCTION

Sorghum is an indigenous cereal of Africa. Unlike maize, it is relatively tolerant to low rainfall conditions. Maize promotional programmes have led to its adoption as a staple food in the Southern African region at the expense of sorghum and millets. Studies on sorghum have been at a much slower pace than with other cereals. The major cause of the slow pace has been lack of research money. Sorghum has traditionally been milled to obtain a meal from which various products are made. Research has been initiated to diversify the utilization of sorghum, through modification of grain types and development of processing technologies to provide grains and products which are appealing to the consumer, who now prefers the white, refined, modern cereals (maize, wheat and rice). In Southern Africa, research on the production and utilization of sorghum has been actively pursued since the establishment of the Southern African Development Community/International Crops Research Institute for the Semi-Arid Tropics (SADC/ICRISAT) sorghum and millet improvement program (SMIP) in 1984 at Matopos Research Station in Zimbabwe. The Center has been involved in sorghum and millet improvement through breeding and selection of cultivars with suitable grain quality traits and processing characteristics. This chapter presents the current processing methods of sorghum.

2. SORGHUM PROCESSING

Sorghum production in southern Africa is aimed mainly at the market for human consumption. Surplus sorghum is then channeled to the stockfeed sector. In South Africa alone an average of 195,000 tonnes per annum is processed for human consumption (see Table 1). The processing sectors are, in order of size: malt, meal, brewers grits and edible rice.

Cereals: Novel Uses and Processes, edited by Campbell *et al.*
Plenum Press, New York, 1997

Table 1. Utilization of sorghum (tonnes) in the Republic of South Africa during the period 1987–1995

	Malt			Rice + grits				
Year	Indoor	Floor	Meal	Brew	Human	Other	Feed	Total
1987	40217	78068	13069	986	1646	7622	254264	395871
1988	48914	90839	20879	5985	931	5170	174139	346671
1989	49995	97049	17830	7184	64	2276	167379	341777
1990	56541	102307	27050	6520	124	1115	121696	315270
1991	55104	105776	29437	4140	236	522	50060	245275
1992	53385	103623	38485	2222	946	235	43945	242840
1993	56736	98489	42177	1647	582	66	87114	286811
1994	53444	88931	37817	3594	234	54	226098	410171
1995	39957	94646	43207	3894	603	293	120011	302612

Source: Sorghum Board, Pretoria, Republic of South Africa.

2.1. Malting Technology

Sorghum is grown on a large scale for the brewing of opaque beer; otherwise most sorghum for food is produced by subsistence farmers. An average of 200,000 tonnes of sorghum are malted annually and about 3000 million litres of opaque beer brewed each year, both at home and in commercial breweries. This is in comparison with about 2500 million litres of lager beer produced each year in the region. The three main steps carried out in sorghum malting are steeping, germination and drying.

Steeping sorghum grain in water enables water uptake to about 35% (wet basis) moisture. High tannin sorghum is steeped in a dilute formaldehyde solution (0.04–0.08%) for the first 4 hour steep to inactivate the tannins. The grains have to be rinsed thoroughly afterwards. Surface sterilization of the grain by formaldehyde or hypochlorite helps to reduce mold infection and proliferation.

Germination of the steeped grain takes place either in an open or covered floor for beds up to 0.3 m in depth, or on a perforated floor on top of a ventilating duct for beds up to 1.5 m. On the floor, the steeped grain is maintained at optimum temperature, moisture and aeration for the desired rate of germination. In floor malting, this is achieved by turning, controlling the depth of the malt bed and frequent watering. A period of 4.5 to 5.5 days is normally required to reach maximum quality. In industrial malting, temperature maintenance is achieved by forced aeration with humidified air of the required temperature. The rates of air flow range from 1,000 to 1,200 m^3/hr per metric tonne of grain (Joustra and Field, 1978). Turning and loosening germinating grain is necessary to prevent matting of seedlings, reduce resistance to aeration, and to assist uniform watering of green malt. The process tries to minimise breakage of roots and shoots to prevent unproductive regrowth and also prevent access to toxin-producing fungi. The industrial and floor malting processes have been compared in a laboratory experiment, with the former giving malt of superior enzyme activities (Beta et al, 1995).

Drying of the germinated grains takes place in thin layers in the sun in the traditional process. Industrially, forced-draft dryers with high volumes of air heated to not more than 50°C are used to dry the malt so as to arrest growth and prevent further metabolic losses. Dried malt is not polished, as roots and shoots are important sources of free amino nitrogen (Taylor, 1983) and flavour. The dried malt is usually milled using hammer mills to a fineness specified by the brewer. A 0.5 mm sieve is commonly used.

2.2. Brewing Technology

Opaque beer brewing consists of a number of steps which include souring, cereal cooking, mashing, straining and alcoholic fermentation. A typical industrial process for opaque beer brewing is the Reef Process shown in Figure 1. The order of the steps varies between different brewing processes, and two or more steps can take place simultaneously, especially in home brewing. Home brewing involving simultaneous mashing, souring and fermentation have been described (Faparusi, 1970; Novellie and Schaepdrijver, 1986).

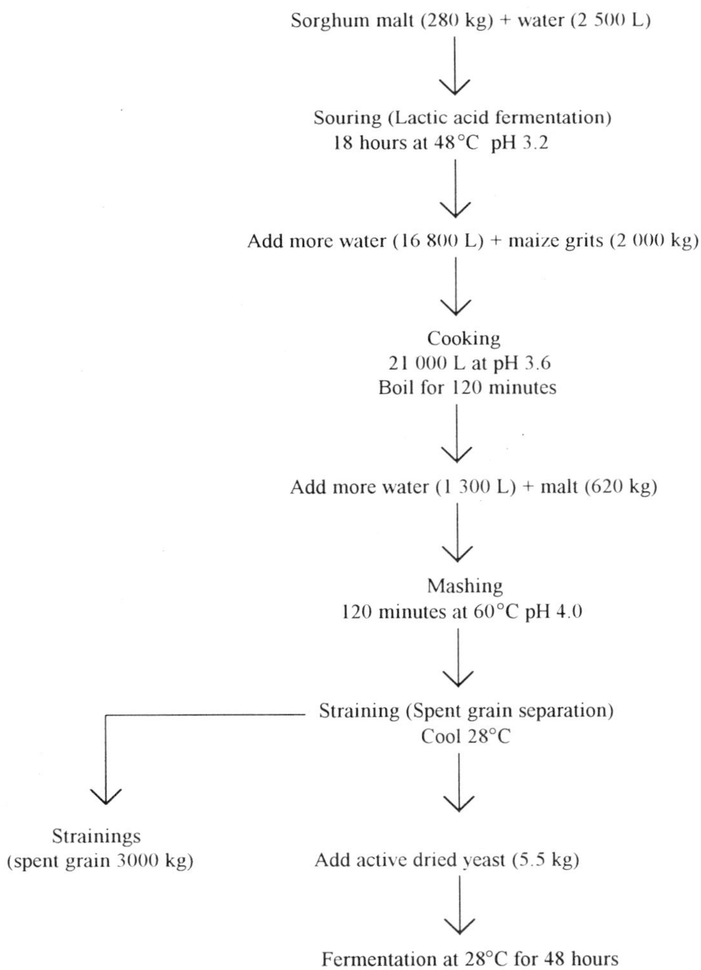

Figure 1. The Reef type sorghum beer brewing process. Source: Taylor, 1993.

Souring. is carried out at 48 - 50°C to produce lactic acid that gives beer its characteristic sour taste and to lower the pH of the beer. The low pH prevents complete hydrolysis of starch into sugars, so that the residual starch imparts an opaque, viscous character to the beer. Two methods of souring are used. Spontaneous souring is dependent on the natural microflora in malt. Lactic acid production will vary between different batches of malt. In industrial brewing and many home-brewing processes, inoculated souring is used. A previous sour of about 10% by volume, containing viable lactic acid bacteria, is used to inoculate new sour. Souring is done in less than two days in industrial practice and leads to lactic acid content of 1–2% and pH of about 3.0. In home brewing, temperature is not strictly controlled, resulting in the flavour being imparted by compounds other than lactic acid.

Cooking. is carried out in all opaque beer brewing processes to gelatinise the starch of the unmalted cereal adjunct. The high altitude of Southern Africa (1,000–2,000 m) lowers the boiling point of water and necessitates long cooking times. The boiling period is 1.5–3 hours in industrial brewing or 2–7 hours in home brewing. In the Reef Process (Novellie and De Schaepdrijver, 1986) (see Figure 1) and in many home brewing processes (Haggblade and Holzapfel, 1989), the sour is cooked with the unmalted cereal adjunct. The development of high temperature short time extrusion cooking has led to the use of extruded grits which can be added to the mash without further treatment.

Mashing involves incubating milled sorghum malt with cooked adjunct at 50–60°C for 1.5–2 hr (Novellie, 1968). This is done to solubilise the components of malt and adjunct so as to produce a fermentable wort. The ratio of adjunct to malt in the mash is about 4:1 in industrial brewing in South Africa (Taylor, 1989). In the Reef process, the mash has 6–7% fermentable sugars, 3% water-soluble dextrins, less than 1% gelatinised starch and about 2% ungelatinised starch. Optimum mashing conditions for free amino nitrogen (FAN) production are about 51°C at pH 4.6.

Straining is done after mashing using solid bowl centrifuges (decanters) in industrial brewing (Novellie and De Shaepdrijver, 1986). All particles larger than 250 μm in diameter are removed. These strainings contain 46% starch and 25% protein. (Van Heerden, 1987). Sugars, dextrins and fibre make up the remainder. The strainings are sold as animal feed. Home brewers use metal screens of appropriate mesh size for straining (Haggblade and Holzapfel, 1989).

Alcoholic Fermentation is brought about by wild yeasts in the wort in traditional home brewing. Commercially produced active dried yeast is used to inoculate wort in the industrial process (Novellie and De Schaepdrijver, 1986). Strains of top fermenting yeast *S. cerevisiae* which rapidly ferment beer wort at ambient temperatures are generally used. The beer is packed shortly after pitching with yeast or fermented at about 25°C in batch-sized fermentors for more than 25 hours. Beer is not filtered, and the yeast is not removed. The draft beer is transported in road tankers to its place of consumption, or dispersed by beer pumps into beakers of 2 or 4 litres capacity from which beer is consumed. The beer is consumed in an actively fermenting state.

A major challenge facing commercial opaque brewing is to develop products with an extended shelf life of up to several weeks. At present the maximum shelf life is five to seven days. New developments in opaque beer include processing to produce "instant" beer powder to which warm water is added and alcoholic fermentation allowed to take place in 24 hours. Pasteurized opaque beer and "pre-brewed" shelf stable wort have been developed by the Council of Scientific and Industrial Research (CSIR) in South Africa. The wort is concentrated to a syrup of high solids content and low water activity or further spray-dried into a powder. The product has been successful in Namibia.

3. MILLING

The potential to utilise sorghum industrially for food products in urban areas has been little developed. Most sorghum for food is grown on a subsistence basis. Cereal-based industries already mill maize from large local farms and wheat, much of which is imported. The purpose of milling is to reduce the grain in such a way that the different botanical parts are separated as cleanly as possible by sifting. A simple milling process involves grinding the whole grain in a stone hand mill or a hammer mill driven by a diesel engine, after which the most coarse bran particles are sieved away. This gives about 95% extraction of product for food use from the raw materials.

3.1. Traditional Milling

Figure 2 illustrates the traditional milling methods. In rural areas, home decortication in a stone or wooden mortar and pestle is common. Decortication of wet grain is done by pounding in mortar and pestle to remove the pericarp. The pestle weighs about 3 kg while the mortar is about 60–70 cm high with a diameter of about 30 cm. The bran is removed by winnowing in woven baskets. Whole grain or decorticated grain is reduced into flour using either the mortar and pestle or a stone mill (quern).

The stone mill consists of a base plate and roller stone which is oval or round in cross-section but becomes flattened after use. The base plate is positioned such that it slopes away from the operator. Flour particle size can vary from very fine to coarse meal depending upon the milling procedure used and the consistency of the product desired. Locally fabricated sieves of various mesh sizes are used to obtain the desired flour particle size. The flour is usually sun-dried and cooked preferably within 48 hours. Processing of small grains is usually time-consuming and labor-intensive, as most of the dehulling and grinding is done by hand.

3.2. Mechanical Milling

All the mechanical mills are much more effective in removing the grain outer layers than the traditional mortar and pestle. The hammer mill is used for crude milling of whole or decorticated grain. Many people choose to dehull their grain manually and then send it to the hammer mill for reduction to a meal.

The other approach to dry milling which has been successful is pearling or decortication. Sorghum kernels, being flattened spheres, work relatively well with decortication equipment. All are designed to remove the outer part of the kernel by abrasion with little breaking of the kernel (see Chapter 32 for a similar process for wheat bran removal). The type of dehuller used for sorghum in Africa is the one originally developed for the Canadian International Development Research Centre (IDRC) by the Prairie Research Laboratory (PRL) (Bassey and Schmidt, 1989). Abrasive action is provided by 13 carborundum stones (30.5 cm in diameter and 3.2 cm thick) or 27 resinoid discs (same diameter but 0.64 cm thick) mounted on a horizontal rotor. Fines are removed by a fan and bagged at a cy-

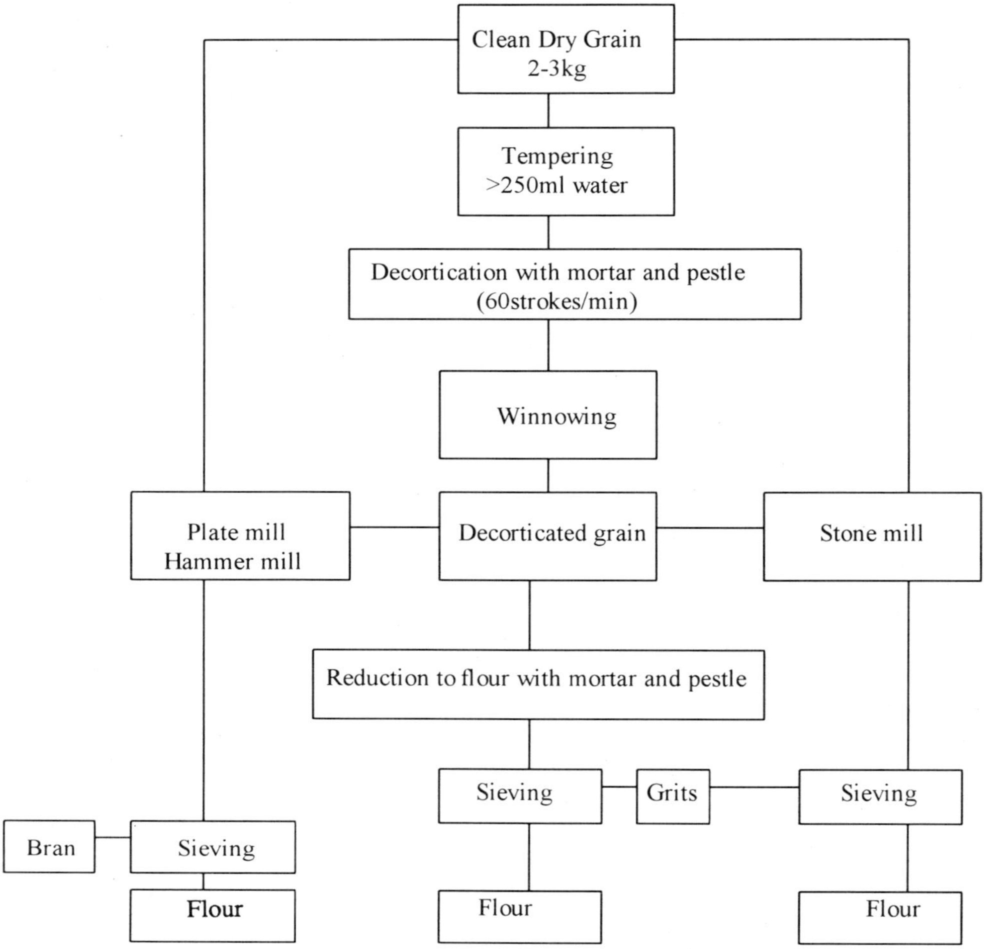

Figure 2. Traditional milling methods. Source: Murty and Kumar, 1995.

clone. The batch of grain (5–25 kg) is discharged when sufficiently dehulled. The Rural Industries Innovation Centre (RIIC) in Botswana successfully scaled down the PRL dehuller to enable to it to dehull batches as small as 5 kg. The significant technical advance of the RIIC dehuller is that it facilitates both batch processing and continuous flow production. The RIIC model has been successfully sold in Southern Africa. Underutilization of the RIIC dehuller has led to the adoption of mini-PRL models, some of which have no aspirating systems. An example is the mini-ENDA dehuller introduced by Environment and Development Activities (ENDA) in Zimbabwe. The dehullers are being sold by local engineering firms to large manufacturing firms, small-scale processors and service millers.

Sorghum roller milling has been developed to a lesser extent and few institutions are involved in this technology. Sorghum grains are more difficult to reduce to a flour fineness than is wheat, giving a coarser flour. Complete separation of the bran and endosperm is not achieved using the conventional roller milling systems thereby limiting its use (Hahn, 1970). A semi-wet milling procedure has been recommended in which added water amounts to 20% instead of the normal 2–3% (Cecil, 1992). Roller milling technology may soon gain importance as better products are realized.

Further research is needed on traditional processing methods and possibilities of applications on an industrial scale. Traditional food processing needs to be optimised with regard to the degree of conditioning for maximum milling yield, period of presoaking and maximum removal of tannins. Differential quality requirements of sorghum for traditional and commercial food processing into porridge meal and baking flour and other milling fractions, such as semolina and grits may be needed.

4. SORGHUM PRODUCTS

Meal. Dehulled, fine or coarse textured sorghum meals are available in shops in some countries. The most common and simplest food prepared from sorghum meal is porridge. Thick porridges are prepared by adding flour to boiling water in increments accompanied by vigorous stirring. The flour is cooked until it forms a thick, homogeneous and well-gelatinized mass devoid of lumps. The difference between thick and thin porridges is the concentration of flour.

Breakfast Cereal. Malted and toasted sorghum meal is used as a breakfast cereal. Thin porridges mentioned above are also used as breakfast meals.

Sorghum Rice. Whole grains, dehulled grains, cracked grains and grits are boiled to give rice-like products.

Malt. Coarsely ground sorghum malt is used for home brewing of sorghum beer with a mild alcoholic content. Finely ground sorghum malt is used to produce nonalcoholic beverages. Mahewu (mageu) is a liquid sour product prepared from sorghum malt and maize.

Beer Powder. This is a blend of malted grain sorghum, pre-cooked maize meal and brewers' yeast specially designed to obtain a fermented and nutritious sorghum beer after a 24-hour fermentation period.

Extruded Products. Sorghum flour is used to a lesser extent to produce blended nutritious foods and sorghum snacks using extrusion. In Mahewu products sorghum levels are as high as 35%. Maisoy-sorghum puffed flakes have been developed with levels of sorghum limited to 20%.

5. CONCLUSION

The development of new and improved processing technologies of sorghum should be actively pursued if the grain is to gain importance as an industrial raw material that can compete with maize. Production technologies should not be carried in isolation with utilization techniques. Promotional programmes on sorghum should take a systems approach encompassing production, processing and marketing strategies.

REFERENCES

Bassey MW and Schmidt OG (1989) "Abrasive Dehullers in Africa - from Research to Dissemination." IDRC, Ottawa. 98

Beta T, Rooney LW and Waniska RD (1995) "Malting characteristics of sorghum cultivars." Cereal Chem. 72(6), 533–537

Cecil JE (1992) "Semi-wet milling of red sorghum-a review." In "Utilization of sorghum and millets." MI Gomez, LR House, LW Rooney and DAV Dendy, eds. International Crops Research Institute for the Semi-Arid Tropics, Patancheru, 23–26

Faparusi SI (1970) "Sugar changes during the preparation of Barakuti beer." J. Sci. Food Agric. 21, 79–81

Haggblade S and Holzapfel WH (1989) "Industrialization of Africa's indigenous beer brewing." In "Industrialization of Indigenous Fermented Foods." KH Steinkraus, ed. Marcel Dekker, New York. 191–283

Joustra SM and Field A (1978) "Technological factors which influence malt quality." BB 118. Sorghum Beer Unit. Council for Scientific and Industrial Research (CSIR), Pretoria, SA. 1–20

Taylor JRN (1993) "Sorghum malt its current use and future potential for brewing in southern Africa." In "Cereal Science and Technology Impact on a Changing Africa." JRN Taylor, PG Randall and JH Viljoen, eds. ICC International Symposium, Pretoria, SA. 413–431

Taylor JRN (1983) "Effect of malting on the protein and free amino nitrogen composition of sorghum." J. Sci. Food Agric. 34, 885–892

Murty DS and Kumar KA (1995) "Traditional uses of sorghum and millets." In "Sorghum and Millets: Chemistry and Technology." DAV Dendy, ed. American Association of Cereal Chemists, Minnesota. 185–221

Novellie L and De Schaepdrijver P (1986) "Modern developments in traditional African beers." Industrial Microbiology, Vol 23. MR Adams, ed. Elsevier, Amsterdam. 73–157

Van Heerden IV (1987) "Nutrient content of sorghum beer strainings." S. Afr. J. Anim. Sci. 17, 171–175

34

CEREAL PROCESSING IN NEW ZEALAND

Inversion, Diversification, Innovation, Management

T. N. Lindley[1] and N. G. Larsen[2]

[1]Grain Foods Research Unit
Crop and Food Research
Private Bag 4704, Christchurch, New Zealand
[2]Crop and Food Research International
PO Box 7, North Ryde, NSW 2113, Australia

1. INTRODUCTION

New Zealand has many advantages as a producer of arable crops. The temperate climate is reliably cool in winter and warm to hot in summer, and the soils are naturally fertile. Farmers are well educated and have access to modern technology. There are some disadvantages, however. The country is geographically diverse, which means that crop quality varies considerably between locations, and the principal cereal cropping areas are separated by sea from the largest population centres. Perhaps the major constraints on cereal cropping are the better returns for less risk in other forms of farming and the high cost of land.

2. THE NEW ZEALAND WHEAT BOARD

Between 1965 and 1987, the New Zealand wheat growing, flour milling, and baking industries were tightly controlled by the New Zealand Wheat Board (for a detailed history see Bushuk, 1995).

According to the Wheat Board Act, the primary function of the Board was *"to promote and organise the orderly development of the wheat growing and flour milling industry and to promote greater efficiency in those industries."* The Board set out to achieve this goal by controlling all sales, purchase and distribution of wheat and flour. Growers and millers were guaranteed a market, while bakers and other flour users had to make do with whatever was provided, regardless of quality.

In effect, this amounted to a lack of confidence in the ability of the New Zealand industry to compete on even terms with the rest of the world, even in the domestic market.

Cereals: Novel Uses and Processes, edited by Campbell *et al.*
Plenum Press, New York, 1997

The attitude of the Board reflected the general political climate of the day and in fact became a self-fulfilling prophecy. At the end of the period, New Zealand wheat, flour and baked goods were all of inferior quality, and innovation or development within the industry was a near impossibility. The single exception was that New Zealand's bakers became highly proficient at adapting their processes to cope with the variability of their raw material.

3. A DE-REGULATED WHEAT INDUSTRY

Disbanding of the Wheat Board in 1987 initiated a period of very rapid change, not all of it beneficial. Within three years, the main effects of de-regulation had been:

- a rapid rise in the quality of wheat, flour and baked goods
- fast uptake of new cultivars (especially Otane) by growers
- rapid decline in production and increase in imports
- mill closures and a crisis in the wheat growing/flour milling industry

Was the old system better? Or was the decline an inevitable pain that must be suffered as part of the cure?

3.1. Inversion

By 1991 there was real concern that the NZ wheat growing industry would disappear altogether. Production was less than half that in the era of regulation and, in fact, was so small that the infrastructure was becoming unsustainable. Not only were wheat imports increasing rapidly, bulk importing of flour was being proposed as an option.

To reverse this situation the millers and growers formed a strategic partnership which used industry workshops to identify the collective actions necessary to rescue the industry. The overarching theme in the resulting industry plan was that *"the future of the NZ wheat industry will be driven by requirements of the NZ domestic and export markets and the ability of farmers through millers to meet these requirements"* (Growers and Millers, 1991, 1993).

The most surprising feature of this plan was that they meant it. Within two years the industry was operating from a completely different, market focused standpoint. Walter Bushuk, professor of Food Science at the University of Manitoba, worked in New Zealand in 1986 and in 1994. On his second visit he noted that New Zealand had "completely re-invented its wheat industry" in the intervening years. And as research managers, we found that conversations with wheat growers were more than likely to be about technical matters.

Since then, the wheat growing and milling industry has steadily recovered. The baking industry, relishing the opportunities afforded by the quality materials now available, developed a burgeoning export trade. Cereal-based food has been by far the fastest growing earner of foreign exchange in New Zealand's food sector since 1991. However, the domestic consumer has benefited most from the revitalised industry.

3.2. Diversification

Most of New Zealand's bread is produced with high-speed, Tweedy-type mixers using the mechanical dough development (MDD) process with work input optimised to match flour properties. Potassium bromate is not widely used and it is expected that this ingredient will be banned. Millers may not add anything to flour, including gluten or enzymes, unless the flour is sold as a 'premix'.

Post-deregulation, the first requirement was for wheat varieties suited to the new commercial environment. Fortunately, wheat breeders had anticipated the need and appropriate material quickly became available.

The mainstay of the bread wheat crop became "Otane", a spring wheat with an interesting biochemical composition and appropriate functional properties. Otane has an unusual profile of high molecular weight glutenins. It scores only as a medium quality bread wheat on analysis of the electrophoresis bands, but performs much better than that in the bakery. This apparent anomaly can be explained by the fact that Otane has an exceptionally large proportion of its HMW glutenins as sub-unit 7 (Sutton, 1991). The high performance of Otane is therefore not obvious from electrophoresis but clearly shows up under reverse-phase HPLC (Figure 1). Otane's most valuable properties are that it provides high baking volumes at modest total protein levels (10.5 - 12%) and at low work input requirements relative to other cultivars. Otane is now being replaced with agronomically-improved cultivars with similar protein composition.

The new flours based on Otane and other high performance wheats provided the baking industry with completely new capability, in that carrying capacity, in terms of loaf volume, increased by more than 25%. This capacity has been used to produce a remarkable variety of breads to attract the consumer. Coloured wheat varieties, other cereal grains, added fibre components, and most recently, nutraceuticals such as omega-3 oils are commonly included in New Zealand bread.

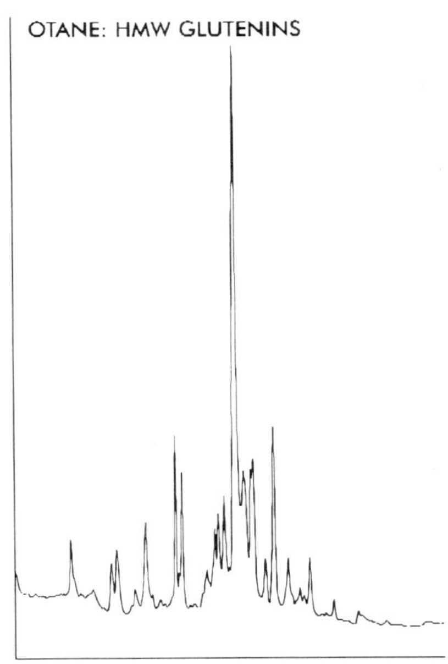

Figure 1. Reverse phase HPLC chromatogram of High Molecular Weight glutenins of the wheat cultivar Otane.

A typical large manufacturer makes up to fifteen types of sliced pan breads, including:

- white sliced breads in three weights and two thicknesses
- open-textured, unlidded, seed topped loaves
- light mixed grain breads containing added meals and kibbled or whole, coloured (purple) wheat and rye
- similar mixed grain breads that may contain either barley, oats, or corn grits
- a range of dense 'health' loaves featuring rye, and high proportions of added seeds or nut pieces
- straight wholemeal loaves
- high fibre white loaves

A rapidly diversified product range was not restricted to the large bakeries, however. Medium and small businesses, often completely new operations, have generated similar growth in new products. While the popularity of Middle Eastern flat breads has grown spectacularly, many other new breads have been successfully introduced to New Zealand life including French, Italian and Gaelic styles, and specialities such as bagels and rewena (a NZ Maori bread).

The main beneficiary of all this activity has been the consumer, as volume sales appear to have grown more rapidly than dollar sales. Bread is now the top supermarket key account category. It was not in the top five a decade ago.

This growth in business has not only affected the domestic market. A new export industry in cereal-based foods has emerged in response to opportunities afforded by the freer market. More than 10% of cereal-based foods produced in New Zealand are now exported (from 2–3% five years ago) and all of this production represents new growth. A particular feature has been the linking of the cereal industry with New Zealand's high-tech dairy industry. Most of the exports feature butter or milk in products such as:

- unbaked pastry,
- butter croissants,
- butter biscuits and biscuit doughs,
- roux bases,
- milk sauces.

3.3. Innovation

The reform of the cereal processing industry was a boon to cereal processing researchers. Along with the opportunities provided by deregulation came unanswered questions and the will to provide solutions. However, the times were also turbulent for scientists. In 1992, Government research providers were re-organised to convert two departments (Department of Scientific and Industrial Research, and Ministry of Agriculture and Fisheries) into eight Crown Research Institutes. These compete for funds from a 'Public Good Science Fund', in an attempt to establish a 'market economy' for science.

The reforms brought about many benefits including: eliminating 'black hole' programmes into which scarce resources disappeared without hope of payback; introducing sound project management practices to research; and renewing the relevance of science. The reforms also introduced confusion in the scientific community, an overemphasis on short term expediency and wastefully expensive systems for obtaining funds. Nevertheless, now that things have settled down, the future looks hopeful for science in New Zealand.

The main thrusts of cereals research at Crop & Food Research are:

- identification and provision of quality in grains for processing,
- understanding the materials as they pass through processing and providing means for monitoring and optimising processing,
- understanding cereal foods post-manufacture and measuring relevant textural properties,
- solving industry problems.

In new cereal production, considerable effort has gone into developing purpose-specific cultivars and introducing novelty, especially thorough the use of coloured grains. Purple-grained wheats are now well established and used, both to add visual interest in kibbled grain mixes, and to produce wholemeal loaves that look like dark ryebreads but have a light texture.

Process optimisation research covers several products including biscuits, extruded products and corn foods. The main emphasis has been on bread and pastry doughs. In bread the frustration in being able to research dough development during laboratory scale mixing but not on an industrial scale led to the invention of the 'Dough Probe' for Mechanical Dough Development mixers (Wilson, 1992). Dough probes have proved effective in optimising industrial processing, and units have been installed in bakeries in Australia, New Zealand and the UK. Similarly, our researchers have been able to quantify the rheological properties of laminated doughs by applying rheometrical measurement techniques based on biaxial extension (Morgenstern et al, 1995) and are having success in predicting the properties of the baked product. Allied to the physical studies have been research programmes on the effect of enzymes, notably amylases and ascorbic acid oxidase (Every et al, 1995; Every and Ross, 1996) and proteins during dough development.

Textural studies on baked goods have been supported by a raft of instrumental tests for measuring physical properties, using the Instron Universal Testing Machine (in the laboratory) or the Crop & Food Research Textron (for factory measurements). The Textron is a versatile, portable, pc-linked, precision compression/extension machine built within the Institute for cereal and seafood texture studies. These tests include measurement of crumb softness, strength, stickiness, and resilience (recoverable energy). The latter two tests have recently been applied with success in an industry-commissioned study of the relationship between grain quality tests (Falling Number, α-amylase) and incidence of bread crumb faults (stickiness, doughiness).

The two major problems facing the New Zealand bread industry over the last five years have been how to obtain sufficient dough oxidation in the MDD (Chorleywood) process without bromate, and how to obtain stable, strong doughs without requiring high levels of work input. An extensive series of research projects funded by the industry have led to a good functional understanding of dough oxidation and dough development. Much of this work is continuing in a Crop & Food Research-led project being undertaken in collaboration with the Australian Cooperative Research Centre for Wheat Quality Products and Processes.

3.4. Management

An unusual feature of the New Zealand wheat growing industry is that individual farmers produce small quantities of wheat that are usually stored on-farm until called up by the flour mill. This preserves a high level of segregation and with it the capacity to specialise for quality attributes on-farm. This facility is valued by millers, who have resisted suggestions of early bulking without their control.

This feature provides an opportunity to manipulate more closely wheat quality for specific characteristics targeted to an end use, without fear that distinctive benefits will be lost during consolidation.

The application of this concept is exemplified in a potential solution to the work input problem. In 1992, baking industry representatives discussed with scientists of the Grain Foods Research Unit the difficulties they were having with dough mixing. Millers had encouraged an increase in wheat protein content through pricing mechanisms, and although this wheat provided some advantages, resulting doughs required much higher levels of work input to achieve development. This resulted in excess dough temperatures, gaps in production, or under-developed bread doughs. What could we do about it?

Technical solutions could be applied at several levels. We could:

- build better mixers
- monitor mixing more closely
- control protein streaming during milling for improved mixing characteristics
- control farm management to maximise baking quality without raising work input
- provide cultivars with the desired properties

All of these aspects are now being worked on, but the most interesting results have come from on-farm management.

On reviewing the changes in farm management practice following deregulation, it became obvious that farmers were applying much more nitrogen, mainly in the form of urea, to the crop. Some New Zealand soils are relatively sulphur deficient. Subsequent research (Wooding et al, 1993) showed that while applying nitrogen pushed up loaf volumes, provided better texture, and increased work input requirement, applying sulphur retained the benefits, while reducing the work input (Figure 2).

An interesting aspect of the work was that, while the importance of sulphur on work input was identified on our 125g MDD test bake system and confirmed on an industrial scale trial using a Tweedy mixer, it did not show up on Farinograph or Mixograph measurements. This demonstrates the importance, where possible, of confirming experimental results obtained in the laboratory with experiments on industrial scale equipment, rather than relying on confirmation from other laboratory instruments.

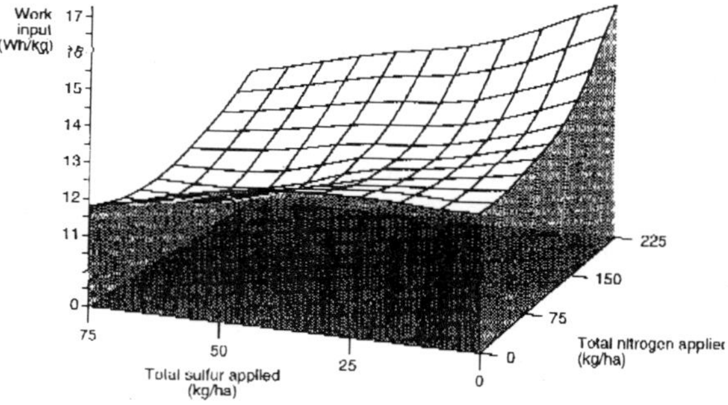

Figure 2. Effect of on-farm nitrogen and sulphur application on work input requirement to achieve dough development.

This result, and its application, underscores the importance of having both the right crop genetics in place and also appropriate crop management. It also demonstrates the fact that crops can be managed for process efficiency as well as food quality.

4. CONCLUSION

The last decade of cereal processing and cereal science in New Zealand has been one of dramatic change. The perspectives and operating frameworks of the cereal processing industry and scientists serving the industry have been overturned. The result has been a rejuvenation of the domestic market, which shows consistent growth in consumption of cereal foods, and the establishment of a significant new export sector.

Crop & Food Research has a vigorous programme of research operating at every point along the production chain to assist the industry supply a wide range of high quality products to the consumer.

REFERENCES

Bushuk W (1995) "Wheat New Zealand - 1769–1994." FoodInfo report No.12. NZ Institute for Crop & Food Research, pvt bag 4704, Christchurch, New Zealand

Every D, Gilpin M and Larsen NG (1995) "Ascorbate oxidase levels in wheat and their relationship to baking quality." J. Cereal Science, 22, in press

Every D and Ross M (1996) "The role of dextrins in stickiness of bread crumb made from pre-harvest sprouted wheat or flour containing exogenous *alpha*-amylase." J. Cereal Science, 23, in press

Growers and Millers (1991) "Strategic Partnerships - Achievement of a sustainable New Zealand Wheat and Flour Milling industry for the '90s." United Wheatgrowers Association, NZ Flour Millers Association, April 1991

Growers and Millers (1993) "Strategic Partnerships - the next step." United Wheatgrowers Association, NZ Flour Millers Association, June 1993

Morgenstern MP, Newberry MP and Holst SE (1995) "Biaxial extension of dough sheets." Proceedings of the 45th Cereal Chemistry Division, Royal Australian Chemical Institute (RACI) conference, 372–375. Eds YA Williams and CW Wrigley. Cereal Chemistry Division, RACI, Melbourne

Sutton KH (1991) "Qualitative and quantitative variation among high molecular weight subunits of glutenin detected by reversed-phase high-performance liquid chromatography." J. Cereal Sci 14, 25–34

Wilson AJ (1992) "Measurement of work input in industrial mixers." Proceedings 42nd RACI Cereal Chemistry Conference. Ed V.J. Humphrey-Taylor. Cereal Chemistry Division, RACI, Melbourne, Australia

Wooding AR, Martin RJ and MacRitchie F (1993) "Effect of sulphur-nitrogen treatments on work input." Proceedings of the 43rd Australian Cereal Chemistry Conference, ed CW Wrigley. Cereal Chemistry Division, RACI, Melbourne

INDEX